# The 2nd International Symposium on Rail Transit Comprehensive Development (ISRTCD) Proceedings

## ISRTCD 2013



Beijing, China
January 11–12, 2013

Hosted by Urban Planning Society of China

In Cooperation with Beijing Jiaotong University, CCDI Group

https://www.easychair.org/conferences/?conf=isrtcd2013

Haishan Xia • Yunan Zhang
Editors

# The 2nd International Symposium on Rail Transit Comprehensive Development (ISRTCD) Proceedings

悉地

Springer

*Editors*
Haishan Xia
Yunan Zhang
School of Architecture and Design
Beijing Jiaotong University
Beijing, China

Supported by NSFC Project 51078022.

ISBN 978-3-642-37588-0 ISBN 978-3-642-37589-7 (eBook)
DOI 10.1007/978-3-642-37589-7
Springer Heidelberg New York Dordrecht London

Library of Congress Control Number: 2014931405



Printed on acid-free paper

Springer is part of Springer Science+Business Media (www.springer.com)

# Foreword I

## Promoting the Guiding Function of Urban Planning Over Rail Transit

Under the positive guidance of the national macro policy, the construction of urban rail transit in our country has achieved rapid development. The average construction scale has reached to 190 km. According to relevant documents commissioned by National development and Reform Commission, the average annual construction scale will reach to the astonishing number of 500 km by 2020. Actually the construction scale will probably be faster than this. Regardless of scale or speed, the construction is unprecedented in the whole world, which will be an opportunity as well as challenge for professionals who work in Chinese urban planning, traffic engineering or other relevant areas.

Meanwhile, new mode of urbanization, after the industrialization, has also become our country's new development policy. President Xi recently in Central Economy Conference stated that: "New mode of urbanization is currently the historic mission of Chinese modernization construction. It is the potential to expand domestic demand, which will not only enhance the quality of urban development, but also to be considered as a practice of scientific outlook." Nowadays the level of urbanization in our country is about 51 %, which has a long distance to reach 60 %. During this process, the biggest problem that we need to face is transportation mode of supporting the development of urbanization. We have gain a lot experience and lessons from this. It is proved that to purely reply on the development of private cars and road traffic is hard to be sustained.

Therefore, on 10th October 2013, there are eight key tasks to be promoted on the State Council Executive Meeting to develop public transportation. In September 2012, National development and Reform Commission also commissioned many relevant projects. Combined with the above two points, we should realized that on this unprecedented urbanization new development stage, we need to carry out planning and construction at fast and high quality level, as well as preventing and solving relevant problems such as traffic congestion, environmental pollution and etc.

Therefore, scientific planning methods should be employed in order to promote and coordinate the development of environmental-friendly transportation system and the urban planning. On one hand, we should expand and seek for potential domestic demand; on the other hand, domestic demand should be adjusted. This is the task that this era endowed for us.

State Council Executive Meeting gives priority to the development of the public transportation with speed, large passage flow, low pollution and high efficiency. Therefore, it is necessary to develop rapid construction of urban transit, which has currently been included into urban planning of many cities. Hence, how to construct rail transit to be fast, safe and high efficient? According to relevant experience in Chinese Hong Kong, Taiwan, and other overseas cities, I think it is very important to integrate new mode of urbanization and urban transit in the areas of investment, planning, construction, operation and management. The target can be achieved through promoting urban transit system towards urban development, as we as improve and utilize the investment structure, construction environment, passenger flow and using experience of urban transit.

Therefore, to plan and develop urban rail transit should focus more on urban planning. Based on this foundation, valuable research projects on rail transit can be carried out around the issues of urbanization and rail transit. In this era, the above issue naturally becomes the central issue to be discussed by professionals and experts from urban transit and urban planning areas. My personal opinions on this will be discussed as follows:

Firstly, urban planning is the foundation of urban transit construction. It is obviously that superstructure relies on the foundation. During the process of urbanization, planning is the foundation, which is also the premise of the following construction. Hence, adjustment of planning will certainly affects urban transit system, which will even leads to collapse between city and transportation system. In China, urban planning and urban transit are made by various people, which generate conflicts. Therefore, planning should be set out scientifically, as well as be integrated with urban planning and complex transportation planning effectively. To conduct planning properly will guarantees the construction of transportation.

Secondly, bearing capacity of the city should be measured. In China, population is a serious issue in many cities, which also has close relationship with environment and resource issues. Water, electricity, or transportation are important components consisted of urban resources. Therefore, principles of the balance between population and environment, harmony between economic society and ecological returns are promoted. Population scale and distribution will have fatal impact on investment scale, construction style, passengers' character and operation safety of rail transit system. Various versions of urban planning in many cities of China have big difference with real condition, particularly in areas of prediction, planning, managing and control of population, which directly or indirectly worse the rail transit construction. Hence, on the overall level of urban planning, how to predict, plan and manage city's population as well as arrange population distribution and urban space should be considered as the important premise of rail transportation development.

In addition to the above two points, the process of urbanization should be properly constructed in order to comprehensively promote resources saving. The land resources in China, particularly in big cities, are limited. Hence, to planning should consider both ground and underground space. Urban planning needs to be carried out in three dimensional levels. In other words, development along the areas of rail transit, underground space of the station, and spaces along subway should also need to be considered.

Overall speaking, in order to achieve the goal of integrating urban planning and rain transit and their interaction, "the complex development of urban transit", the title of this conference, is an excellent research thought. It offers us great opportunities to carry out in-depth discussion and research from various perspectives. It is a great honour to be invited to attend "The Second Conference of Urban Transit Development Theory and Practice". Many experts and honored scholars share their precious experience of complex development on urban planning, transportation, investment, architectural design and engineering. Therefore, this assembly can be considered as the summarized essence of many experts' academic outcome and experience. I believe that this will make great contribution to the sustainable development of trail transit and urban planning. Meanwhile, I also hope there will be more academic papers and research materials produced by experts, scholars and engineers from various fields, to enrich the future decades' theory and practice in this professional field.

Academician the Chinese Academy of Sciences 施仲衡
Beijing Jiaotong University Zhongheng, SHI
July, 2013

# Foreword II

Our country is experiencing an unprecedented urbanization. During this process, on one hand, as the fundamental facilities supporting urban development, rail transit has gradually become important means for solving urban traffic problems in many cities. On the other hand, with the increasingly improvement of urban rail transportation network, rail transit has exerted great influence on urban development, which has become more and more complex and in detail. Therefore, the construction and operation of large-scale rail transit calls for new requirements in the fields of urban planning, construction, and investments. There are several common issues need to be confronted in many cities, such as how to seek for positive interactions between urban rail transit and urban complex development under the big framework of urban development and urban transportation.

It is obvious that China has achieved great developments of rail transit. Meanwhile, there are immense difficulties and challenges. Therefore, in-depth research of theory, methods and experiences in preliminary strategy, planning design, architectural design, development construction, institutional guarantee, management supporting, within which will be involved in the lands' complex using and utilization of surrounding area, could be regarded as highly realistic issues. Because this relates to topics such as urban land use right, rail transit investment and financing system, rail transit operating mode, as well as the selection of rail transit route, and land utilization policy. Followed by the first session of "International Seminar of Transportation and development – Rail Transit Complex Development", which was held by China Planning Academy in Shanghai in 2011, "The Second Session of International Seminar of Rail Transit Complex Development – Complex Development and Planning along the Rail Transit Line" was organized and held by the Academy on 11th–12th January 2013. The purposes of all these seminars were to discuss the above issues.

During the process of organizing the seminar, we carried out a special investigation on the cities with constructed rail transit and on-going constructing rail transit. From the investigation, we discovered a series of new problems under the current situation of intensive construction of rail transit, such as: rail transit occupies high percentage of the urban fiscal expenditure, how to use rail transit invest capital in a sustainable way? The construction of rail transit not only exerts positive influence in solving transportation problem, but also has great impact on the development of urban structure, particularly the urban expansion and the development of urban central area. Therefore, from the perspective of the selection of rail transit line and station location, how to integrate with urban planning? It is well known that the developed countries have gained rich experience in the land development along the rail transit line. Hence, under our country's current land and investment policy, how to realize the positive interaction between rail transit construction and urban development? How to connect and mediate the planning work between rail transit and urban at different levels and phases? How to coordinate conflicts between various works and technical standards? And etc.

All these issues must be solved by expects coming from areas in urban planning, urban rail transit, and investment. We hope this seminar could become the platform for expects from the related areas to communicate and to exchange ideas, within which everyone could reasonably discuss the issues on urban complex development and seek for answers for these common questions. The selection of the theme of "The development and planning of Rail Transit along Line" actually represents this significance.

The following leaders and expects attended this seminar: China Urban Planning Academy Honorary Chairman Zou, Deci, Academician of the Chinese Academy of Sciences SHI, Zhongheng, China Investment Association Secretary-General Zhang, Yonggui, President of Beijing Jiaotong University NING, Bin, Chairman and general manager of CCDI Zhao, Xiaojun. Meanwhile, there are more than 300 people attending the seminar, including managers and technicians from various provincial railway investment companies, various cities' Development and Reform Commission, urban management department, city investment company, urban rail company and etc. This 2-day seminar comprehensively reflects every expert's research outcome and practical experience, which exerts a positive influence on the development of further research, the impetus of mediation of urban rail transit planning and urban planning, the promotion of rail transit construction and the complex development of land along the line.

Currently, papers have been called for in a national range. Expects from related institutions such as universities, planning bureau, subway company, and investment association submitted papers. After the seminar, we rearrange these writings into the assembly documents of the Second Session of the International Seminar of the Complex Development of Rail Transit, in order to summarize the phase outcome of research in this area. I do hope from now on, there will be more and more scholars, expects, managerial staff and investment-decision making members will focus on research in this area, with more research outcome and outstanding papers produced.

I also hope China Urban Planning Academy will make a great contribution to the development of urban rail transit.

I would also like to take this opportunity to present my sincere gratitude to the Seminar's undertaker – CCDI and Beijing Jiaotong University.

China Urban Planning Academy Vice Chairman and Secretary-general Nan SHI
International Urban and Regional Planner Academy Vice President
Beijing, April, 2013

# Contents

# Organizing Committees

**Honorary Chairman**

| | |
|---|---|
| Deci Zou | Urban Planning Society of China |
| Zhongheng Shi | Beijing Jiaotong University, China |
| Bin Ning | Beijing Jiaotong University, China |
| Xiaojun Zhao | CCDI Group |
| Zengliang Shan | CCDI Group |

**General Chair**

| | |
|---|---|
| Shi Nan | Urban Planning Society of China |

**Program Chairs**

| | |
|---|---|
| Changhong Qu | Urban Planning Society of China |
| Haishan Xia | Beijing Jiaotong University, China |
| Mu Wang | CCDI Group |

**Publication Chairs**

| | |
|---|---|
| Haishan Xia | Beijing Jiaotong University, China |
| Yunan Zhang | Beijing Jiaotong University, China |

**Financial Chairs**

| | |
|---|---|
| Xinyu Liang | CCDI Group |
| Jinfeng Dong | Beijing Jiaotong University, China |

**Publicity Chairs**

| | |
|---|---|
| Yunan Zhang | Beijing Jiaotong University, China |
| Yinghua Bao | Beijing Jiaotong University, China |
| Tiejun Gao | CCDI Group |

**Organization Chairs**

| | |
|---|---|
| Guobiao Zhang | Urban Planning Society of China |
| Yuan Lu | Beijing Jiaotong University, China |

| | |
|---|---|
| Yunan Zhang | Beijing Jiaotong University, China |
| Xiao Bo | CCDI Group |

**Secretary**

| | |
|---|---|
| Qingfeng Dong | Beijing Jiaotong University, China |
| Xia Zhao | Beijing Jiaotong University, China |
| Zitao Zhang | Beijing Jiaotong University, China |

# Program Committees

| **Program Chair** | |
|---|---|
| Shi Nan | Urban Planning Society of China |
| **Executive Chairman** | |
| Changhong Qu | Urban Planning Society of China |
| Haishan Xia | Beijing Jiaotong University, China |
| Mu Wang | CCDI Group |
| **Program Committees** | |
| Deci Zou | Urban Planning Society of China |
| Zhongheng Shi | Beijing Jiaotong University, China |
| Nan Shi | Urban Planning Society of China |
| Guoyong Li | National Development and Reform Commission |
| Yonggui Zhang | The Investment Association of China |
| Bin Ning | Beijing Jiaotong University, China |
| Xiaojun Zhao | CCDI Group |
| Gerald Olivier | The World Bank |
| Mu Wang | CCDI Group |
| Haixiao Pan | Tongji University, China |
| Huiye Zhu | The Investment Association of China |
| Kao Akira | CFK, Japan |
| Yan Wang | Urban Renewal Authority, Singapore |
| Jian Zhao | Beijing Jiaotong University, China |
| Yuan Lu | Beijing Jiaotong University, China |
| Kubota Koji | Hanshin Electric Railway CO., LTD, Japan |
| Dandan Yu | CRIC CO., LTD, Shanghai, China |
| Zhengmin Feng | National Chiao Tung University, China |
| Kai Wang | China Academy of Urban Planning & Design |
| Steven Highfill | WAI, USA |

| | |
|---|---|
| Chungfan Yuen | MTR Corporation (Shenzhen) Limited, China |
| Ke Qin | CCDI Group |
| Mingchun Luo | ARUP, UK |
| Zhang Ming | Wuhan University, China |
| Gang Meng | CCDI Group |
| John Bilmon | PTW CO., LTD, Australia |
| Liaoyuan He | The Investment Industry and Financial Commonwealth of China |
| Yu Zhou | Urbanseal CO., LTD, Beijing, China |
| Yida Xu | Jiading New City CO., LTD, Shanghai, China |
| Xinhai Yu | CCDI Group |
| Haishan Xia | Beijing Jiaotong University, China |
| Linfei Han | Beijing Jiaotong University, China |
| Yunan Zhang | Beijing Jiaotong University, China |
| Yuxiang Dong | Beijing Jiaotong University, China |
| Wei Sun | Beijing Jiaotong University, China |
| Yingdang Hu | Beijing Jiaotong University, China |
| Yinghua Bao | Beijing Jiaotong University, China |
| Kaiyu Zhang | Beijing Jiaotong University, China |

# Part I
# Theory of Comprehensive Development

# Chapter 1
# The Structure Research on Comprehensive Development and Integrated Use of Land and Space Around Metro Stations for the Sustainable Rail Transit Development

**Yuan Lu and Ke Qin**

**Abstract** Urban rail transit system is now playing a key role in utilizing and expanding city's scale and space structure, and is the key part of the construction of urban transport infrastructures. With the growing demand of introducing and pushing the intensive utilization of urban rail transit station area, the article summarizes the issues and challenges which occurred during the practice of developing rail transport in city, and proposes a research structure for the theory and policy research on comprehensive development and integrated use of land and space around metro stations.

**Keywords** Rail transit • Surrounding land • Comprehensive development • Structure research

## 1.1 Rapid Development of Urban Rail Transit in the New Period

### *1.1.1 Three Stages in the Development of Urban Rail Transit Construction*

Since the middle 1960s, the urban rail transit construction has been activated in mainland China. As an initial period, the urban rail transit construction was restricted by the economic and social development, during which the speed of

Y. Lu (✉)
School of Architecture and Design, Beijing Jiaotong University, 100044 Beijing, China
e-mail: luyuanup@foxmail.com

K. Qin
CCDI Group, Rail Development Business Sector, Train Transit Comprehensive Development Research Centre, 100013 Beijing, China
e-mail: willqinke@163.com

H. Xia and Y. Zhang (eds.), *The 2nd International Symposium on Rail Transit Comprehensive Development (ISRTCD) Proceedings*,
DOI 10.1007/978-3-642-37589-7_1, 

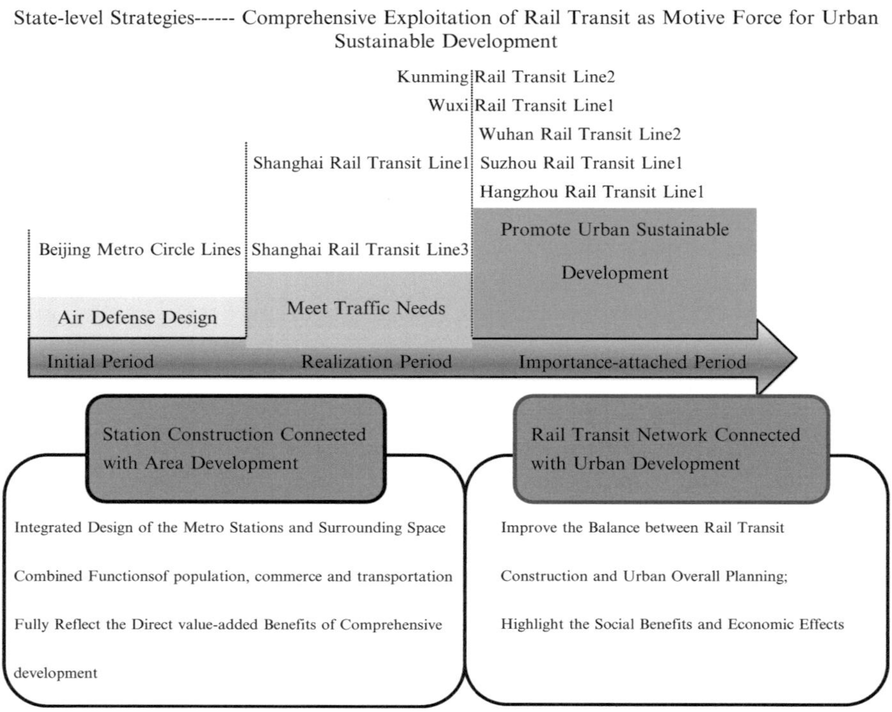


**Fig. 1.1** Three developmental stages in rail transit construction and exploitation strategies

construction was relatively slow. Throughout the 20 years of the 1960s–1980s, three lines were constructed of which the total length was only 47 km. Under special historical background, the construction of urban rail transit gave priority to combat readiness and civil air defense; tackling daily traffic difficulties only served as a byproduct. From 1985 to the end of twentieth century, after the reform and opening up, cities like Beijing and Shanghai initiated a second-round urban rail transit construction with the main purpose of public transit system construction. During these 10 years, a total of three lines were built; a new mileage of 54 km was added. At present, the rail lines built during this period plays a key as well as fundamental role in the urban public transportation system (Fig. 1.1).

A decade after entering the new century, cities in China have gone into a highly-motorized stage. The explosive growth of the vehicles leads to the occurrence of traffic jams frequently. As a response to the phenomenon, from 1999 to 2009, the country's first-tier mega-cities have initiated a large-scale urban rail transit construction plan. The new mileage adds up to 832 km over the 10 years. By the end of 2009, the operated lines of Shanghai, with the longest urban rail transit in mainland China, have been running for nearly 300 km, followed by Beijing, with a mileage of nearly 250 km. Other cities like Guangzhou, Nanjing, Wuhan and Shenyang also enter the era of rail transit.

Since 2008, in order to cope with the global financial crisis, China has proposed a macroeconomic development strategy that takes urban rail transit as "precedent areas, advanced planning and timely construction" [1]. It vastly stimulated other local cities into urban rail transit construction, forming a new trend of the rail transit pervasion. Urban rail transit has transferred from the first-tier cities to second-tier cities and even third-tier cities, from the eastern coastal areas towards the central and western underdeveloped areas. Based on reliable statistics, up to 2010, nearly 30 cities and regions in China had carried out rail transit construction and planning, a total of more than 110 lines involved. Among the 25 recently-ratified construction plans of urban rail transit, the new lines which are coming into operation reach a mileage of 2,610 km. It is estimated that during the "12th Five-Year Plan", more than RMB 700 billion will be invested into rail transit construction [2].

### 1.1.2 Issues and Challenges Caused by the Rapid Development

The urban rail transit of China has entered an unprecedented period. With more big cities and mega-cities trying to join the rail transit club, the comprehensive influence of rail transit on space structures [3], function organization patterns and industry layouts is expanding rapidly in breadth and depth. However, the public investmental risks, financial risks and financial burdens caused by the large-scale construction are also accelerating and accumulating. In this case, how to deal with the complex effects on urban integrated system caused by rail transit, namely: integrate rail transit construction and urban development strategy based on Comprehensive Development Theory; how to raise the fund of rail transit construction, reduce the operating loss and increase the public revenue by means of efficient usage of the land and space around metro stations; how to adjust the selection of metro stations and sites as well as design ideas, as to maximize the public benefits and development performance by means of comprehensive development; how to establish a policy environment in accordance with the new pattern of construction and development, optimize incentive mechanism and system guarantee by means of comprehensive development pattern... All of these topics have gradually transcended technical matters, turning into major issues in the development of urban rail transit construction.

### 1.1.3 Research on the Necessity of Comprehensive Development

As successful domestic and foreign rail transit construction experience shows: urban rail transit as the core, fully integrate transportation and urban functions, realizing a comprehensive development of multi-agent, multi-mode and assembled model is an essential means to deal with the above challenges and achieve rail transit and the urban sustainable development. Therefore, while recent urban rail transit faces significant opportunities, discussions on theories, routes, methods, innovations and guarantees of Comprehensive Development are extremely realistic, necessary and urgent.

## 1.2 The Structure Research on Comprehensive Development and Integrated Use of Land and Space around Metro Station

### *1.2.1 Coordination Between Rail Transit and Urban Overall Planning*

#### 1.2.1.1 Development Tendency

With the large-scale development of rail transit construction and improvement of network coverage level, rail transit is exerting comprehensive influence on the urban system which is or has transcended transportation field and extends further to comprehensive fields such as urban space and urban industries [4]. Rail Transit has to some extent changed the life style of the residents by means of changing the transportation means of the surrounding areas; affect the functional organization of the city and the tendency of industry-location selection; propose new requirements on the land use of surrounding areas, industrial function layout, service network and corresponding transportation system [5].

#### 1.2.1.2 Relevant Phenomena and Issues

Faced with new opportunities and challenges, rail transit network planning and other relevant city planning, such as city space planning, comprehensive transportation planning, industrial development planning and housing construction planning, are lacking in mutual communication and integration. It directly hinders the achievement of urban comprehensive development and effectiveness of rail network operation.

#### 1.2.1.3 Research Focus and Field

Therefore, we are in urgent need to apply relevant theories of transit-oriented development (TOD), theories of rail transit with city integration (TOD) theory, as well as the domestic and international experience to the study of rail transit and the future integration strategy; research the method of achieving the strategy of urban comprehensive development with the assistance of rail transit; research and integrate rail transit with city planning, coordinate rail transit and urban transportation planning, optimize the series model of city industrial layout planning under the guidance of rail transit.

#### 1.2.1.4 Case—Line No. 1 of Wuxi and Spatial Structure [6]

Wuxi put forward the target of building "Five Centers" and "Five Famous Cities", of which the major functions of recent downtowns will be undertaken by 16 functional carriers. The 16 major functional areas are the major carrier to realize the macroscopical development strategy of Wuxi. At the same time, the under-construction Rail Transit Line 1 connects 7 functional carriers, namely: Yanqiao District, Xizhang District, commercial and business areas of central cities, ancient canal-featured area, Taihu new town CBD, Taihu Science and Education Industrial Park and Taihu International Science and Technology Park. In the mean time, the Line No. 1 connects Huishan New Town and Central area and Taihu New Town in series from north to south, undertaking major commercial, business, administrative, productive services and residential functions. The rails will enhance the functional level, increase strength of development, and intensify mutual connections. Whether the rail transit can integrate with the surrounding territory and functions is a key factor in achieving the macroscopical development strategy.

Based on the adjustment of spatial form, as the Wuxi Rail Line No. 1 suggests, modify the functional structure of the land and space around metro stations, making the land of highly-efficient use, particularly public infrastructure use, set along the rail transit. Through adjustment, increase the utilization value of space, the effectiveness of investment, at the same time, increase the passenger flow and earnings, even beneficial to the development of the land and space around the rail transit. Based on relevant research of Wuxi Rail Transit Line No. 1, the following adjustments are conducted to the surrounding land and space: Adjustment 1: make the land occupied by residents and public facilities approach and concentrate nearby the rail transit. Increase the percentage of high-correlational land use in the land and space around metro stations. Adjustment 2: On the premise of the same total quantity, optimize the layout of public facilities; give priority to the high-grade surrounding site combining with the available land lots. Adjustment 3: consummate the functional attributes of land use, increase the corresponding traffic for the land use of comprehensive development—develop land of hybrid functional use, guarantee the land control for comprehensive exploitation. Increase commerce-residency and commerce-office mixed functional land for the key regions, reserve land with flexible layout for comprehensive exploitation.

After adjusting the land use planning, forecast results of the passenger flow show: In 2020, the daily passenger flow of Line No. 1 is about 378,000, with 20,000 more people after adjustment; the total travelling volume of the 500-m station area of influence rises by 11.6 %, accounting for 60.5 % of the volume of rail travel, rising by 6.3 % than the existing regulatory plan; the effects are remarkable, reflecting the advantages of the new land use plan in improving rail-using efficiency and ameliorating the service of residents' travelling.

## 1.2.2 Rail Transit and Investment and Fund-Raising of Infrastructures

### 1.2.2.1 Development Tendency

Large-scale rail transit construction has put forward higher requirements to the city's financing and investment and fund-raising system of public services. On one hand, public financing needs to raise more capital fund for the construction of new rail transit lines; on the other hand, it also needs to take on the consistent losses of the operation of the existing rail transit [7]. The experience of the cities which have already built rail transit shows: if rail transit fails to hold the surrounding properties and lacking in comprehensive development with the surrounding properties, it will be quite difficult to achieve the operating fund balance only by the ticket income [8]. Moreover, along with the expansion of the rail transit network, the loss will continue to expand as well.

### 1.2.2.2 Relevant Phenomena and Issues

In reality, the developed first-tier cities are relatively wealthy and there are a variety of policy supports. As a result, they can bear the loss to a certain extent. Therefore, in the current rail transit cities, the surrounding space and land cannot be reasonably and fully utilized. The urban space and land cannot achieve proper, sustainable and recyclable development benefits. The rail transit system construction cost and the loss of operation and management of a lot of developed first-tier cities are taken on by public finance, becoming a heavy burden as well as a "special case" which cannot be emulated for the second-tier or third-tier cities who want to join the rail transit club. As rail transit construction enters the second-tier or third-tier cities, from the east coastal cities to western underdeveloped cities, the rail transit losses will increase greatly in proportion with the finance of the cities. Accordingly, this kind of losses will lead to a serious, long-term continuing financial pressures and financial risks.

The successful experience of Hongkong metro and Japanese rail transit shows: It's a necessary guarantee for the rail transit investment and fund-raising system to combine the construction of rail transit with the surrounding properties and realize the effective integration and comprehensive utilization of the transportation and the city. Both the sustainable rail construction and sustainable rail operations depend on the effective integration of rail transit and surrounding properties. On one hand, through the highly-efficient land use, achieve the efficiency and commercial value of the space and land around the metro stations. On the other hand, by means of rail transit subjects, participate in a variety of ways in the Level 1, Level 2 and Level 3 development of the aboveground and underground space and land around the metro stations. This can also make the public sector and subjects of construction and operation obtain more rail transit land value and premium efficiency. Enhance rail transit capital by means of properties development as well as make up the operating loss [9]; guarantee the sustainable development of the rail transit system whose investment and operating costs are highly (Table 1.1, Fig. 1.2).

**Table 1.1** Annual capital fund plan of Nanning Rail Line No. 1 and No. 2

| Projects | Total investment (100 million RMB) | Investment plan (100 million RMB) | | | | | | | | | |
|---|---|---|---|---|---|---|---|---|---|---|---|
| | | 2008 | 2009 | 2010 | 2011 | 2012 | 2013 | 2014 | 2015 | 2016 | 2017 |
| Line No. 1 | 111 | 0.52 | 16.02 | 16.79 | 18.13 | 18.99 | 19.39 | 10.58 | 10.58 | 0 | 0 |
| Line No. 2 | 80.5 | 0.00 | 0 | 0.37 | 11.62 | 12.18 | 13.14 | 13.77 | 14.06 | 7.68 | 7.68 |
| Total | 191.5 | 0.52 | 16.02 | 17.16 | 29.75 | 31.17 | 32.53 | 24.35 | 24.64 | 7.68 | 7.68 |
| Financial income of Urban Central Areas (UCA) | | 92.89 | 120.46 | 156.10 | | | | | | | |
| Percentage in the financial income of UCA | | 0.6 % | 13.3 % | 11.0 % | | | | | | | |

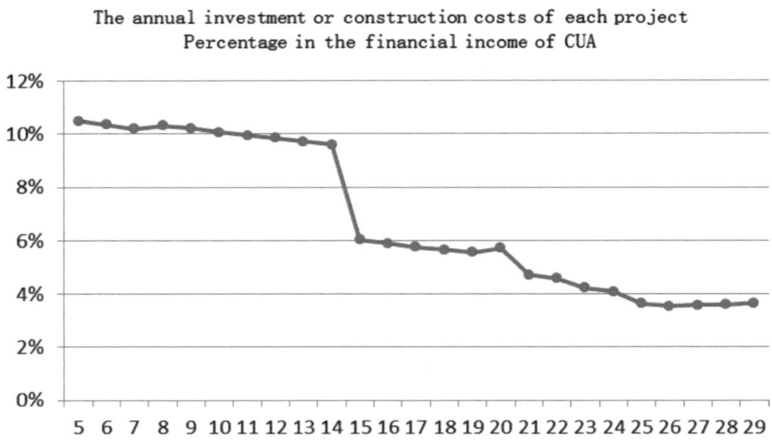


**Fig. 1.2** Investment of Suzhou Rail Line No. 2

#### 1.2.2.3 Research Focus and Field

Taking the successful experience of the rail transit development and construction in Hongkong and Japan as reference, we desperately need to apply relevant domestic and foreign practical experience into our rail transit development, namely, integration of rail transit and the investment and fund-raising, comprehensive development of rail transit and surrounding properties; how to raise the massive amount of funds for rail transit development by exploiting the land and space around metro stations; how to make up for the rail transit operation losses by exploiting the land and space around metro stations; discuss the variable ways of land fund-raising as well as the ways in which private capital get involved in the rail transit construction, for example, Model PPP, Model PFI and Model BOT; examine and draw up the strategies and methods for reserving and transferring the land and space around metro stations; examine the institutional obstacles, model renovation and practice during the city's investment and fund-raising process.

#### 1.2.2.4 Case Analysis—Benefit Evaluation of the Surrounding Land and Space of Nanning Line No. 1 [10]

According to the relevant research on Nanning: Despite the fact that Line 1 mainly passes through the built-up area of Nanning central district, still through thorough classification, modification, feasible analysis and cost evaluation, we can attain exploitable land with a certain amount of profits. Particularly, the outer area of the central district which connects with the rail transit is the main part where the most benefits can be attained. Also, it's the major area which can be used to supplement the capital of rail transit construction. Provided effective reserve is conducted on the major surrounding land and space during the proper time of rail transit development, the added-value of the surrounding land and space caused by the rail transit will be recycled to the maximum, even laying a solid foundation to reserve the capital for the long-term and sustainable rail transit construction.

## *1.2.3 Comprehensive Use of Rail Transit and Land and Space around Metro Stations*

### 1.2.3.1 Development Tendency

In theory: after the introduction of rail transit, it exerts variable influence on the urban land utilization and spatial development through the variability of the land accessibility [11], including: in the aspect of land utilization, the increase of public service facilities will grow from relying on the artery to rail traffic corridor and the surround site distribution; in the aspect of traffic organization, the urban transport system will transfer from road transport dominance to the , multi-model integrated system in which pedestrian and rail transit function as the core; in the aspect of urban open space, the introduction of rail transit will greatly promote the development of city pedestrian system and open space system, creating the possibility of the construction of city open space network [12]; in the aspect of the intensity of land development, urban land development intensity will change in accordance with the layout of the metro stations, accumulating around them; in the aspect of land development model, the introduction of urban rail transit has injected new vitality to the development of node areas, making the spatial integration of urban transit node areas an inevitable outcome [13].

### 1.2.3.2 Relevant Phenomena and Issues

At present, the development and construction of the land and space around metro stations don't quite achieve the basic goal of comprehensive development and integrated use at home. Most of the built-up rail transit and relevant facilities at the moment are installed separately from surrounding land. There into, certain metro stations deliberately evade from urban development and other functional systems in order to avoid cross property right and management disputes. Consequently, the metro stations cannot perfectly achieve the seamless connection and integrated development with the surrounding urban space. Under some circumstances, metro stations can't even well transfer or connect with other matching transport means. Rail transit system has become an independent system isolated from urban comprehensive transit network in aspects of space, function and transit system. The consequence is that consumers are suffering from this terrible design: walk by themselves through the systemic gaps which shouldn't have existed and waste the consumers' precious time. The difficulty of change and connection and the last 1 km obstacle have become a common phenomenon and problem of rail transit facilities. It tremendously reduces the appeal of rail transit to the residents, in reverse, boosting the urban mechanization and overuse of automobiles.

Simultaneously, there exists another tendency in the cases which attempt rail transit comprehensive development and transfer payment of property benefits: due to the lack of systematic development and construction plan, the exploitation of the land and space around metro stations focuses on commercial and service functions, presuming that comprehensive development based on rail transit is equal to station construction plus commercial and property development. On one hand, as the commercial service based on metro stations fails to interact or coordinate with relevant urban planning, hindering the standard layout of urban public service facilities, even the city's overall functional structure. In some cases, owing to the lack of overall coordination, the planned business circles and the ones which are built based on metro stations compete with each other rather than coordinate. It damages the proper urban supply and demand relationship, obstructs the role of land efficiency and lessens the spatial value of the development of metro stations. On the other hand, without an urban planning guaranteed by the force of law, the service property based on the metro stations has trouble in coordinating development and functions with surrounding land of a wider range, even more trouble in connecting and communicating with surrounding business and service facilities. In some cases, the isolated development based on the metro stations fails to form a business district effect due to cramped land and limited scale. The level of commercial facilities is fairly low and the due value has been weakened. Moreover, when some construction bodies are developing the service property based on metro stations, they overlook the integrated construction of supporting transport functions connected with rail transit. It leads to the difficulties of rail transit connection and at the same time, the reachability of facilities and commercial passenger flow are also affected.

The formation of these phenomena in some cases is caused by the interests-led differentiation between construction bodies and development bodies. In some first-rated regions, "traffic" is the main goal of the construction of rail transit. Therefore, the rail construction bodies rarely have the pressure of capital and operating costs. The construction and operating bodies of rail transit usually consider simple construction and operation of transport facilities as the main target, paying less attention to the connection and integration with other systems, nor do they increase the cost of construction and operation of the land and space around metro stations or the interests incentives of collaborative development risks; in other cases, it is due to the lack of systematic theoretical guidance, pre-planning and technical analysis on the development of the metro stations.

#### 1.2.3.3 Research Focus and Field

Consequently, we are in urgent need to base our research on the comprehensive development and proper use of rail transit and the land and space around the metro stations, combine with relevant domestic and international practical experience, explore the development process of rail transit, guarantee the comprehensive output mode of public benefits and development benefits in the process of comprehensive development, explore and configure the land functions of land and space around metro stations and the types of operation, TOD all-fronts development and planning metro stations as well as Hub Integrated Positioning Method, a matched form between differentiated types of operation and integrated transit, develop in a multi-model way on the land control, land reserve and transfer strategy; explore and confirm the transportation functions of the land and space around metro stations and the method of matching scale, explore the methods to effectively control the development of the land and space around the metro stations.

#### 1.2.3.4 Case—Land Planning Adjustment and Comprehensive Station Positioning of Nanning Line No. 1 [14]

During the comprehensive evaluation of Nanning Rail Transit Line No. 1, aiming at the 436 key plots in the influential area along the line, combining with the Basic Premium Rating Chart of Nanning Land, using Nanning real estate price current-time dynamic data of commerce and office etc. of the areas along the line from 2003 to 2009 provided in the database of real estate transactions, short-term and long-term evaluation on the development potential of the commercial service functions of the available land resources along the line are conducted. Further clarify the possibility of the development of business and service along the line in the short term, middle term and long term based on the outcome of evaluation, thusly, providing significant support to the optimization of the developmental function direction for the land and space around the major metro stations.

In order to better achieve the transfer efficiency of the rail transit in comprehensive transfer system, relevant research of Nanning Line No. 1 integrates and optimizes the grades and construction model of the rail transit hubs. For example, the layout of the modified hub distributes as follows: First-class hub: four, namely, Xixiangtang Passenger Station, Nanning Railway Station, Liangdong Passenger Station and Nanning Eastern Station. Second-class hub: seven, namely, Shibu Station, Chencun Station, Guangxi University Station, Village Ma Station, Five-elephant Square Station, Dongmeng Commercial District Station, Yunjing Road Station. Third-class hub: fourteen. The hub operational strategy of the land and space around metro stations after modification includes: Coordinated development of transportation and commerce (T+C): 7, namely, Xixiangtang Passenger Station, Village Chen Station, Zoo Station, Xinwei River Station, Guangxi University Station, Yunjing Road Station, and Nanning Eastern Station. Coordinated development of transportation and residency (T+R): 2, namely, Shibu Station and Gaopoling Road Station.

### *1.2.4 Rail Transit and Integrated Development of Underground Spatial Resources*

#### 1.2.4.1 Development Tendency

Systematic development and multi-functional utilization of underground space is a significant characteristic of the land and space around rail transit underground metro stations. Rail transit has become a tremendous opportunity for the development of urban underground space, in which commercial value and transport value get promoted unprecedentedly [15]. Metro stations as the core, there arises a developmental base for underground commercial network and transport network layout, for the underground rail attracts large quantities of consumers, enabling the underground space to acquire, even to replace the aboveground as a major source of urban functional passenger flow, promoting greatly the commercial value of the underground space. Simultaneously, the rail transit space which is located in the road intersection and plot connection can effectively integrate scattered underground commercial walking system, even aboveground walking system, exerting the underground commercial network and constellation effects, as well as the aboveground. As a result, it has become a significant principle to integrate the surrounding land and space by seamlessly connecting the underground spatial network of central business area with the comprehensive transfer space of the rail transit metro stations through unified pathways and connections.

Along with the construction of underground rail transit, sections, stations and every other kind of corresponding lines tend to be the self-owned space and direct development object of rail transit bodies as well as an essential helper of the development and utilization of the underground space along the line. Among the new upsurge of rail transit construction, the main construction bodies all fully realize the value of this underground space, expecting to achieve development benefits and operation compensation through the utilization of underground space. Therefore, under the double promotion of interest incentives and practicability guarantee, the development and construction of the underground space along the line increases rapidly. The development and use of the underground space provides significant opportunities for the network development of urban underground space.

#### 1.2.4.2 Relevant Phenomena and Issues

At present, as systematic development and commercial development of the urban underground space are still at the exploratory stage, multi-aspect matching strategies are still unsound, such as corresponding planning theories, design methods, planning tools, ownership configuration model and collaboration mechanism, incapable of fitting into underground spatial network, integration and comprehensive utilization resulting from rail transit construction, making the achievement of proper use and effective configuration of resources quite difficult.

For example, in some cases of rail transit underground spatial construction, the lacking in a full understanding of the inside and outside space, public plots and development plots leads to the result that during the rail transit, particularly underground lines development and construction, underground spatial resources cannot be utilized, some high-value underground space fail to achieve proper functional configuration.

Take another example: in some other cases, without scientific and systematic methods to forecast development demand and supply, the development scale, functional types and layout of types of operation of underground space all lack the market demand and supply relation as analytical foundation as well as rational and scientific forecasting methods as analytical basis. Phenomena like blind development and over use emerge in the process of massive rail transit construction and development. It leads to the decrease of the actual usage efficiency of underground space, at the same time, the underground space developed with high costs are in idle state. Space value fails to come into actual effects and investment performance of construction slides tremendously. In some areas, certain underground development projects which were supposed to bring about benefits and compensate for the operational losses become heavy financial burden.

In the mean time, the more prevalent phenomenon in underground space development is: each underground spatial development body, including rail transit, usually operates on its own and performs independently, failing to achieve effective connection and communication; underground space is fragmented and underground spatial network cannot be formed as well. In this way, the business circle functions or network effects cannot be formed, nor can the target of making up for the urban functions and transport network through underground space be achieved, highly reducing the economic and social benefits of underground spatial development. This is exactly the result of absence and insufficiency of management, institutional guarantee, pre-planning and effective communication in the underground spatial development and construction.

#### 1.2.4.3 Research Focus and Field

Consequently, based on the relevant domestic and foreign practical experience of research on the multi-functional and integrated development of rail transit and surrounding properties, explore scientific methods of underground spatial need forecast, types of operation analysis and network planning in the rail transit developing process; explore the issues like coordination and benefits-distribution of underground space development and relevant institution establishment under the condition of multi-agents; explore how to conquer the distribution with development benefits as the subject and form an integrated underground spatial development network with metro stations at the core; explore how to achieve the staged construction and integration in aspects of technology and planning; explore how to solve the collaboration and benefit distribution mechanism of multi-agent underground space development; explore how to resolve the technical issues (air defense and fire control etc.) in the comprehensive development and use of underground space; exploring the institutional barriers and innovation in the long-term underground space development will become the focus and hot issue in the field of urban planning and engineering design.

#### 1.2.4.4 Case: Integrated Use of Underground Space in the West Railway Station Area of Jinan Metro Line No. 1 [16]

The region between the West Railway Station and Cultural Center Station in Jinan Rail Transit Line 1 is the core area of the western sub-center in Jinan. To this end, the region formulates special regulatory plan for the development and use of underground space. The planning program, through the Sub-index Prediction Method, Location Model Method, Case Analysis Ratio, Inverse Market Value Checking Method etc., predict developmental needs of the underground space in the region. Multiple indicators are predicted like the total size of development, functional categories and format layouts. Developmental experience shows: on one hand, the prediction results fully reflect the location superiority and land value, on the other hand, they also demonstrate the relation between demand and supply as well as the reality of development and construction. In the mean time, as the subject of the underground space development in the local area is very complex, including the Railway Bureau, orbital companies, government public sectors, developers, air defense and municipal administration, the plan particularly emphasizes underground space interconnection and integration under the condition of multiple subject. By means of detailed and specific multi-interface, multi-node and multi-period docking control, realize the business circle function and the network value of underground space, greatly enhancing the economic and social benefits of underground space development.

## *1.2.5 Station Design Optimization Under the Target of Integrated Development*

### 1.2.5.1 Development Tendency

Urban development strategy implementation based on the rail transit; rail transit financing balance based on the development and utilization of land and space resources; optimal utilization of land resources and effective development of the underground space resources, with integrated transfer system as the core, depend on the line selection and optimized station layout of rail transit network, and the optimized design of lines, stations and facilities. Therefore, station optimization design based on comprehensive development and integrated utilization is an important foundation for the realization of rail traffic micro, meso and macro economic and social benefits. The TOD and DOT oriented station optimization design task is an important procedure to realize the goal of integrated development system.

Therefore, comprehensive development of multi-target, multi-body and multi-system raises new requirements and challenges to the traditional design method of station selection and design. In the practical level, only by optimizing station selection plan and station design can rail transportation space resources get fully excavated; only by increasing the value effects of space resources can we guarantee the goal of comprehensive development through engineering and technical means.

### 1.2.5.2 Relevant Phenomena and Issues

However, the station selection in reality not only pays attention to traffic volume, more attention is paid to interval setting, consideration of construction cost, consideration of land use and feasibility of engineering; station design in reality is more dependent on the mode design of standard stations, ignoring the land use feature of the surrounding space, ignoring the actual demand of comprehensive development to facilities layout, docking requirements, and access settings. As a consequence, under the condition of comprehensive development, station and developmental function are very difficult to achieve seamless connection. Reduce the value of developmental space of to a certain extent, forming subjective obstacles to integrated development and functional integration.

### 1.2.5.3 Research Focus and Field

As a consequence, we are in urgent need to combine with relevant international and domestic experience of rail transit and the comprehensive development of surrounding property, and explore the method of optimizing station design from the perspective of urban development; explore technological method to optimize station selection and the means of station optimization, interval vertical design, layout, access setting, equipment and layout of ancillary facilities; explore key technical problems of comprehensive development and urban functional construction by means of parking and vehicle depot as well as turn line intervals.

#### 1.2.5.4 Case: The Optimized Design of Guangda Station of Changsha Metro Line No. 2 [17]

Guangda Station of Changsha Metro Line No. 2 is the easternmost station of Line 2, located east of Changsha South Station of the Wuhan-Guangzhou high-speed railway. It's the only recently-implemented rail transit station of the southeastern peripheral of Changsha Center City and southern area of Changsha County. It's also the most important public transit node for these areas to enter the core area of Changsha by rail transit. However, the existing Guangda station is a typical shallow station located within the land. The transfer between the rail transit and other public transport is conducted by the ground and in the open air. In the mean time, at the important node, the use of upper space is not taken into consideration, which shows poor land-use efficiency. As a consequence, the research and analysis of Guangda Station firstly analyzes its functional positioning and clarify its responsibility as an important transport node and developmental needs as large commercial complex of the suburb. Next, the program optimizes the vertical design through processing standard elevation of each layer and increases the comprehensive transfer layer of the basement floor, achieving a tridimensional seamless transfer of rail transport with other modes of transport. Furthermore, the program sets development of upper space as a starting point and optimizes various equipment of the station, channels, entrances, exits and construction method, which realizes the combination of large commercial complex with metro stations. Achieve the philosophy of comprehensive development of rail transit and integrated design with engineering practice.

### *1.2.6 System Innovation and Guarantee in Integrated Development*

#### 1.2.6.1 Relevant Phenomena and Issues

Institutional building is the foundation and important guarantee of the comprehensive development of rail transit and the comprehensive utilization of the aboveground and underground space along the line. However, constricted by the current system of land use, urban planning system and building use system, the current comprehensive development work is still in discussion and practice in various cities within a small area [18]. In some cases, the high commercial value generated by the rail transit makes various development subjects have comprehensive development needs to the land and space around the metro stations driven by interests.

Nevertheless, there are some conflict and limitation due to law and related regulations to integrated development based on stations, especially the integrated development of underground and aboveground. At present, facing the development impulse, a majority of the development cases take on the way of "each discussion to each issue", resorting to the planning leaders for approval on specific projects. This is obviously a mode of operation lacking in security and long-acting mechanism, bringing more uncertainty and risks of system for comprehensive development in the future.

#### 1.2.6.2 Research Focus and Field

Therefore, based on the international and domestic experience of system guarantee and management innovation in rail transit and comprehensive surrounding property development, integrated utilization of underground and aboveground space, we are in urgent need to explore the principle of optimizing the system of land management under the condition of the comprehensive development; explore the condition-setting form of public interest expression under the condition of comprehensive development in the procedure of Land Bid, Auction and Listing; explore how to connect the public with the private and the effective implementation method to protect the public interests in the construction of specific projects; explore the mode and experience of joint management and common maintenance in the process of comprehensive development, combined construction and public-private participation; explore specific measures and suggestions on optimization of land ownership, management supporting with public and private collection, solve the final 1 m matching, layering development and ownership optimization, coordination of multi-interests subjects and development incentive innovation with public interest orientation and other similar aspects.

## 1.3 Conclusion

To sum up, in order to achieve the coordination and sustainable development of the city and rail transit, interdisciplinary research and in-depth exploration of the two specialized areas is very necessary and urgent. Research of this field is not only the important research field of city planning, but also the important content of architectural design, engineering design, city policy, project management and other research fields. More important, in this special interdisciplinary research, all kinds of close and complex relations of mutual influence and restriction exists between the different research fields. Accordingly, the research of this field needs more multidisciplinary collaboration.

Overall research framework diagram of the comprehensive development of rail transit:

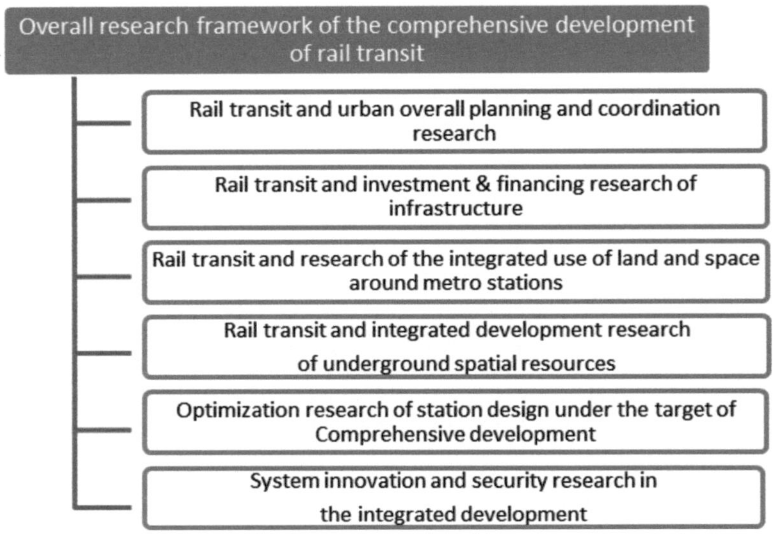


**Acknowledgement** Supported by National-sponsored Funding Program for Social Sciences of China (13&ZD026), Supported by National Natural Science Foundation of China (51108018).

## References

1. The outline of the eleventh five-year program for the national economic and social development of People's Republic of China [R]. http://www.gov.cn/gongbao/content/2006/content_268766.htm. March, 2006
2. Xinhua Daily (2010) Experts advise that urban rail transit pay less attention to image, more to service [N]. Xinhua Daily, 19 March 2010
3. Pan Haixiao, Rail transit and optimization of spatial structure in metropolitan. Collected papers and international forum on urban development and planning in 2008, pp 25–34
4. Cai Guotian, Chen Zhongnuan (2004) The influence of rail transit on the spatial layout of Guangzhou retail commerce [J]. Mod City Res (4):65–67
5. Qin Yingbing (2000) Influencing factors analysis of urban rail transit on the urban structure [J]. J Southw Jiaotong Univ (3):284–287
6. Yuan Lu etc, TOD research of the land and space around Wuxi Metro Line 1, Line 2. Research report 2008, pp 72–73
7. The Beijing News (2011) Beijing subsidized bus and subway industry 13.5 billion to maintain low fares last year [N], 2 January 2011. http://news.qq.com/a/20110102/000057.htm
8. Shen Lixia, Tan Guotian (2006) Analysis of the influence of urban rail transit on surrounding properties [J]. Urban Rail Transit (4):49–55

9. Wang Yuguo etc (2004) Comparison and evolution of the investment & fund-raising mode of urban rail transit [J]. J Beijing Jiaotong Univ (Soc Sci Ed) 3(4):16–21
10. Spatial development and utilization planning of the underground space in Nanning, 2009
11. Pan Haixiao, Ren Chunyang (2005) Spatial coupling relation between Rail transit and the urban public activity center system – a case study of Shanghai [J]. J City Plan (4):76–82
12. Li Guodong (2009) Research on the design of subway urban space in the central areas of mega-cities [D]. Xi'an University of Architecture and Technology, Xi'an
13. Song Minhua, Feng'aijun (2005) Underground space development relying on urban rail transit [J]. Urban Rapid Rail Transit (1):21–25
14. See Ref. [11].
15. Lai Dongfeng (2007) Underground commercial space development combined with subway construction [J]. Urban Rapid Rail Transit (2):77–79
16. Utilization of underground space planning in Jinan West Railway Station, 2010
17. Planning and architectural design programs of the space around Guangda Station in Changsha, March 2011
18. Luohan Bin, Qi Shenjun (2007) Study on the joint development policy mechanism between rail transit and real estate [J]. Constr Econ (10):42–45

# Chapter 2
# Explore the Impact of the Metro Station to the Surrounding Area

**Lingru Yang**

**Abstract** The development of rail transit for the impact of urban land use has become increasingly prominent, and researching the influence of rail transit on urban development has become one of the hot topics of the moment. In the city level, the metro is the artery of urban development, so the reasonable selection of metro station becomes an important part for the regional development and construction. Well, what is on earth the interactive relationship between the station and the surrounding area? This study is based on the thinking of the question, and selects the Shengtai Road Station on the Southern extension line of Nanjing Metro Line 1 as an example, analyzing the impact of the metro station to the development of the surrounding area from different levels and scale on the basis of spot investigation and data analysis. Trying to draw the law of how the rail transportation guide urban development, as well as exploring that what factors should be considered to the construction of the station surrounding areas and which road the regional development should take.

**Keywords** Metro station • Metro line • Urban development • Surrounding area • The land use

## 2.1 Background

Urban rail transit is a kind of urban passenger transportation which is featured as quick, efficient, energy saving, environmental protection and high-capacity, along with the development of rail transit, its impact on the development of urban land is becoming increasingly prominent. Southern extension line of the Nanjing Metro Line1 (southern extension line for short) starts from Andemen station stretching

L. Yang (✉)
Urban Planning Major, Southeast University, NO. 2 Si Pailou, 210096 Nanjing, China
e-mail: ylrg@foxmail.com

H. Xia and Y. Zhang (eds.), *The 2nd International Symposium on Rail Transit Comprehensive Development (ISRTCD) Proceedings*,
DOI 10.1007/978-3-642-37589-7_2, © Springer-Verlag Berlin Heidelberg 2014

south to Dongshan new urban area, through the Nanjing station which is under the construction of the Beijing-Shanghai high-speed railway station in Nanjing, across the Yuhua District, Jiangning District and ends at the China Pharmaceutical University Station. The length of the line is 25.08 km. Shengtai Road Station area is an important node between Dongshan Town and the Development Zone, and it was once an important carrier of the Jiangning economic leaps and bounds. With the rapid development of urbanization in Nanjing, Shengtai Road Station area is facing the challenge of transforming. So the opening of the Southern extension is undoubtedly a best opportunity.

## 2.2 Research Summary

This study will give a comprehensive understanding on the basis of spot investigation and data analysis of the situation of Shengtai Road station and its surrounding areas. From the statistical analysis of the data, we will get further understanding of the relationship between Shengtai Road station and its surrounding areas and rail transit.

### *2.2.1 Research Object*

As the circle effect of rail transit station, the main object is a range that uses Shengtai Road Station as the center of the circle, and 1,000 m for the radius of the circle, which is divided into four layers from the radius of 100, 200, 500 to 1,000 m to comparative study. The main focus of the survey is the scale and structure of the land use.

### *2.2.2 Research Framework*

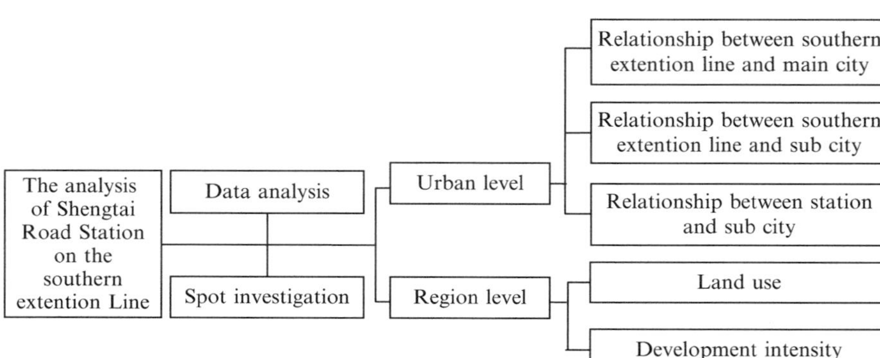

## 2.3 Relationship Between the Metro and Urban Development

### *2.3.1 Study of the Relationship Between Southern Extension Line and the Development of the Main City*

Nanjing is experiencing the process of transforming from the big city to the mega-city, so it has new requirements for Dongshan Sub City in aspects of the urban population, industrial development, center system formation and transport facilities construction. The construction of southern extension line is under this background, expecting to ease the stress of development of the main city by having close contact with the Dongshan Sub City.

The new round of city planning of Nanjing (2010) forecasts that by 2030, Nanjing urban population will reach 11.3 million, which is about twice that of the urban population in 2007. The Main city population density is about 15,000 people/km$^2$, and the Sub City population density is 10,000/km$^2$, the size of the population is more than 500,000 people. The construction of southern extension line can solve the problem of the population movements between the main and sub city.

According to the theory of economic location and distribution characteristic of the modern urban industry, the industry within the city range will show “Circle” layout, while the Dongshan sub city is located at the junction of the first circle and the second circle, so it should focus on the development of modern service industry and high-tech industries. Southern extension line can better promote alternating the evolution of the modern service industry and high-tech industries.

According to the new general plan for Nanjing, the new city center system will change from the original basis of “one main city with multi-sub cities” to “city center – city sub-center – new city center – groups” (regional) city center system. In Dongshan Sub City there are a city center of south Nanjing, the Dongshan sub city center, as well as multiple groups centers. Southern extension line acts as the spine to support the entire Dongshan sub city center system.

It is in this context of development, according to the Nanjing rail transportation network diagram (2050), in Dongshan sub city there will be line 7 (Southern extension line), line 5, and yudai line rail transit as well as high-speed rail hub – Nanjing South Station. This will greatly enhance the locational advantages of the Dongshan sub city (Figs. 2.1, 2.2, 2.3, and 2.4).

### *2.3.2 Study of the Relationship Between Southern Extension Line and the Development of Dongshan Sub City*

Since the end of last century, the development of Jiangning focus on the West Development Zone, and the Dongshan Town centers in the east. The former is the

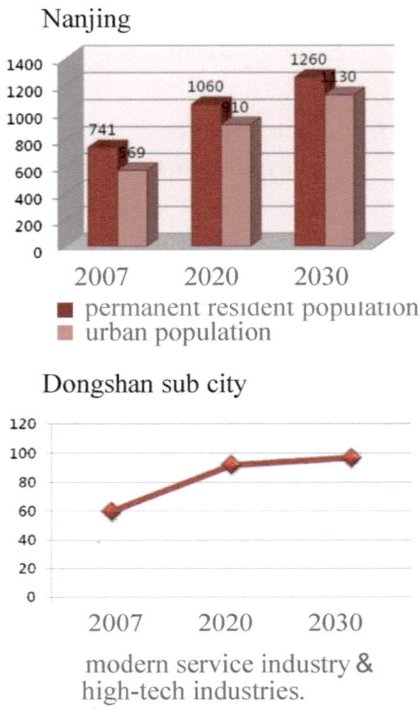


**Fig. 2.1** Urban population diagram

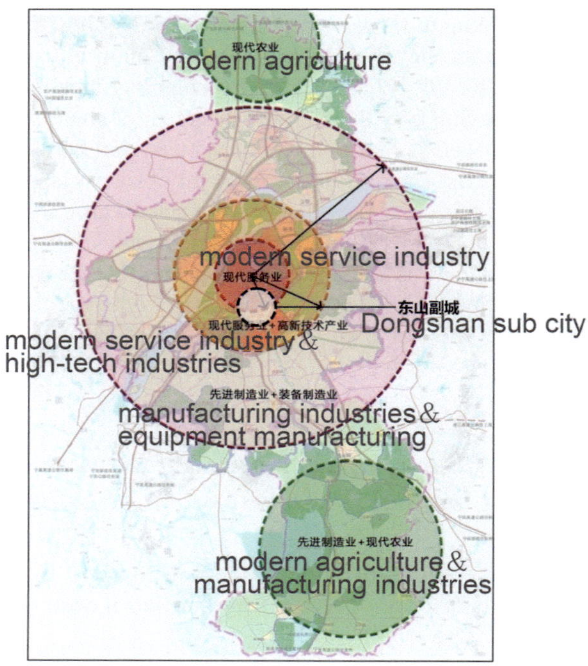


**Fig. 2.2** Layout of industrial development diagram

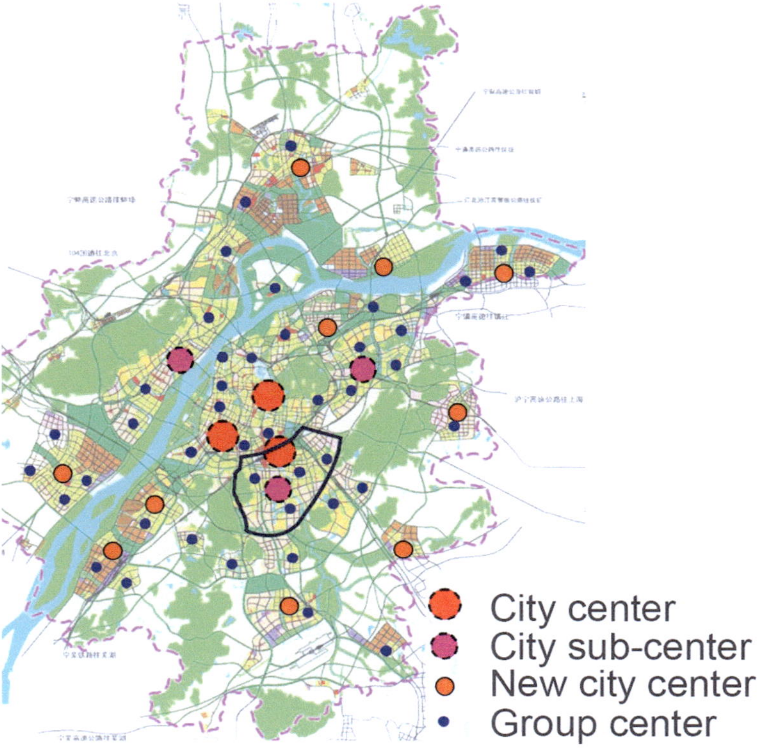


**Fig. 2.3** City center system diagram

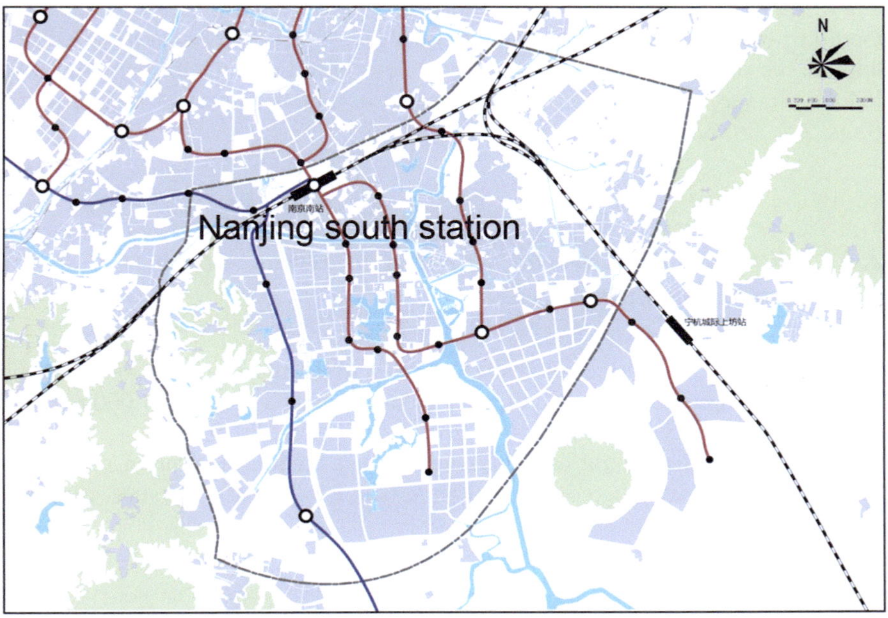


**Fig. 2.4** Dongshan sub city traffic facilities diagram

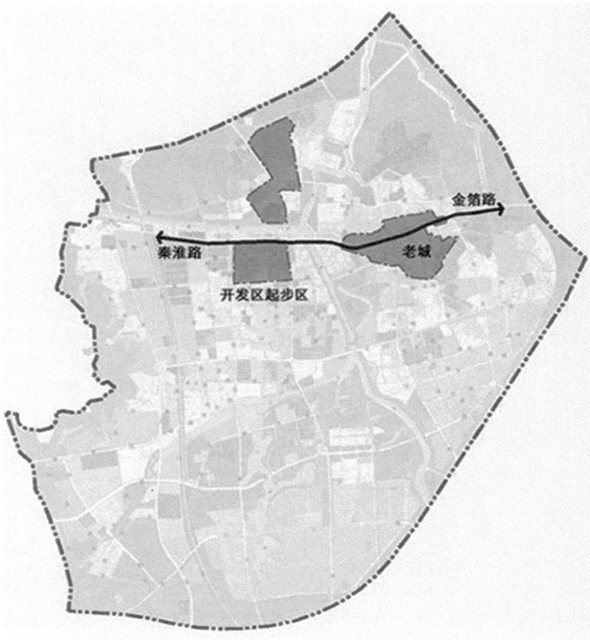


**Fig. 2.5** Period of industrial land use (1992)

important industry pillar in Jiangning district, while the latter is the living supporting center in Jiangning. The situation that the industry and life supporting are separated has been restricting the development of Jiangning City. According to a new round of Jiangning city general planning requirements, the area which is to the north of Baijia Lake, between Shuanglong Avenue and Liyuan Road will be built as the new city center area, and coincides with the southern extension line radiation range, and Shengtai Road is located in a more important node position (Figs. 2.5, 2.6, 2.7, and 2.8).

### 2.3.3 *Study of the Correlation Between Metro Station and Dongshan Sub City*

Shengtai Road Station is located in the northwest of the sub center of Dongshan sub city, near the center of the circle of three groups. Radiation stations 1,000 m range overlaps the sub city hub range.

Besides, near the central of the Shengtai Road Station there are a number of public facilities, including commercial land, land for finance, insurance, commercial comprehensive land and so on.

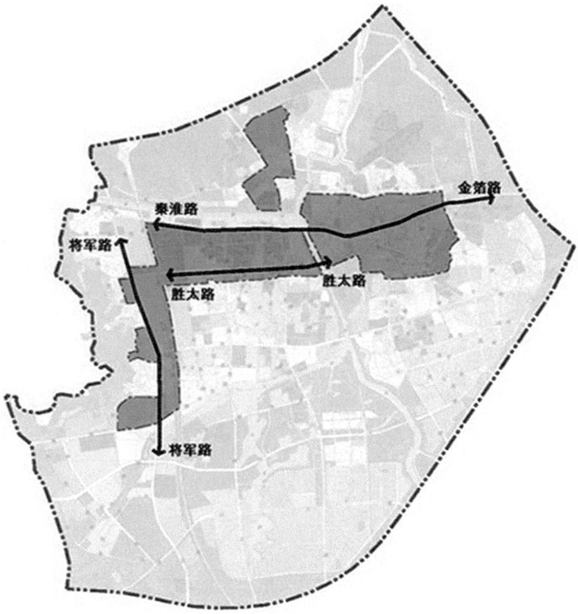


**Fig. 2.6** Period of Jiangning development (1998)

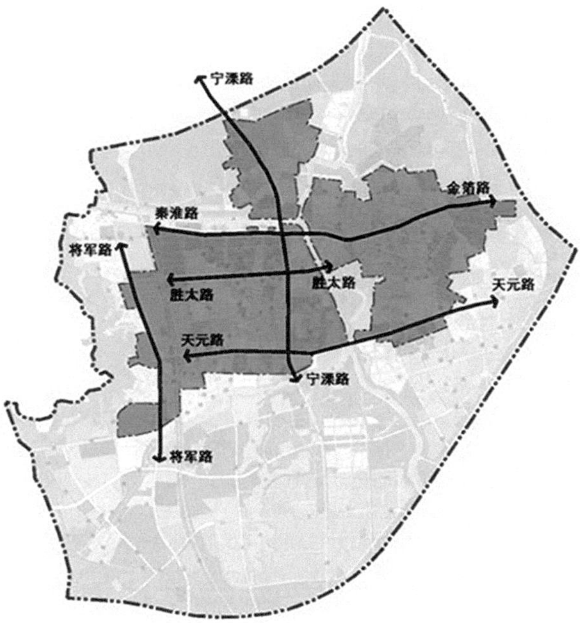


**Fig. 2.7** Period of national development zones

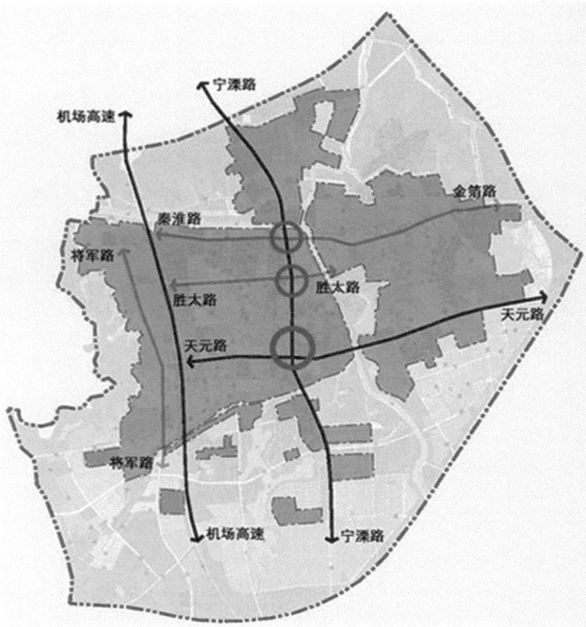


**Fig. 2.8** Period of new city

Within 1,000 m of Shengtai Road Station is the distribution of a number of large settlements, including Lakeside Apartments, Shengtai new apartment, Wangfu garden, Baijia Lake villa garden, artistic apartment and so on.

But in the same context, it is also distributed a large number of industrial land, mainly concentrated in the North of Shengtai Road, West of Shuanglong Road, south of Chitian Road, and East of Shuanglong road (Figs. 2.9, 2.10, 2.11, and 2.12).

## 2.4 The Impact of the Metro Station to the Surrounding Area

### *2.4.1 Impact to Land Use*

For more specific reflection of the impact of the construction of the metro stations to the condition of the surroundings, this study will use the research methods of a combination of vertical and horizontal.

The vertical method will use time as units, setting the characteristics years by comparing the change of land development around the metro stations before or after the construction of metro stations and analyze the trend of the change. As the Nanjing southern extension line opened on May 28th, 2010, this study sets the land use in 2009 which was before the metro station was built as the research object, the

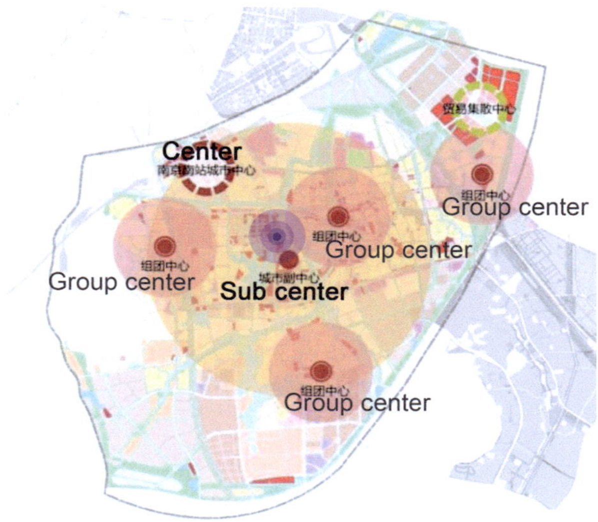


Fig. 2.9 Stations and city center system

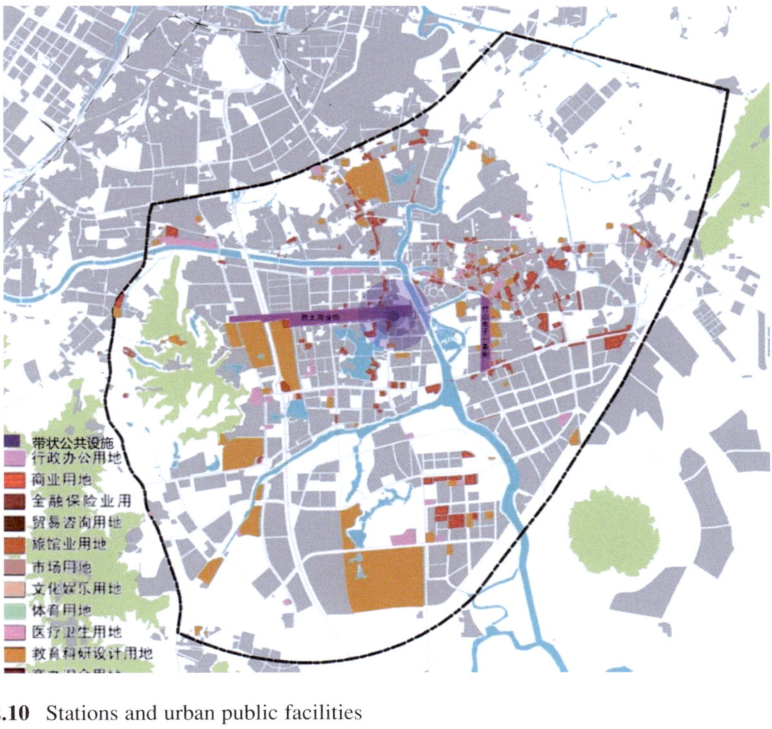


**Fig. 2.10** Stations and urban public facilities

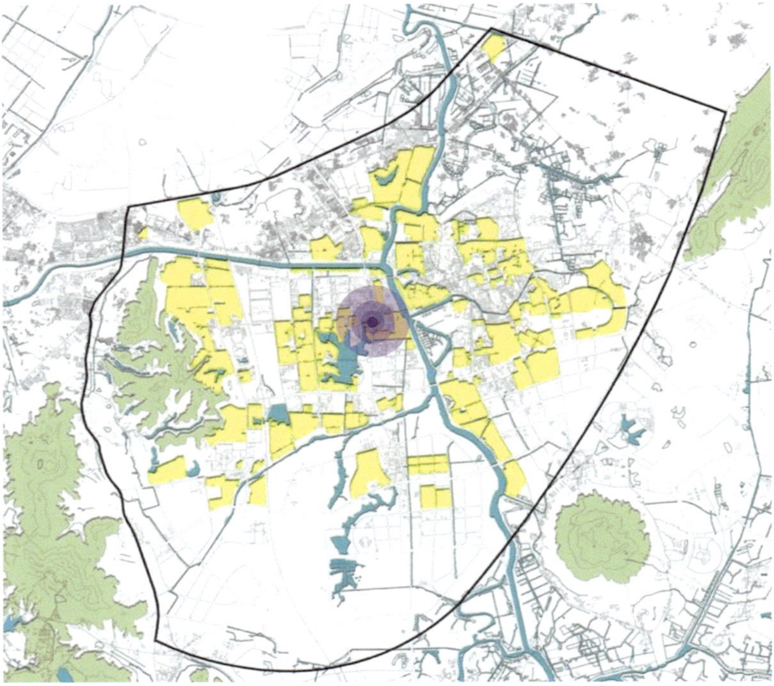

**Fig. 2.11** Stations and urban settlements

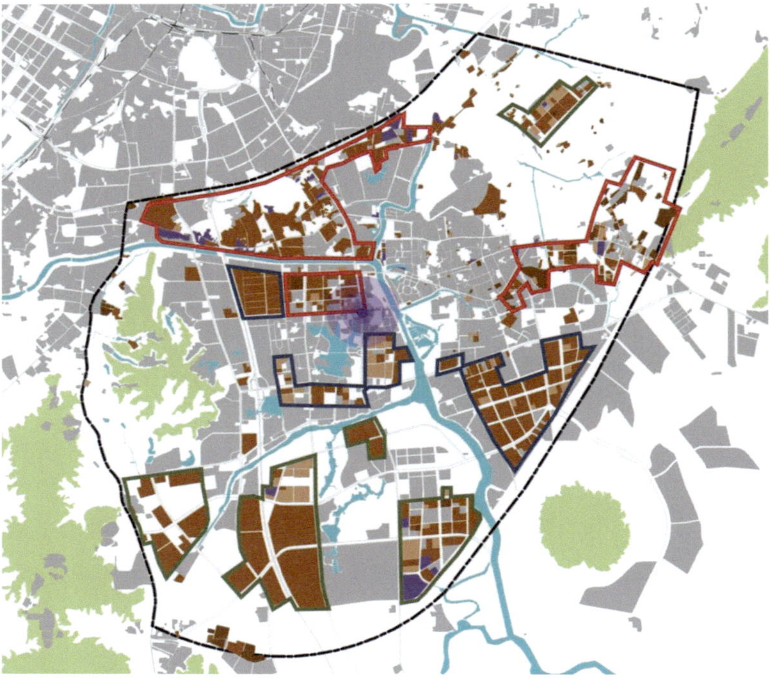

**Fig. 2.12** Stations and urban industrial areas

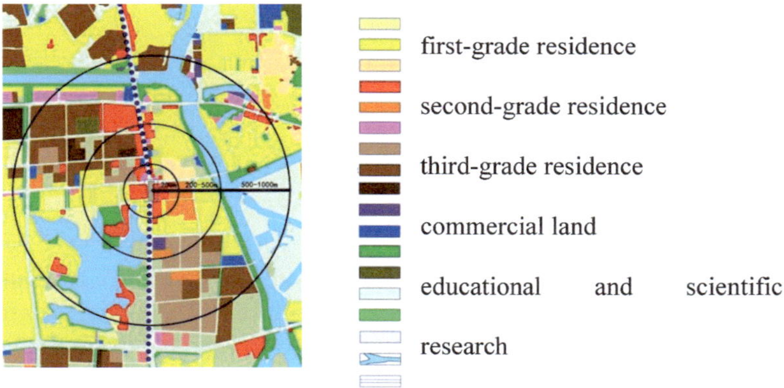


**Fig. 2.13** Present land use situation of the surrounding area

current land use is based on field research, and future land development of the surrounding area around the metro stations will use the controlled detailed planning drawings of the area between 2010 and 2030 years for reference. And through the statistical analysis of the hypothesis to test and verify whether the construction of metro station has significant effects on surrounding land use.

The horizontal method will research by dividing the circle of ring layer. According to the transportation planning theory, the adjustment of the metro station on the nature of the surrounding land can be divided into less than 200 m, 200–500 m and more than 500 m. This study is a comparative study, takes Shengtai Road metro station as the center, and divides outside area into three layers, core layer 0–200 m, inner layer 200–500 m and the outer layer of 500–1,000 m (Fig. 2.13).

#### 2.4.1.1 Land Use of the Surrounding Area

Shengtai Road station is located at the crossroad between Shengtai Road and Shuanglong Street, and it is a commercial street, there is a centralized commercial land existing by the Shengtai Road station. According to land use distribution of the surrounding area, we can see that to the east of Shuanglong Road, most of the residential land located in this area with living vitality; to the west of Shuanglong Road and the south of Shengtai Road, as it's surrounded by the Baijia Lake, it has excellent environmental conditions. There is not a large number of better green beauty of the landscape, but also the distribution of large tracts of residential areas, the construction of the villa area, international hotels, clubs and other recreational architectures; and to the west of Shuanglong Road and the north of Shengtai Road is a central area of science and technology zones. Thus, land use of the station surroundings has various functions, and many kinds of people mixed together, so the metro station as the form of public transport can be full used by the crowd surroundings (Table 2.1).

**Table 2.1** Land use function structure (within 1,000 m ring layer)

| The main land use types | Residence | Commercial | Manufacture | Green space | Transportation | Water | Others |
|---|---|---|---|---|---|---|---|
| Area (ha) | 94.7 | 41.1 | 27.3 | 9.8 | 81.6 | 41.9 | 17.6 |
| Structure (%) | 30.2 | 13.1 | 8.7 | 3.1 | 26 | 13.3 | 5.6 |

#### 2.4.1.2 Distance and Land Use

From the vertical and horizontal comparison and analysis, we can conclude that the station of public buildings in the inner ring is higher than the outer ring, showing the characteristics that public facilities are centralized distribution around the station, and residential land is mainly distributed in the outer ring, this phenomenon reflects the characteristics that the commercial areas are usually located in the intensive rail transit and walk zone, because only businessmen can pay the higher rent, and service radius is about 200 m. Secondly, in the reasonable area of rail transportation, jobs are relatively less, so the main land use is residential land, and the service radius is about 500 m or so. In the periphery of the reasonable pedestrian transport more than 500 m range, there are city park, industrial land, and form the secondary closely related area of rail transit station. According to the research to the land use of the Shengtai Road station and the surrounding area, we can analyze that the metro station really has certain impact on the land use of the surrounding area (Tables 2.2, 2.3, and 2.4).

### *2.4.2 Impact to Land Development Intensity*

#### 2.4.2.1 Land Development Intensity of the Surrounding Area

Shengtai Road Station surroundings is mainly used as residential and enterprise factory, the scenic Baijia Lake accounted for almost a quarter of the area of the scope of the study, which makes the region overall volume rate is not high. Only in the commercial center of Shengtai Road station surrounding area, the complex volume rate is relatively high, and new residential and business office building in the surrounding area of the stations usually have a height of 20 or more layers. We can clearly see the land use development trend of the station surroundings (Fig. 2.14).

#### 2.4.2.2 Distance and Land Development Intensity

The station is the center, radiating outward and dividing the surrounding area into three layers, it can be found by comparing the land development intensity of the kernel of the station to the outer ring: the proportion of the low-intensity

**Table 2.2** Land evaluation of the surrounding area (500–1,000 m ring layer)

| Characteristics years | Before the metro station built (2009) | After the metro station built (2010) | According to the control detailed planning (2030) |
|---|---|---|---|
| Land use map | 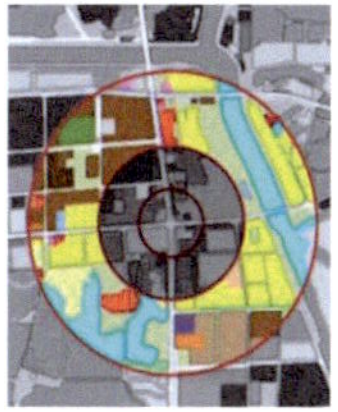 |  | 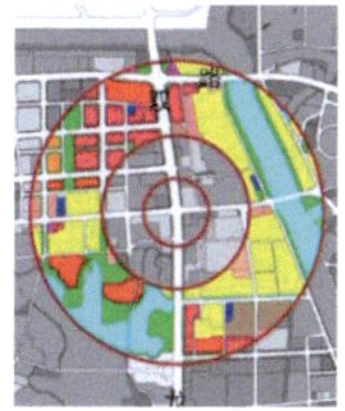 |
| Residence (ha) | 68 | 82 | 84 |
| Commercial (ha) | 7.1 | 12.3 | 21.2 |
| Manufacture (ha) | 36.7 | 21 | 7.5 |
| Education & scientific research (ha) | 3.6 | 3.6 | 3.6 |
| Green space (ha) | 4.7 | 9.8 | 13.8 |
| Street (ha) | 56 | 60 | 64 |
| Water (ha) | 39.6 | 39.6 | 39.6 |
| Others (ha) | 19.8 | 7.2 | 2.2 |
| Comprehensive study | 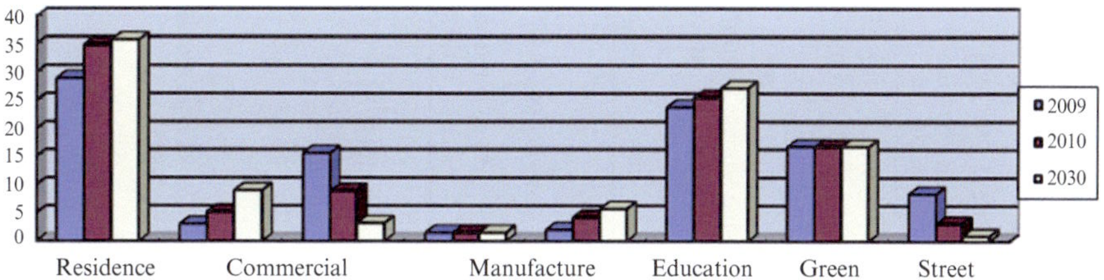  Judging from the outer layers of data and statistics, the weighting of residential land is large, and volume growth trend of development increases, and construction of infrastructure facilities trends also increases, the intensity of land development increases, and the amount of industrial land reduces, environmental construction of green space are valued | | |

**Table 2.3** Land evaluation of the surrounding area (200–500 m ring layer)

| Characteristics years | Before the metro station built (2009) | After the metro station built (2010) | According to the control detailed planning (2030) |
|---|---|---|---|
| Land use map | 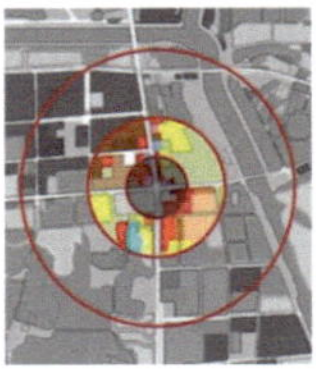 |  | 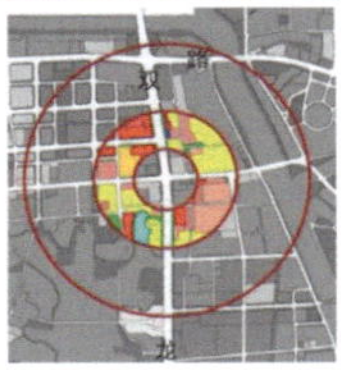 |
| Residence (ha) | 4.4 | 12.1 | 13.4 |
| Commercial (ha) | 11.4 | 12 | 16.4 |
| Education & scientific research (ha) | 8.6 | 8.6 | 8.6 |
| Manufacture (ha) | 7.1 | 6 | 2.9 |
| Street (ha) | 15.6 | 17.6 | 18 |
| Water (ha) | 2.3 | 2.3 | 2.3 |
| Others (ha) | 16.6 | 7.4 | 4.4 |
| Comprehensive study | According to the above statistics, within layers 200–500 m, commercial development decreases, and residential land development intensity is strong, type of land increased, and with the future advancement of the land utilization rate will continue to increase | | |

**Table 2.4** Land evaluation of the surrounding area (0–200 m ring layer)

| Characteristics years | Before the metro station built (2009) | After the metro station built (2010) | According to the control detailed planning (2030) |
|---|---|---|---|
| Land use map | 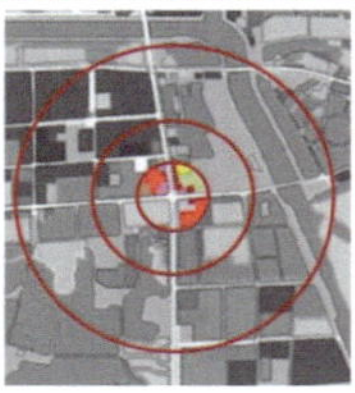 | 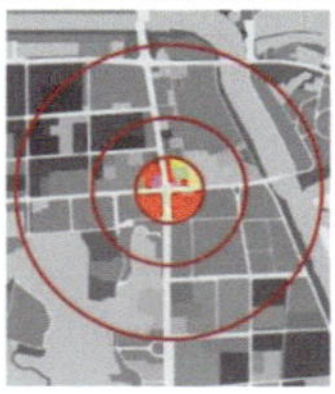 |  |
| Residence (ha) | 0.53 | 0.62 | 0.53 |
| Commercial (ha) | 4.1 | 5.6 | 7.2 |
| Manufacture (ha) | 0.45 | 0.3 | 0.25 |
| Street (ha) | 3.8 | 4.0 | 4.0 |
| Utilities (ha) | – | – | 0.4 |
| Others (ha) | 3.68 | 2.04 | 0.18 |
| Comprehensive study | In the core circle, Residential land and industrial land have a small proportion, public services land is dominated, and has a rapid growth before and after the construction of the metro station | | |

development (FAR $<$ 1) and medium-intensity development (1 $<$ FAR $<$ 3) increased, and the number of high-intensity (FAR $>$ 3) declined. From the current situation of Shengtai Road, the development intensity of the inner ring is higher than that of the outer ring. As the time of Shengtai Road Station construction is

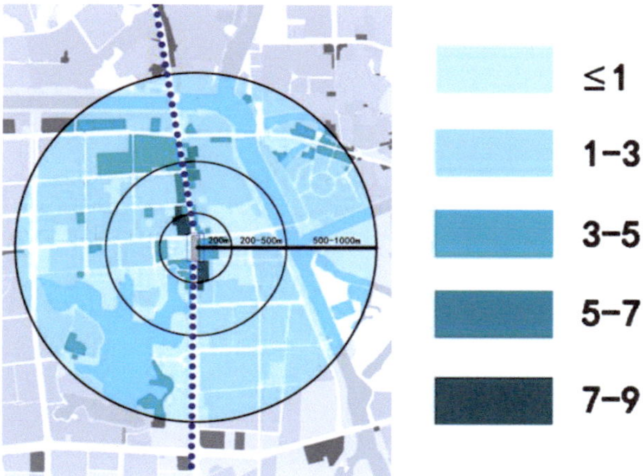


**Fig. 2.14** FAR of the surrounding area

**Fig. 2.15** 0–200 m ring layer

short, we do not have enough data to easily get this conclusion, but from the land use situation of the construction of the station surroundings, this phenomenon is indeed consistent with the principle of land bid-rent curve and reflects that the impact of the transportation to the land use is showing the law of diminishing, in the future it will be more clear in front of us (Figs. 2.15, 2.16, 2.17, and 2.18).

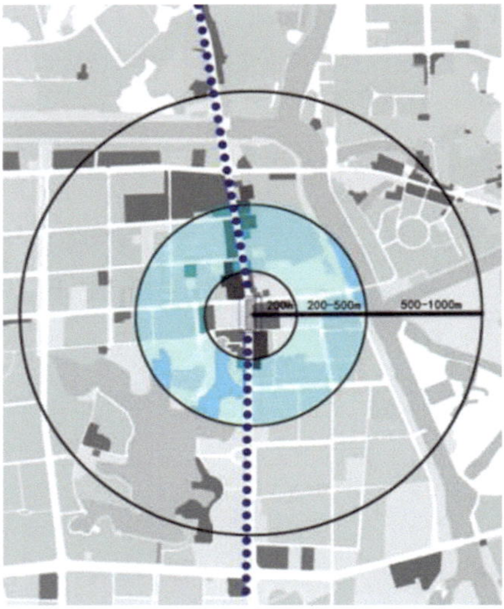


**Fig. 2.16** 200–500 m ring layer

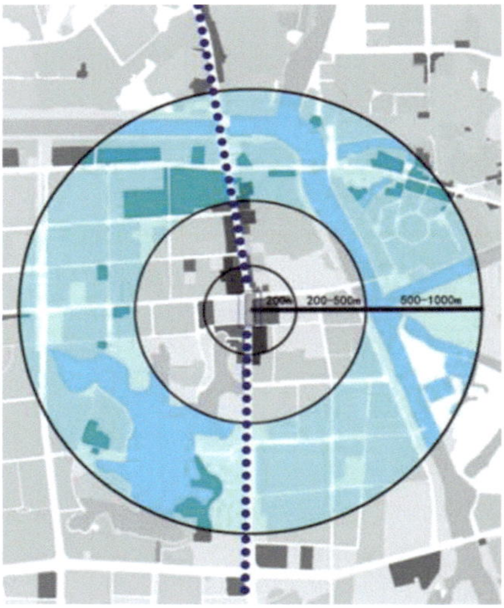


**Fig. 2.17** 500–1,000 m ring layer

## 2.5 Conclusion

Metro is the main artery to the development of the city, the metro stations are important points, the construction of the stations will vibrant the station surrounding areas and promote the value of land and enhance the intensity of development. Through the investigation of Shengtai Road, analyze the interaction between station and the city and region from the city level and the station level, to further

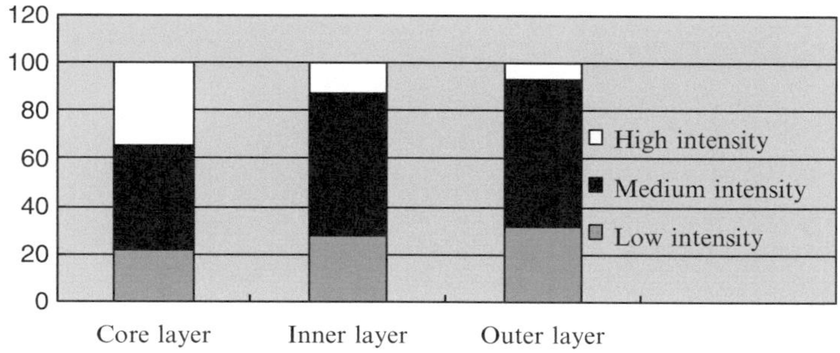


**Fig. 2.18** Distribution gradient of development intensity in surrounding area

understand the influence scope and degree of the station. This has important practical significance for the promotion of public transportation, city function and underground space development, and balance the benefits of the public, environmental and economic interests to ensure that the development and construction of the region is in a healthy, harmonious, and sustainable direction.

## References

1. Luo Lan (2008) Domain underground space development in the metro station with traffic – land feature studies. China's outstanding master's degree thesis full-text database, Tsinghua University
2. Tian Li (1999) Study of the rail traffic along the land-use. Urban Stud (3):26–29
3. Chen Jie, Lu Feng, Cheng Changxiu (2007) Accessibility measure methods and applied research of recent development. Prog Geogr 100–103

# Chapter 3
# Research of Urban Land Use Based on TOD Mode

**Juan Zhang, Liang Chen, and Xinliang Jiang**

**Abstract** TOD in urban planning improve the efficiency of land use, and promote the rationalization of urban space layout, ease the unrestricted spread of the city due to urbanization and motorization, and then solve the traffic congestion, pollution and automobile exhaust pollution, provides an effective way for the sustainable development of cities. This paper elaborates the basic theory of the system TOD concept, discusses the relationship between urban land use and transport system in detail, analyzes the main problems of China's development of TOD, and gives some suggests about our cities TOD in accordance with our transportation, land use layout and socio-economic characteristics, Finally, analyze of the land use situation of a project of Shenzhen.

**Keywords** Urban planning • TOD • Land use • Transport system

## 3.1 Concept of TOD

TOD (Transit Oriented Development) refers to the mode of public transport-oriented urban development. It advocates that urban planning should have a compact layout, the public transport as the support system of city, public transport nodes

---

J. Zhang (✉)
School of Traffic and Transportation, Beijing Jiaotong University, 100044 Beijing, China
e-mail: 12125667@bjtu.edu.cn

L. Chen
Nanjing Chemical Industrial Park Railway Transport Co., Ltd., 211500 Nanjing, China
e-mail: hzsxcl0530@163.com

X. Jiang
China Academy of Railway Sciences, 100044 Beijing, China
e-mail: jiangxinliang7@163.com

H. Xia and Y. Zhang (eds.), *The 2nd International Symposium on Rail Transit Comprehensive Development (ISRTCD) Proceedings*,
DOI 10.1007/978-3-642-37589-7_3, © Springer-Verlag Berlin Heidelberg 2014

as the basis of urban land use layout, urban service facilities lay around public transport sites [1]. And TOD has three basic elements:

- Transit, Large-capacity public transport. And transport carrier for subway, light rail or bus rapid transit and other high-capacity public transport system.
- Oriented, Guidance. As the goal, the pursuit of the high strength profitable real estate development projects and to guide rational and orderly development of the urban space through the development of high-capacity public transport.
- Development, exploit. Characteristics of the development object is close to public transport facilities, compact land layout, mixed land use and high quality of the environment for walking.

## 3.2 Analysis of the Relationship Between Urban Land Use and Transportation System

Relationship of modern urban land use and urban transport systems generally can be attributed to three aspects: the impact of urban land use for the urban transport system, the impact of the urban transport system for land use, and comprehensive coordination of the interaction.

The impact of land-use for transport system can be attributed to: (1) characteristics of land use affect the transportation system from different angles; (2) density of urban land use affect the transportation system mode; (3) land use affect the characteristics of the traffic trips.

Transport system of land use can be attributed to: (1) urban transport system has a profound impact on the urban spatial form; (2) urban transportation systems affect land use layout; (3) construction of urban transport has an important influence on urban land prices.

The study of interaction between urban land use and transport system, in the early many scholars study the theoretical relationship preliminarily, but the form a special issue, the modern relationship between the two theories is very rich, but there are still some limitations.

### *3.2.1 Basic Theory of Relationship Between Urban Land Use and the Transport System [2]*

Urban transport systems and urban land use have a relationship, a cycle of a mutual contact, mutual restraint and mutual feedback. First of all, the urban land use is the root of the needs of the urban traffic, it decides the urban traffic sources, traffic and traffic access. See from the macro level, it provides the structure and foundation of the city traffic, different urban land use situation requires a different mode of urban transportation. Second, actual operation level of urban transport system has impact to urban spatial structure and urban development, thus affecting the urban land-use, especially accessibility of urban traffic has a decisive role for urban economic, commercial and cultural activities.

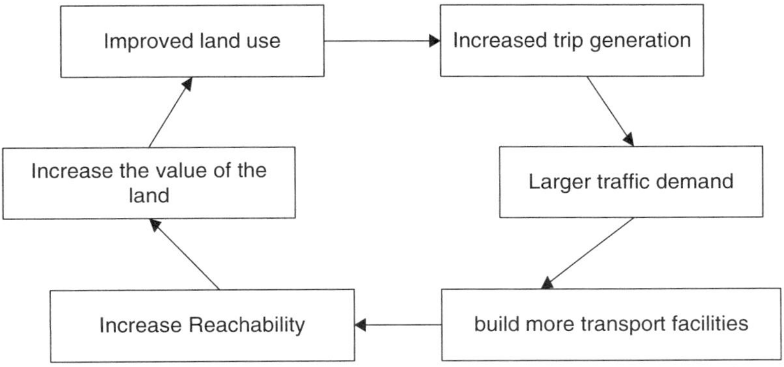


**Fig. 3.1** The mutual influence of land use and positive environment of transportation systems (Source: Quoted from the research and development of public transport-oriented urban land)

Figure 3.1 show the mutual influence of land use and positive environment of transportation systems [3]. Land use is a major factor in travel generated activities, the level of trip generation activities within the study area and the propensity to travel will determine the demand for transport facilities, the supply of these facilities can change the accessibility of land use, while accessibility re-determine the value of the land, the value of the land is a major determinant of land use. In this ring, the changes of any part will result in changes of their own or other components.

### *3.2.2 Analysis of Correlation Between Urban Land Use and Transport System*

The main elements affect the relationship between land use and transportation are urban mobility, traffic relative reachability and directional.

Relations of all kinds of motorized transport (mode structure) to determine the mobility of the city, divided into two categories, car mobility and public transport mobility, which are represent two typical traffic structures. Considering car traffic patterns, changes of spatial relations is to promote the city of expansion. Because of car mobility, Suburbs have a lot of development. For the public transport system, the scales of operation need compact and high-density development. The urban mobility determines mode of the urban axial space development, form a different extension shaft, mobility determines the relative reachability of land, mobility influence the evolution of urban space mode by the relative reachability.

Changes in the structure of urban space largely influence city reachability. The meaning of reachability is wide, sociology and psychological, consider optional of space-time in the transport sector. Transportation accessibility refers to arrive at a place conveniently. Travel time or distance to measure the objective optional from one location to another location. Relative reachability means degree of reaching a place relatively. The examples of expansion of urban space show that, the higher

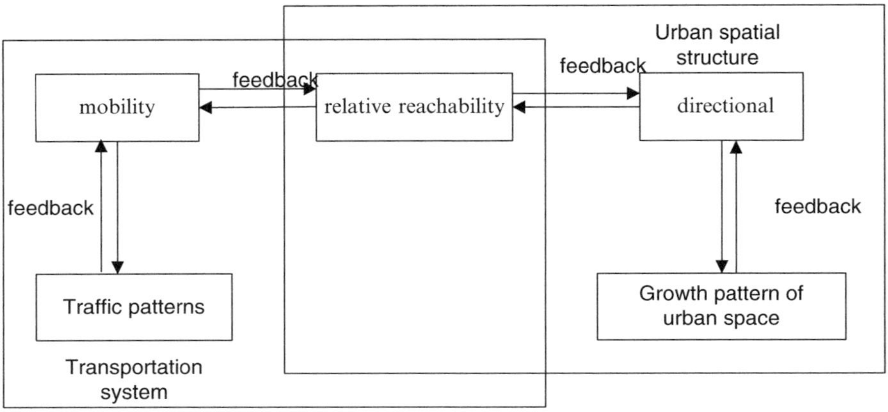


**Fig. 3.2** Mutual use schematic diagram of mobility, relative reachability and directional (Source: Quoted from the research and development of public transport-oriented urban land)

relative reachability necessarily located in direction of the least resistance of the expansion of urban space.

The traffic point concept earliest appears in "location theory" by geographers, in regional development and urban construction, t development of the location influenced by traffic point., urban form and mastery time of transportation systems are related., if the main transport in one direction faster than others, it is easy to form constantly axial expansion of land stretching axis by traffic point, and promote axial development of urban land. Scale of urban land use does not exceed the size of the main mode of transport, which has distance of 46 min.

The above analysis shows that the expansion of urban land use and development performance as a result of traffic directional, relative reachability decides traffic directional, relative reachability depends on urban mobility (Fig. 3.2) [4].

## 3.3 China's Urban Development Problems About TOD and Measures

### *3.3.1 China's Urban Development Problems About TOD*

Many cities in China have begun to develop TOD mode, but there are still a lot of difficulties. (1) the transportation system planning could not consider very well in the overall planning, urban development was an aimless spread state, TOD theory in China is still not widely understood. (2) considers of transportation structure, many cities in China have just begun to implement bus priority development programs. Currently, most of the urban public transport system is not yet developed, and the number of car traffic has a rapid growth, low bus-sharing, low service level, and the lack of large traffic volume has caused great difficulties of public transportation for development of TOD. (3) in terms of land use, city's land use planning cannot be combined with the transportation system planning, many projects attracted a lot of traffic volume are concentrated on a road leading to severe traffic problems [5].

### *3.3.2 Propose of China's Urban TOD Implementation*

In order to solve the problems of TOD in China, we should not only to learn successful experience of the world, but also to get the strong support and cooperation of various government departments. At the same time, efficient operation and financing mode is essential for TOD whether it can transfer from theory into practice from theory into practice.

1. the introduction of TOD concept in the overall planning of cities is widely publicized and learned successful TOD cases and advanced experience. The urban land-use planning combine with transportation system planning reflect the mutual benefits.
2. vigorously develop public transport, improve the public transportation network density and service levels, moderately develop rail transportation, bus rapid transit (BRT) and large capacity public transport system, adopt congestion pricing and other measures to curb the growth of the demand for car travel, and optimize the urban transportation structure, increasing the bus sharing rates, to provide better support for the implementation of TOD.
3. TOD and urban planning coordinated development

Practice has proved that the urban space based on TOD planning is a continuous feedback loop process, as shown in Fig. 3.3. Specific planning should be based on urban spatial structure, then determine the initial program of urban spatial structure and track network, analysis for space growth and TOD strategy coordination, proposed adjustment and optimization for the layout of the bus network into the next cycle, until the space growth and the degree of TOD coordination strategy to meet the planning target, so as to get a bus network coordinated development and orderly growth of urban spatial structure.

China's urban space development has three basic modes, the axial development mode, sector development mode, multi-core development model. Combine the characteristics of urban land development and characteristics of bus travel, cities should adopt a form of the urban continuity axial expansion plus balanced low-cost multi-center urban land layout. Continuous axial expansion mainly rely on various large capacity of highway system to set urban settlements and jobs. Multi-center equalizer mode focuses on land partition, highlight the nature of the key sites, which also have tour production, work, life, living, entertainment and other buildings and facilities, within the district also has its own commercial and cultural center as near as possible, to solve their daily problems between the various tours that are relatively independent, closely linked, also form an organic whole.

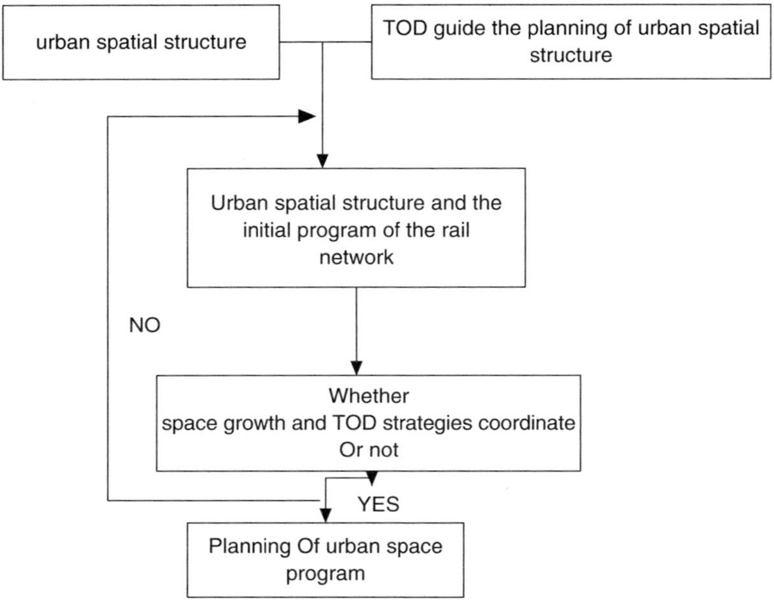


**Fig. 3.3** Planning of urban space layout based on TOD

Now our major cities' land layout mostly is centralized, [6] for the current situation of China, urban planning model should be carried out in the city center, district center, the center of the residential area, the hierarchical structure of the cell center, form multi-center in cluster, to adapt low cost implementation of the strategy to limit traffic; also consider the relationship between the place of residence and place of work, adjust industrial layout and staff accommodation; balance of production and living place, reflects the travel convenience and economy; multi-level center services for the production and living, alleviate downtown burden, to maximize the use of public transport commuter.

4. public-private partnership financing of development model

First, government departments should actively promote the development of TOD land, become a successful regional urban development. Second, invest public land surrounding the site to reduce the risk of developers, and scientifically assess overall return of TOD project.

5. support of national and local policy

In the implementation of TOD, when many agencies involved in the management of land development and service delivery, coordination and effectiveness of the system is particularly important, so a lot of system security and authorization legislation needed.

## 3.4 The Land-Use Study of Shenzhen Project [7]

Shenzhen has entered a rapid development period of rail transportation, it will be completed 400 km of backbone network in 2020, formed 542.8 km vision networks in 2030, in order to reach the city center in 1 h in the Pearl River Delta, and half an hour to the main development areas in the SAR, 45 min to the urban sub-centers, the targets greatly improve the efficiency of public travel.

For this reason, there is an urgent need to study how to have a positive effect by track construction of urban development, including the upgrading of the potential land along the track, guiding of optimization model of the land use, the value-added benefits to maximize feedback public. Paper analyzes rail traffic on the utilization of underground space and urban land use function.

### *3.4.1 Promotion of Rail Traffic Along the Underground Space Utilization*

The project promotes the re-development of the areas along the rail, central business district and transportation hub area have frequent activities, greater improvement in the surrounding environment. Rail traffic driven the underground space development, Table 3.1 show main functions in underground space. However, the use of underground space development is single function, lack of co-ordination, the overall quality needs to be improved. According to public evaluation of the underground public facilities, connectivity, identification and comfort need to be improved.

### *3.4.2 Analysis of Urban Land Use Along the Rail*

In 2005, housing, road infrastructure, public facilities commercial, government associations and greenbelt along the rail are main land use, accounted for more than 10 % in the proportion of urban construction land, land use indicators of the reside (28.14 %), commercial public facilities (15.76 %), government (11.67 %) and green (11.62 %) in line with the Shenzhen and national standards. Land for road facilities accounted for 21.9 % beyond the standard of the national standard, mainly because the city main road in the track range and red line wide between 100 and 150 m.

From 1999.1 to 2005.9, urban new construction and transformation have increased a total of 910.5 ha, which represents approximately 1/3 of the land for construction; including new construction is about 384.42 ha, the transformation is

**Table 3.1** The main features list of underground space development

| Function | Total area (10,000 m$^2$) | | Proportion (%) |
|---|---|---|---|
| Parking in basement | 77.04 | 149.54 | 65.00 |
| Parking in story underground | 72.5 | | |
| Commercials in the basement | 29.63 | 57.52 | 25.00 |
| Commercial in story underground | 27.89 | | |
| Other functions in basement | 11.85 | 23.01 | 10.00 |
| Other functions in story underground | 11.16 | | |
| Total | 230.07 | 230.07 | 100 |

526.10 ha of land, with an average 76.9 and 105.22 ha in annual. Commercial public facilities, large living and government and community site increase of 7.70, 6.97 and 3.47 %; unused land and industrial land greatly reduced, the number is 11.22 and 9.79 %; unused land transformed into residential and commercial land for public facilities. The changes in 5 years represent a meaning of attract and opposite to the rail transportation public facilities for commercial, residential, government and community site as well as industrial land.

The present land use situation of Shenzhen track project comply with the major mode of high-density housing business and retail center, which meet the laws of economic development and the effectiveness of urban land; meanwhile the nodes of the track with other transit lines should be different levels of city.

## 3.5 Conclusion

The ultimate purpose of the urban land development is to promote the optimal allocation of land resources and the most economical use, TOD mode can effectively use our limited land resources, change from extensive to intensive, and avoid consumptive land and traffic problems caused by urban construction. Urban planning, land use and TOD strategy closely linked, complete perfect combination of public investment and private investment development mechanism, curb disorder spread of land, to form "green, efficient and economic" integrated transport system, to further promote the city sustainable and harmonious development.

## References


1. Bernic M, Cevero R (1997) The transit village in the 21st century. McGraw Hill, New York
2. Gu Kedong (2004) Public transport-oriented urban land development study. School of Architecture Southeast University, Nanjing
3. Xiao Qiusheng, Xu Weici (1990) Urban transport planning. China Communications Press. Beijing.
4. Cao Muhua, Zhang Lu (2003) Studies of rail transport and urban space orderly growth. Urban Rail Transit Study

5. Song Jingxing (2010) Research of urban land public transport oriented development (TOD) mode. Sci Technol Innov Herald (36):4–5
6. Feng Siqing (2004) Research of interactive relationship qualitatively between urban transport and land-use. Hefei University of Technology (Natural Science)
7. Li Zhe (2010) TOD mode under the guidance of urban rail traffic along the land-use study. Huazhong University of Science and Technology

# Chapter 4
# The Evolvement Mechanism of Land Use and Urban Space Structure Along Rail Transit

**Zhihua Xiong and Zhisheng Yao**

**Abstract** Previous attempts to obtain the qualitative interpretation that rail transit affects the land use and urban space structure are lack of effective quantitative evaluation model. In this paper the evolvement mechanism of land use and urban space structure along rail transit can be proposed to objectively reflect the relationship and the complexity between rail transit and land use structure. Firstly, the land development intensity in suburb when rail transit opened in Beijing can be obtained. Then, the information entropy model can be used to evaluate the land use structure along the rail transit in three development stages that is 2003, 2007 and 2010 respectively. Finally, the land use structure along different type of rail transit station can be proposed. It shows that the passing by rail transit plays a significant role in space structure of suburb. Meanwhile, the residential, financial and official land development may be attracted along rail transit and industrial land repelled according to their information entropies. Especially the financial and official land development is closer from rail transit station, its intensity is higher and the scope of attraction is expanding.

**Keywords** Rail transit • Land use • Information entropy • Space structure

---

Z. Xiong (✉)
School of Traffic and Transportation, Beijing Jiaotong University, 100044 Beijing, China
e-mail: zhhxiong@bjtu.edu.cn

Z. Yao
Beijing Municipal Institute of City Planning and Design, 100045 Beijing, China

H. Xia and Y. Zhang (eds.), *The 2nd International Symposium on Rail Transit Comprehensive Development (ISRTCD) Proceedings*,
DOI 10.1007/978-3-642-37589-7_4, 

## 4.1 Introduction

The rapidly construction of rail transit in Beijing has brought opportunities and challenges to the development of land use and urban space structure. There was only a line in 1969 which was 23.6 km long. It took 34 years that rail transit operation mileages broke 100 km in 2003 and took only 6 years that the mileages arrived 228 km in 2009. In 2011, the operation rail transit lines have reached to 15 which were 372 km long finished within only 2 years. Rail transit system has transferred from some lines to a network that will be 561 km long in 2015 and about 1,000 km for the longer term. As large-capacity, fast passenger transport system by which the trip proportion undertaken will be transferred from 6 to 23 % [1], rail transit network will play a key role in the pattern of urban space development.

The main concentration that the rail transit affects the pattern of urban space development is the urban job/housing space pattern. The existing studies show that the convenience of rail transit already has become the second influencing factor to make their residential choice after real estate price and the proportion achieved 25 % [2]. More than 50 % respondents took the rail transit as the import or very important factor as their employee choice.

Meng (2008) discussed the variance of land use and population distribution after the operation of Shanghai UMT Line 1 and the results showed population evacuated in the urban city and aggregated in suburb [3]. Dong (2008) discussed corridor effect and TOD mode changed the range of living spaces [4]. Liu (2009) analyzed the effects of the metro line on land use structure around the station and the aggregated business [5, 6]. In this paper, the evolvement mechanism of space structure along the rail transit can be obtained by the information entropy model to provide the guidance to the coordinated development between the rail transit and land use.

## 4.2 Rail Transit Development and Space Structure in Suburbs

The existing studies show that the degree of separation from the job and house is less in urban and the effect of rail transit on the distribution of job/house is no significant [2]. The degree of separation from the job and house is more obvious in suburbs for the residential choice and the effect is significant.

According to Beijing urban master plan, there are ten suburbs which are Qinghe, Beiyuan, Jiuxianqiao, Dongba, Dingfuzhuang, Fatou, Nanyuan, Fengtai, Shijingshan and Xiyuan. To support the development of suburbs, many rail transit lines were built in recent years. The average floor area ratio of suburbs changing with passing by rail transit is shown in Fig. 4.1. Average floor area ratio is the ratio of total floor area to gross floor area, which shows the land development intensity.

The results show that the effect of rail transit on the land development in suburbs is significant. Average floor area ratio of Beiyuan, Qinghe and Jiuxianqiao increased rapidly after M13 was operated in 2003 and the land development

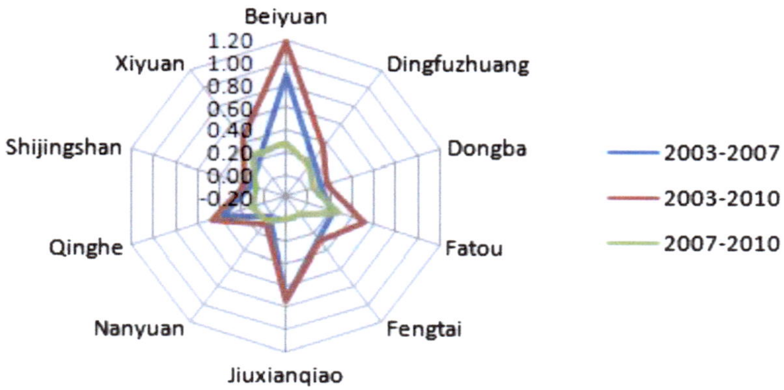


**Fig. 4.1** Average floor area ratio of suburbs

intensity improved significantly. Dingfuzhuang, Dongba and Fengtai continued to grow and Beiyuan further develop after M5 was completed in 2007, Xiyuan developed for M4 was operated in 2010. It appears quite different than urban city. The average floor area ratio of urban city generally increased at first, but decreased later. It shows that the land development intensity begins to decrease and some functions of land use transfer from the urban to the suburbs. Therefore, rail transit system will play a key role in the pattern of urban space development. The suburban area along the rail transit shows a growing trend of highly concentrated development and the land use function migration to the suburb can be accelerated.

## 4.3 Dynamic Evolvement of Land Use Along the Rail Transit

The structure of land use can be described by information entropy model for urban land use system is a complex and open system [7]. There are N types of land use and each land use area is $A_1$, $A_2$,$\cdots$, $A_N$ respectively. If the land use area is A, then $A_1 + A_2 + \cdots \ldots + A_N = \sum_{i=1}^{N} A_i = A$. The proportion of each land use area can be obtained by $P_i = A_i/A = A_i/\sum_{i=1}^{N} A_i$ and $\sum_{i=1}^{N} P_i = 1$. The information entropy of land use structure can be proposed according to the information entropy as follow:

$$H = -\sum_{i=1}^{N} P_i \log P_i \quad (4.1)$$

Where H is the information entropy of land use structure, $H \geq 0$.

The degree of land use structure diversity can be described by the value of information entropy. Higher the value is, more the types of land use are and smaller the distinction of different land use area is [5, 8]. When $A_1 = A_2 = \cdots \ldots = A_N$,

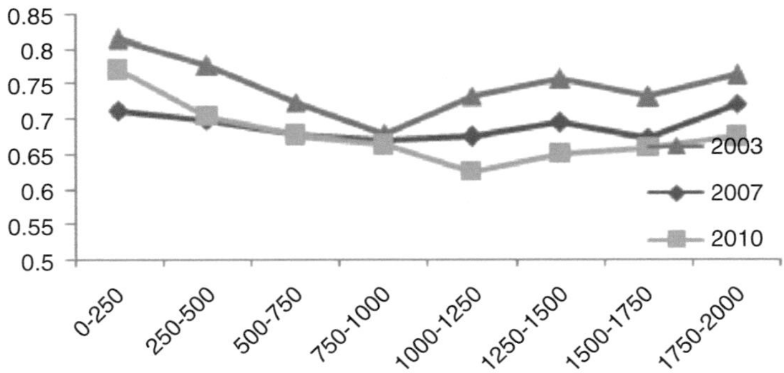


**Fig. 4.2** The information entropy of land along rail transit

H arrives the maximum and can be obtained by $H_M = \log N$, $P_1 = P_2 = \cdots \ldots = P_N = 1/N$ and the balance of each land use area arrives which shows the pattern of land use structure. The entropy of each land use type is $H_i = -P_i \log P_i$, $i = 1 \ldots\ldots N$, which shows the proportion of each land use area. Bigger the entropy is, higher the percentage of land use area is.

The existing studies show that the range with a radius of 500–800 m is generally known as the scope which is affected by rail transit significantly [9]. The expanding range is used as the effect of rail transit is strengthening in this paper. There are 250, 500, 750, 1,000, 1,250, 1,500, 1,750 and 2,000 m respectively. The information entropy of seven types land use such as residential, commercial, official and industrial land can be obtained in Fig. 4.2.

It shows that the information entropy is decreased from 2003, 2007 to 2010. The information entropy is lowest in 2010 that shows the diversity of the different land use area is growing along the rail transit and the attraction to each land use type is quite different. The information entropy of land use structure first decreases then increases slowly. The lowest point of the information entropy in 2010 locates in 1,250 and 1,000 m in 2003, 2007, and then begins to rise. It shows that the range that the effect of rail transit on land use development is expanding.

The entropy of each land use can be obtained according to the ratio of residential, financial, official and industrial land area to total floor area shown in Fig. 4.3.

It shows that the entropy of residential, financial and official land goes up, that of industrial land goes down. The area of residential, financial and official land decreases with the distance father from rail transit station and industrial land area increases. Therefore, residential, financial and official land is attracted around rail transit and industrial land is repelled as the entropy shows.

The area of residential and financial land is larger along the rail transit according to their higher entropies. The entropy of financial land goes down significantly with the distance father from rail transit station and residential and official land is not significance. It shows that rail transit system had a distinct impact on the distribution of financial land.

In conclusion, the land use distribution is imbalance for the information entropy is lower along the rail transit, especially residential, financial and official land may

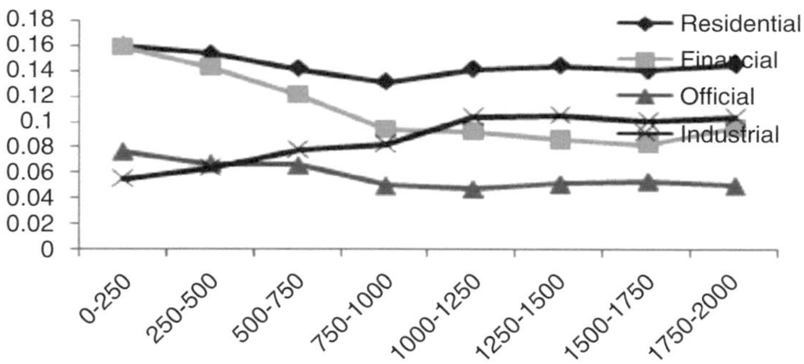


**Fig. 4.3** The evolvement of four land use types' entropy in 2003

be attracted by rail transit and industrial land repelled. The attractive range of rail transit is expanding and the distance about 1,000 m can be as the reasonable scope. With the rapidly decrease in the entropy of financial land, rapidly increase of industrial land and the slowly decrease of residential and official land when the distance is farther from rail transit, the distribution space structure can be obtained as follow. The land development will give priority to financial land within 250 m from rail transit station. When the range is from 250 to 500 m, the area of residential and official land can be increased as appropriate. The residential land is applicable to be developed within the scope from 500 to 1,500 m. Other types of land use with lower density can be developed in father distance.

## 4.4 The Evolvement of Land Use for Different Types of Rail Transit Station

The pattern of land development is distinct for different type of rail transit station has different geographical location or different traffic function. The rail transit station can be divided into two-line transfer and three-line transfer according to their intersected line number. The three-line transfer station is the comprehensive hub and two-line transfer as common hub. The information entropy of land use structure along rail transit station can be shown in Fig. 4.4. The information entropy of common hub is higher than the comprehensive hub. The industrial land is rejected by three-line transfer station and its proportion nearly is zero.

As we all known, residential, financial and official land are mainly developed along the rail transit. Therefore, the entropy of these three-type land uses is taken to reflect the evolvement of land use structure shown in Fig. 4.5.

For the variance of the entropy in two-line transfer station is less than three-line transfer, the proportion of these three-type land uses is more balance in two-line transfer station than three-line transfer.

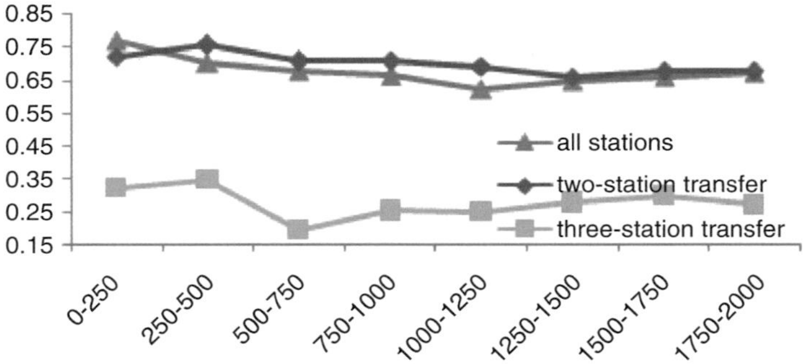


**Fig. 4.4** The information entropy of land rail transit station

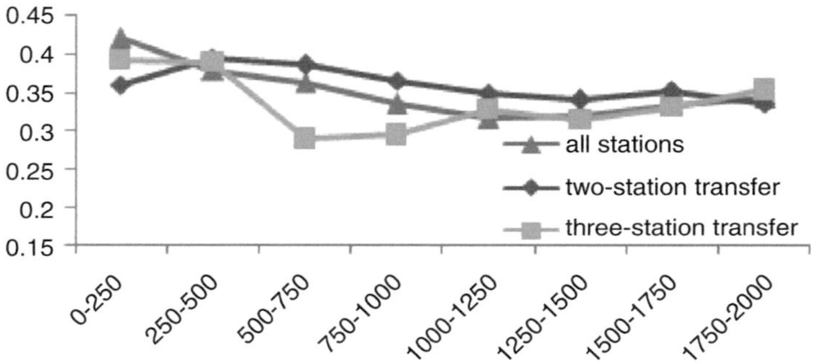


**Fig. 4.5** The entropy of three types land use along rail transit station

The entropy can be stability at 1,000 m far from the station whether two-line transfer or three-line transfer. So the scope that rail transit affects on residential, financial, official land is within 1,000 m.

The proportion of residential land in three-line transfer is higher than two-line transfer and its entropy from 500 to 1,000 m is outstanding. The proportion of each land area is distinct shown in Fig. 4.6. The proportion of residential land can be arrived 60–70 % when the distance is from 500 to 1,000 m in Fig. 4.6a. But the diversity in each land area of two-line transfer is more than the three-line transfer within 250 m from rail transit station. The proportion of financial land can be arrived 50 % higher than the three-line transfer within 250 m in Fig. 4.6b.

The conclusion shown in Fig. 4.6 can be obtained as follow. The proportion of financial land along transfer station is more than common station among which the proportion of financial land along two-line transfer is the maximum. The proportion of residential land along common station is generally more than transfer station, but the proportion of residential land along three-line transfer is the maximum when the distance is from 500 to 1,000 m. The proportion of official land along transfer station is more than common station among which the proportion of official land is

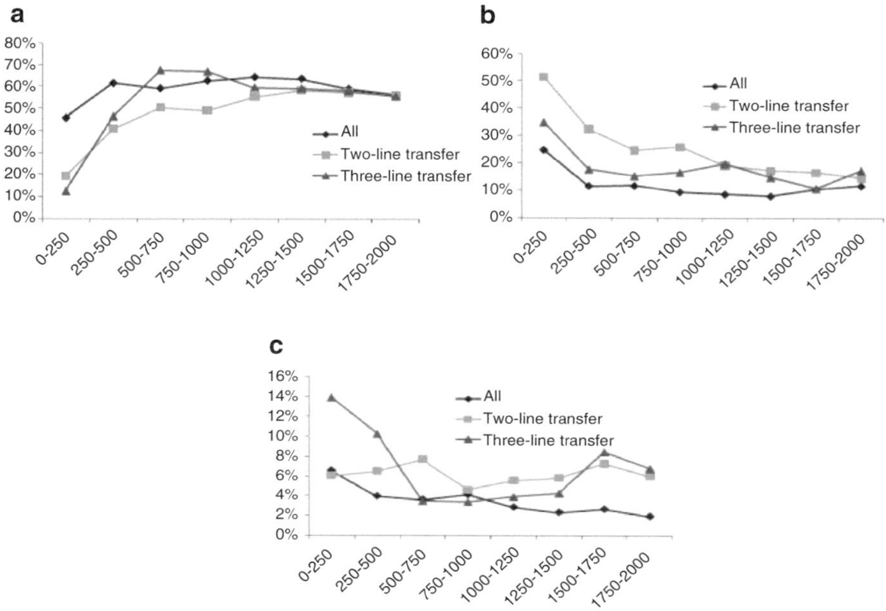


**Fig. 4.6** The proportion of three types land use along the rail transit. (**a**) Residential land. (**b**) Financial land. (**c**) Official land

the maximum when the distance is within 500 m along two-line transfer and three-line transfer is the maximum from 500 to 1,500 m.

Therefore, the land development is mainly in high-density financial and official land within 250 m from comprehensive hub. Moreover, the proportion is higher when the distance is closer from the hub. The land development is mainly in residential, financial and official land along the common hub among which the proportion of residential land is higher along all station for the trip convenience.

## 4.5 Conclusion

The integrated plan and design or comprehensive development along the rail transit can be suggested according to the evolvement mechanism of land use and space structure along the rail transit. The mechanism of coordinated development between rail transit and land use can be constructed. It shows that high-density land use will be developed along the rail transit and the intensity is weakening with the distance farther from station. Moreover, the residential, financial and official land should be preferentially developed along the rail transit according to their suggested distance to promote the sustainable development between rail transit and land use.

**Acknowledgement** This research is supported by the Fundamental Research Funds for the Central Universities (No. 2011JBM257).

## References

1. Lai-fang S, Yang-kun W (2012) The effect of rapid development about rail transit network on urban space [J]. Compr Transp 5:12–17
2. Hui Z, Jun Y, Chang-ping L (2011) Research on characteristics of the Job/housing distribution and the changing commuting along subway lines – a case study from Beijing [J]. Econ Geogr 31 (9):1445–1451
3. Qingyan M, Haolan X (2008) The impact of UMT on population redistribution – a case study of Shanghai UMT line 1[J]. Urban Mass Transit 3:7–12
4. Dong Shen, Shi Qixin (2008) Impact of corridor effects of rail transit on urban living space [A]. In: The 6th national civil engineering graduate student academic BBS [C]. Tsinghua University, Beijing
5. Baokui L, Changchun F (2009) Research of impact of urban railway transportation on land use structure based on GIS and information entropy [J]. Urban Stud 16(4):149–155
6. Zhongnuan CGC (2004) The impacts of rail transport on retailing spatial pattern in Guangzhou, China [J]. Mod Urban Res 19(4):65–67
7. Yongzhong T, Cifang W (2001) The laws of the information entropy values of land use composition [J]. J Nat Resour 18(1):112–117
8. Zhi C (2006) Analysis of the evolvement in information entropy of land use structure and its driving forces in Wuhan [J]. J Hubei Univ (Nat Sci) 28(3):317–320
9. Xingdong D, GanYonghua LJ (2012) Land use development of rail transit terminal in Guangzhou [J]. Urban Transp China 10(2):15–20

# Chapter 5
# Frequency Setting of Urban Mass Transit Based on Passenger Flow Along the Entire Line

**Lu Tong, Leishan Zhou, and Jinjin Tang**

**Abstract** Frequency setting plays a fundamental role in increasing the level of service and the efficiency of operation in urban mass transit system, which needs information of lines, trains and passengers etc. This study presents the space-time relationship between transit frequencies and passenger flow along the transit line, and distinguishes the service frequencies of the entire transit line from that of a certain section. To deal with the lack of carrying capacity in peak hours, a postponement strategy for passengers failing to board is proposed. Furthermore, we develop an approach to calculate frequencies based on the global passenger information and tested it on a real-scale transit line.

**Keywords** Urban mass transit • Frequency setting • Passenger flow

## 5.1 Introduction

The public transit planning process is usually divided into a sequence of five steps [1]: (1) the design of routes, (2) the setting of frequencies, (3) the timetabling, (4) the vehicle scheduling and (5) the crew scheduling and rostering. The first three steps are always treated as fundamental components in public transit planning process. Transit frequencies setting provides frequencies for every transit line for each time period, going right after the network design. Its accuracy concerns the level of service and the efficiency of operation.

Although the optimization of frequency is an important part of the urban transit operation, it had not received relatively scant attention and usually appeared as a sub problem of transit planning process in 1970s–1980s. Constantin and Florian [2] formulated optimizing frequency problem integrated with route choice as a

---

L. Tong (✉) • L. Zhou • J. Tang
School of Traffic and Transportation, Beijing Jiaotong University, 100044 Beijing, China
e-mail: ltong@bjtu.edu.cn

H. Xia and Y. Zhang (eds.), *The 2nd International Symposium on Rail Transit Comprehensive Development (ISRTCD) Proceedings*,
DOI 10.1007/978-3-642-37589-7_5, 

nonlinear bi-level problem. Two popular concepts in recent studies for this problem are minimizing total waiting time of passengers and total expenses of passengers and transit companies. Huimin et al. [3] conducted a model with the object of minimizing waiting time of passengers with constraints of train operation requirements, the number of electronic multiple units and the benefits of operation enterprises. Yanfang [5], using the other concept, developed a multi-objective optimization model for frequencies setting problem on a urban rail transit network to minimize the variable costs, the total passenger cost and the capacity expenses. Yuval [4] built an optimal cost-based frequency setting model, considering two main cost elements: (a) empty-seat driven and (b) overload and un-served demand.

Until now researchers have shown little concern about relationship between frequencies and passenger flow along the entire transit line in the literature [7, 8]. Moreover, it is difficult to calculate costs of passengers (including waiting costs, congestion cost, etc.) and outlay of transport companies. So this paper presents an easy to use method for solving frequency setting problem.

The remainder of this paper is organized as follows. In Sect. 5.2, a conceptual illustration is provided for the space-time relationship between frequencies and passenger flow. After putting forward a calculation method of transit frequencies in Sect. 5.3, Sect. 5.4 tests the method by a numerical experiment. Finally, conclusions are presented in Sect. 5.5.

## 5.2 Transit Frequency Setting

For a long time, the frequency in one time period is supposed to be determined by the passenger volume in that time period. More precisely, the design hour volume $P_d$ divided by the average number of passengers a transit unit (TU) will carry on maximum load section (MLS) gives the required frequency $f_d$. In Eq. (5.2), $\alpha$, n and $C_v$ donates load factor, number of vehicles per TU and vehicle capacity respectively.

$$f_d = \frac{P_d}{\alpha \cdot n \cdot C_v}. \tag{5.1}$$

### 5.2.1 *Space-Time Relationship Between Frequencies and Passenger Flow*

Whereas, as a matter of fact, $f_d$ calculated by formula (5.1) is required frequencies of the MLS but not frequencies at the terminal station. Urban mass transit lines always extend a few dozen kilometers long and it takes tens of minutes to take a one-way trip from one terminal to the other. In addition, the layout of urban mass transit stations is very simple, which usually only contains two main tracks. These characters of urban transit line make the train route relatively fixed. As a result, when deciding the frequencies, both time factor and space factor should be taken into account.

Consider a single route, in which vehicles of the same capacity C are serving the route at a fixed head-way. During period $t_i$ (usually 1 h) the recommendation level of service is $\alpha^{t_i}$. Let $D_{s_j s_{j+1}}$ be the section between station $s_j$ and $s_{j+1}$, $P^{t_i}_{D_{s_j s_{j+1}}}$ the passenger volume during $t_i$ on section $D_{s_j s_{j+1}}$. The service frequency needed by the section $D_{s_j s_{j+1}}$ in $t_i$ is

$$f^{t_i}_{D_{s_j s_{j+1}}} = \left\lceil \frac{P^{t_i}_{D_{s_j s_{j+1}}}}{\alpha^{t_i} \cdot C} \right\rceil. \tag{5.2}$$

$\lceil x \rceil$ presents an integer no less than x.

A train, departing from the terminal station, will travel through all the section alone the transit line to the other terminal. Similarly, $F^{t_i}$, transit frequency in $t_i$ for the whole line, should be decided upon all the $f^{t_i}_{D_{s_j s_{j+1}}}$ of each section. Let $F_r(F^{t_i})$ be the set of all the $f^{t_i}_{D_{s_j s_{j+1}}}$ of sections that covered by $F^{t_i}$. In the service oriented situation, $F_r(F^{t_i})$ should equal to the maximum frequency of section.

$$F^{t_i} = \max\{F_r(F^{t_i})\}. \tag{5.3}$$

Figure 5.1 is a stretch map of a transit line with five stations and four sections. Figure 5.2 shows its time-distance graph of train. The shaded area represents the space-time area covered by trains departing at station $s_0$ in $t_1$, based on running time in each section.

We can see from the graph that trains belong to $F^{t_1}$ travel through seven space-time area. Therefore, set $F_r(F^{t_1})$ contains $f^{t_1}_{D_{s_0 s_1}}$, $f^{t_1}_{D_{s_1 s_2}}$, $f^{t_1}_{D_{s_2 s_3}}$, $f^{t_1}_{D_{s_3 s_4}}$, $f^{t_2}_{D_{s_1 s_2}}$, $f^{t_2}_{D_{s_2 s_3}}$ and $f^{t_2}_{D_{s_3 s_4}}$, and $F^{t_1}$ should equal to the maximum value of these elements.

### 5.2.2 Lower and Upper Bounds of Frequency

Frequencies of urban mass transit are restricted to both technical condition provided by the transit line and service level set by the authority. On one hand, signal system or train control system offers a minimum headway, thus there is an upper bound $f_{max}$ of frequency. On the other hand, in order to maintain a high service level, authorities or transit companies always make regulations specifying the headway in different time period, so there is also a lower bound $f_{min}$.

In off-peak hours, for instance, early morning, due to low passenger volume, $F^{t_i}$ calculated according to formula (5.3) will less than $f_{min}$, so $F^{t_i}$ will equal to $f_{min}$; in peak hours, commuters accumulating at some busy stations, the value of $F^{t_i}$ will

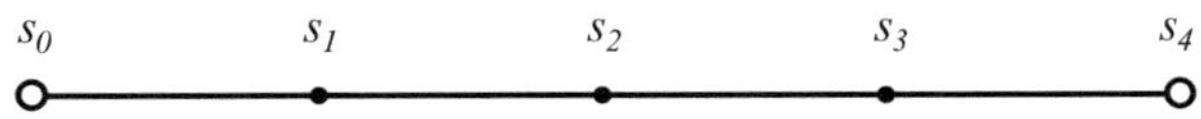


**Fig. 5.1** Stretch map of a transit line

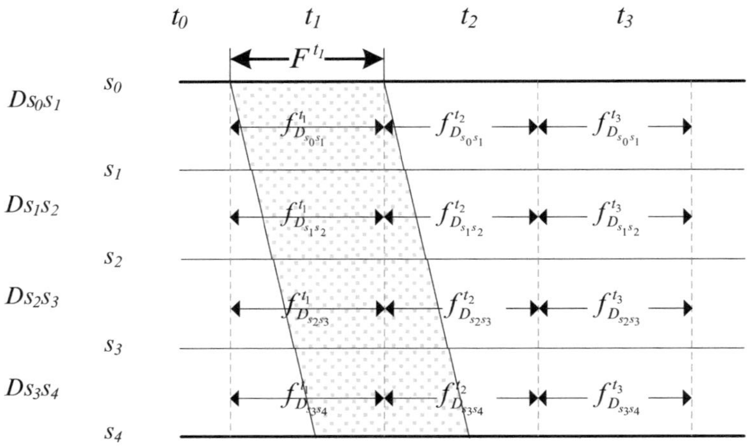

**Fig. 5.2** Time-distance graph of the train-set

larger than $f_{max}$, so under this circumstance $F^{t_i}$ will equal to $f_{max}$. After computing m periods of the whole day, set of transit frequencies F can be derived.

$$F^{t_i} = \begin{cases} f_{min} & (F^{t_i} < f_{min}) \\ \max\{F_r(F^{t_i})\} & (f_{min} \le F^{t_i} \le f_{max}) \\ f_{max} & (F^{t_i} > f_{max}) \end{cases}. \tag{5.4}$$

## 5.2.3 Un-served Passengers

When the third condition of formula (5.4) occurs, $F^{t_i} > f_{max}$, passenger volume is greater than the carrying capacity Q on at least one section. That means passengers at a particular stop will be unable to board a vehicle that is already heavily loaded. This kind of passengers is called un-served passengers. Number of un-served passengers $PS^{t_i}_{D_{s_j s_{j+1}}}$ equals to $P^{t_i}_{D_{s_j s_{j+1}}}$ reducing Q. If there are un-served passengers in $t_{i-1}$, they must also be postponed to this period.

$$PS^{t_i}_{D_{s_j s_{j+1}}} = \begin{cases} \max\left\{P^{t_i}_{D_{s_j s_{j+1}}} + PS^{t_{i-1}}_{D_{s_j s_{j+1}}} - Q, 0\right\} & (m > i \ge 1) \\ \max\left\{P^{t_i}_{D_{s_j s_{j+1}}} - Q, 0\right\} & (i = 0) \end{cases}. \tag{5.5}$$

## 5.3 Process for Solving Frequency Setting Problem

Using the concept proposed in Sect. 5.2, frequencies of urban mass transit can be computed according to passenger volume directly. The process of solving frequency setting problem and flow chart are as follows.

**Step 1:** Initialize frequencies, and input information of vehicles, passengers and level of service.
**Step 2:** Select the first period $t_0$ and frequency $F^{t_0}$ in this period. Let $i = 0$, and go to **Step 4**.
**Step 3:** Select period $t_i$ and frequency $F^{t_i}$. Check whether un-served passengers in $t_{i-1}$ exist. If there are un-served passengers, calculate their volume, and then go to **Step 4**; else, go to **Step 4** directly.
**Step 4:** Establish $F_r(F^{t_i})$, the set of frequencies covered by $F^{t_i}$, and make $F^{t_i} = \max\{F_r(F^{t_i})\}$. On the basis of $f_{max}$ and $f_{min}$, if $F^{t_i} < f_{min}$ let $F^{t_i} = f_{min}$; else, if $F^{t_i} > f_{max}$, let $F^{t_i} = f_{max}$ and deduce the un-served passengers. Go to **Step 5**.
**Step 5:** Decide whether $t_{i+1}$ is still in operation time period. If this is the case, let $i = i+1$ and return to **Step 4**; else, go to **Step 6**.
**Step 6:** Output the set of transit frequencies at the end of process (Fig. 5.3).

## 5.4 Numerical Experiments

The following case study considers Beijing Subway Line 4 consisting 24 stations along its 28.2 km long main tracks. Running times and headways follow an actual train diagram. The minimum headway is set to 2 min. Transit frequencies from 6:00 to 13:00 are examined in this section. Table 5.1 shows the results of calculation.

Restricted space-time area in Table 5.1 indicates the heaviest load section covered by frequency $F^{t_i}$. From the data we come to a conclusion that the restricted space-time area of $F^{t_i}$ is not always in period $t_i$ but in period $t_{i+1}$ sometimes. For instance, the frequency from 6:00 to 7:00 is determined by the passenger volume between stations Caishikou and Xuanwumen from 07:00 to 08:00.

## 5.5 Conclusion

In this paper, we have analyzed the space-time relationship between frequencies and passenger flow. Based on the solution for un-served passengers, an approach to solve frequency setting problem is introduced. Finally, a numerical experiment using a real-world subway line is conducted to further demonstrate the concept.

**Acknowledgment** This work was supported by National Science and Technology Support Program (2011BAG01B01).

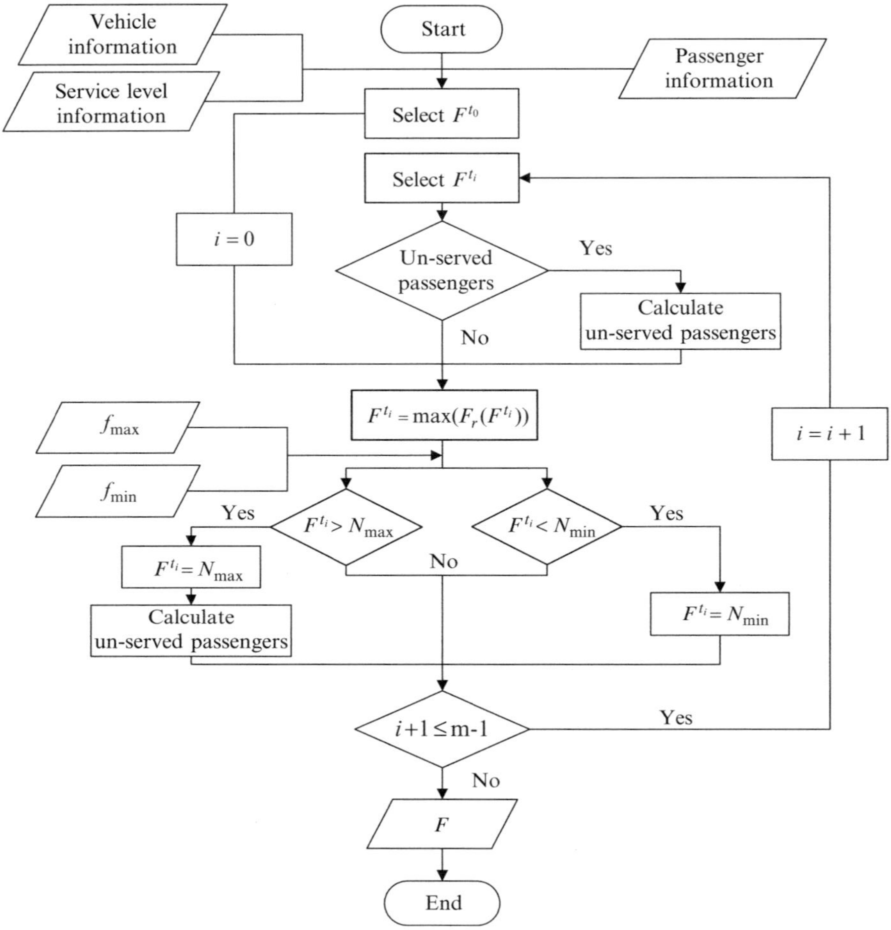


**Fig. 5.3** Flow chart of urban mass transit frequency setting

**Table 5.1** Frequencies for Beijing subway line 4 from 6:00 to 13:00

| Time period | Frequency (train/h) | Headway (min) | Restricted space-time area |
|---|---|---|---|
| 06:00–07:00 | 30 | 2 | Caishikou → Xuanwumen (07:00–08:00) |
| 07:00–08:00 | 30 | 2 | Caishikou → Xuanwumen (07:00–08:00) |
| 08:00–09:00 | 24 | 2.5 | Caishikou → Xuanwumen (08:00–09:00) |
| 09:00–10:00 | 20 | 3 | Caishikou → Xuanwumen (09:00–10:00) |
| 10:00–11:00 | 20 | 3 | Xizhimen → Dongwuyuan (10:00–11:00) |
| 11:00–12:00 | 15 | 4 | Xizhimen → Dongwuyuan (11:00–12:00) |
| 12:00–13:00 | 12 | 5 | Xizhimen → Dongwuyuan (13:00–14:00) |

## References

1. Valérie G, Jin-Kao H (2008) Transit network design and scheduling: a global review. Transp Res Part A Policy Pract 42(10):1251–1273
2. Constantin I, Florian M (1995) Optimizing frequencies in a transit network: a nonlinear bi-level programming approach. Int Trans Oper Res 2(2):149–164
3. Huimin N, Mingming C, Minghui Z (2011) Optimization theory and method of train operation scheme for urban rail transit. China Railw Sci 4:128–133
4. Yuval H, Matan S (2012) Public-transit frequency setting using minimum-cost approach with stochastic demand and travel time. Transp Res Part B Methodol 46:1068–1084
5. Yanfang Z (2012) Study on integration train timetabling theories and method for urban mass transit network. Beijing Jiaotong University, Beijing
6. Grosfeld-Nir A, Bookbinder JH (1995) The planning of headways in urban public transit. Ann Oper Res 60(1):145–160
7. Ceder A (2007) Public transit planning and operation: theory, modeling and practice. Butterworth–Heinemann, Oxford
8. Vuchic VR (2005) Urban transit: operation, planning and economics. Wiley, Hoboken

# Chapter 6
# Facility Configuration Optimization in Subway Stations

**Xianyu Wu and Zhenzhou Yuan**

**Abstract** According to the analysis results of facility configuration and passenger flow line, the fare collection area in subway stations can be divided into nine functional areas, such as ticket office, ticket vending machines, tickets adjustment, etc. In our study, a microscopic simulation model is developed with the division of the fare collection area and walk speed-density model. The model can be used to evaluate the adaptability of the facility configuration in the fare collection systems. Then the simulation optimization method is presented to obtain the optimum number of fare gates with the simulation program of fare collection system and optimization method based on the genetic algorithm. Lastly, using the micro simulation model of fare collection system and the optimization method based on the genetic algorithm, the facility configurations of the fare collection system in the Dong Zhi Men, Beijing rail transit hub are evaluated to get some improved measures.

**Keywords** Facility configuration • Simulation • Optimization • Genetic algorithm

## 6.1 Introduction

Because of the high cost potential of constructing urban rail transit hubs, it is especially important, at the engineering design stage, to assess the suitability of a hub system for handling passenger flows. Many analytical and simulation models are available for evaluating rail transit hub systems. Simulation models have become popular in such applications due to many advantages of simulation.

Hoogendoorn and Daamen [1–3] established simulation models based on pedestrian flow characteristics and developed two simulation tools, named SimPed and

X. Wu (✉) • Z. Yuan
MOE Key Laboratory for Urban Transportation Complex Systems Theory and Technology, Traffic and Transportation, Beijing Jiaotong University, 10044 Beijing, China
e-mail: wuxy@bjtu.edu.cn; zzyuan@bjtu.edu.cn

H. Xia and Y. Zhang (eds.), *The 2nd International Symposium on Rail Transit Comprehensive Development (ISRTCD) Proceedings*,
DOI 10.1007/978-3-642-37589-7_6, © Springer-Verlag Berlin Heidelberg 2014

Nomad. However, the models did not consider the walking behavior of individual passengers. He [4] developed a metro platform simulation system based on a virtual environment for training metro drivers. However, the system was designed primarily for training metro drivers and it could not fit to evaluating facility designs. Ni [5] developed a subway passenger's simulation system based on virtual reality, but it did not consider the characters of different passengers. Li [6] developed a simulation model named MTR-PedSIM using the multi-agent simulation technique; however, it was not suitable for assessing facility adaptability. Liu [7] established a simulation model to analyze queuing issues at metro entry gates; but the model was not enough to consider different passenger types in fare collection systems.

Different studies have shown that travelers are more sensitive to transfer time than in-vehicle time [8, 9]. Therefore, it becomes critical to minimize passenger transfer times by providing optimal design of transfer hubs, platform allocation, and fare collection system. Traffic operations at transfer hubs are complex, partly caused by the interactions among different processes, as well as different services offered by contemporary transfer hubs. Another complicating factor is the heterogeneity in passenger composition with respect to trip purpose, age, gender, etc.

The focus of this research is to analyze the facilities configuration in the fare collection area of transfer hubs. This paper is organized as follows. After the introduction, analysis of facilities configuration and passenger flow line is presented. The simulation optimization method based on genetic algorithms is then discussed. A case study and results are presented. Finally, some conclusions and recommendations are reached based on the research.

## 6.2 Passenger Flow Line Analysis and Simulation Optimization

### *6.2.1 Analysis of Facility Configuration and Passenger Flow Line*

A fare collection system includes a number of necessary devices, including ticket office, automatic ticket vending machines, automatic analyzers, recharging machines, automatic fare gates, tickets adjustment, service counters, and security check machines. According to the relative position between ticket office and automatic fare gates, the device layout of a fare collection area can be classified into three types: paratactic, parallel, and vertical. Their layout and passenger flow line are analyzed as follows.

Total of 15 subway hubs and 85 subway stations of Line 1, Line 2, Ba Tong Line, Line 5, Line 8, Line 10, Line 13 were analyzed the application of the facility configuration types in the Beijing subway hubs and general stations. The statistic

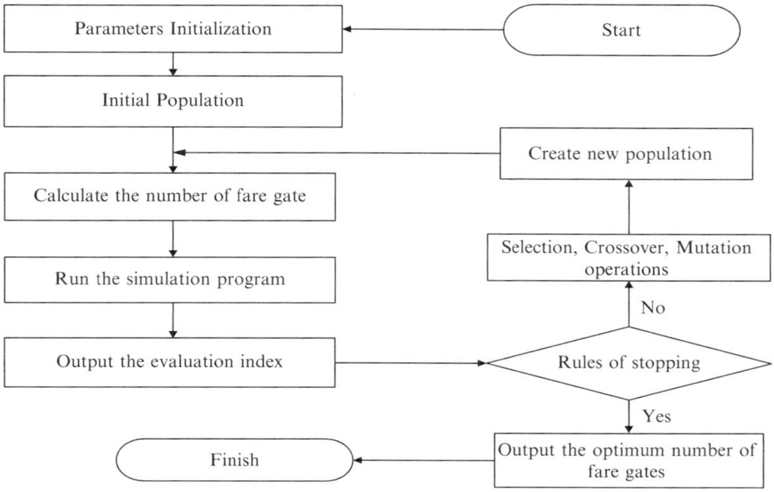


**Fig. 6.1** Flow chart of simulation optimization based on genetic algorithms

results shows that parallel fare gate and ticket office with double exit directions, paratactic fare gate and ticket office with single exit direction are the main facility configuration types of the fare collection areas. Therefore, our main focus of the research is on these two types of configurations.

### *6.2.2 Simulation Optimization Algorithm*

The simulation optimization method is to gain the optimum number of fare gates for an entrance. It has two parts: the simulation program of fare collection system and optimization method based on genetic algorithm.

The passenger flow data and facility parameters of the fare gate are inputted into the fare collection system, a program developed in our research. The evaluation index system can be calculated using the simulation results. The index system mainly includes the proportion of passengers who must stand in a line to get across the fare gates, queue length and average waiting time per passenger. According to the fitness function gained from the evaluation index system mentioned above, the optimum number of fare gates can be calculated using the optimization method based on genetic algorithms. Then the optimized number of fare gates can be returned to the simulation program to get the new evaluation index system. After operating two parts of programs several times, the reasonable and optimum number of fare gates can be obtained as shown in Fig. 6.1.

## 6.3 Case Study

### *6.3.1 Facility Configuration and Parameter Setting*

The passenger flow Line diagram of the main facilities layout in the Dong Zhi Men rail transit hub was drawn in our study. According to a field survey of the hub, the schematic diagram of the main facilities layout in the Northern Fare Collection Area of Line 2 can be drawn as Fig. 6.2 shows.

### *6.3.2 Evaluation Results of Simulating the Fare Collection System*

#### 6.3.2.1 The Input of the Passenger Flow Data for the Simulation System

We have investigated the passenger flow data entering and exiting the fare collection area, from 18:00 to 18:30, using the method of digital video capture. The curve line graph of the time interval (unit: s) serial data can be drawn with the data sorted in descending order, as Fig. 6.3 shows.

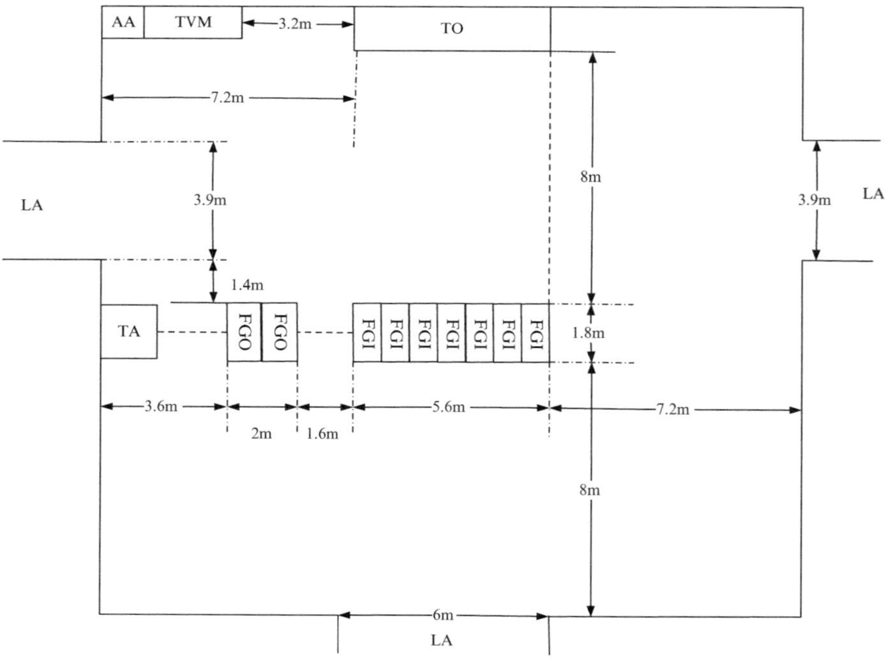


**Fig. 6.2** Schematic diagram of main facilities layout in the northern fare collection area

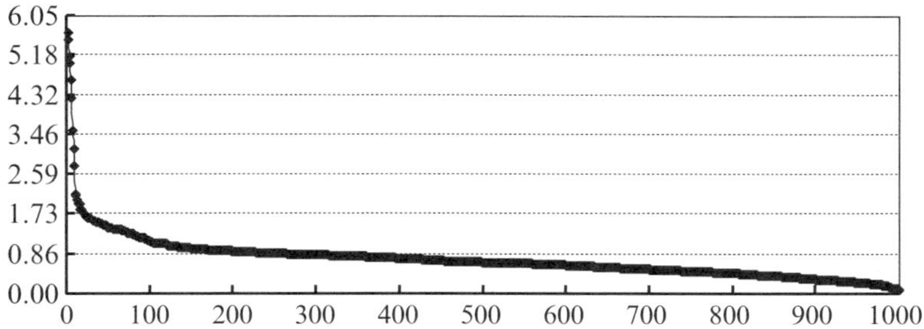


**Fig. 6.3** Time interval of passengers arriving at fare gates for entrance

**Table 6.1** The evaluation index of facilities application in the fare collection area

| Index name | Simulation results in a peak hour (%) | Field survey (%) | Index name | Simulation results in a peak hour (%) | Field survey (%) |
|---|---|---|---|---|---|
| Capacity utilization of FGI | 20.4 | 21.2 | Capacity utilization of FGO | 38.5 | 40.2 |
| Capacity utilization of TVM | 12 | 13 | Capacity utilization of TO | 28.0 | 29.0 |
| Capacity utilization of AA | 7 | 6 | Capacity utilization of TA | 0.33 | 0.30 |

Through curve fitting and probability distribution test, it can be concluded that the time interval of passengers arriving at fare gates obeys the negative exponential distribution with $\lambda = 1.33$ (namely the average time interval is 0.75 s).

In the same way, it can be concluded that the time interval of passengers arriving at fare gates for leaving obeys the normal distribution with $\mu = 1.25$, $\sigma^2 = 0.44^2$ (namely the average time interval is 1.25 s).

#### 6.3.2.2 Evaluation Results of the Simulation System

The evaluation index of facilities application in the fare collection area can be calculated using the micro simulation model and the passenger flow data mentioned above. Table 6.1 shows the simulation results of that.

### *6.3.3 The Simulation Optimization Based on Genetic Algorithm*

The number of population is assigned with a value of 20. When the value of maximum waiting time is less than 90 s, or the max queue length is less than 20, the program will be stopped.

**Table 6.2** Results of the simulation optimizing facilities configuration fare gates-for entrance

| Number of fare gates for entrance | Capacity utilization (%) | Proportion of the waiting passengers (%) | Average queue length (person) | Average waiting time (s) |
|---|---|---|---|---|
| 1 | 99.99 | 99.87 | 18.5 | 81.25 |
| 2 | 89.50 | 75.22 | 7.5 | 21.39 |
| 3 | 58.09 | 25.25 | 3.0 | 2.50 |
| 4 | 44.62 | 20.05 | 1.6 | 1.75 |
| 5 | 36.92 | 17.32 | 1.3 | 1.26 |
| 6 | 29.38 | 14.25 | 0.9 | 1.03 |
| 7 | 20.50 | 10.26 | 0.6 | 0.75 |

**Table 6.3** Results of the simulation optimizing facilities configuration for fare gates-for leaving

| Number of fare gates for leaving | Capacity utilization (%) | Proportion of the waiting passengers (%) | Average queue length (person) | Average waiting time (s) |
|---|---|---|---|---|
| 1 | 60.85 | 67.35 | 8.5 | 16.24 |
| 2 | 38.50 | 40.62 | 4.3 | 7.65 |
| 3 | 29.53 | 22.58 | 2.2 | 2.50 |
| 4 | 17.14 | 10.24 | 0.9 | 1.25 |

The results of the simulation optimizing facilities configuration can be gained using the micro simulation model of the fare collection area and the simulation optimization based on genetic algorithms, as Tables 6.2 and 6.3 show.

The simulation results in Table 6.2 show that the optimum number of fare gates for entrance is 4. In this case, the capacity utilization reaches the value of 44.62. And the average queue length and average waiting time just reach 1.6 persons and 1.75 s, respectively, meaning that the level of service is acceptable for the passenger flow demand. The practical number of fare gates for entrance is 7, which exceeds the optimum number, so it is reasonable to reduce the number of fare gates for entrance to cut down the operation cost.

Alternatively, the simulation results in Table 6.3 show that the optimum number of fare gates for leaving is 3 and the capacity utilization reaches 29.53. In this case, the average queue length and average waiting time just reaches 2.2 persons and 2.50 s, respectively, meaning that the level of service can meet the requirement of passenger flow. According to the field survey, the practical number of fare gates for entrance is 2, which is less than the optimum number mentioned above, so it is recommended to set up another fare gate for leaving. Then, the level of service will be remarkably improved.

## 6.4 Conclusion

This paper documents simulation and optimization models for facility configurations, which can be used for analyzing the adaptability of the facilities, assessing and improving the fare collection system of subway transfer hubs or stations.

Based on the application analysis results of the facility configuration types in the subway hubs or stations, it can be concluded that the parallel fare gate and ticket office is the most important mode of the three types. So the discussions mainly concern micro simulation and optimization methods for this type.

Then the simulation optimization algorithm is presented to gain the optimum number of fare gates with two parts of programs, including the simulation program and optimization method based on genetic algorithm.

Using micro simulation program and the optimization method, the facility configurations of the fare collection system in the Dong Zhi Men rail transit hub are evaluated and improved. According to the simulation results, it is reasonable to reduce the number of fare gates for entrance, and to add the number of fare gates for leaving. In this case, the level of service for the passengers pulling in or out of the fare collection system will be improved.

According to the activity types of the passenger, a transfer hub system can be partitioned into five areas, including access and egress area, fare collection area, transfer activity area, boarding and alighting platform area and assistant activity area. This research addressed evaluating and improving the facility configuration in the fare collection area. As many factors involve passenger interactions, conflict-resolution and interdependent decision-making, new simulation optimization models need to be established for access and egress area, transfer activity area, and platform area in the next phase of this study.

**Acknowledgment** This study is supported by project "The Fundamental Research Funds for the Central Universities (No. 2013JBM049)" and "National Basic Research Program of China (No. 2012CB725403)".

## References

1. Daamen W (2004) Modeling passenger flows in public transport facilities [D]. Delft University, Delft
2. Hoogendoorn SP, Daamen W (2004) Design assessment of Lisbon transfer stations using microscopic pedestrian simulation [M]. WIT Press, Dresden, pp 135–147
3. Hoogendoorn SP, Miklos Hauser, Nuno Rodrigues (2004) Application of microscopic pedestrian flow simulation to station design evaluation in Lisbon Train Stations [C]. In: TRB 2004 annual meeting, 04-3749
4. He HongYun, Su Hu (2008) Metro platform simulation system based on IVE. Comput Simul 20 (15):4073–4077
5. Ni Ju, Zhang Shuquan, Tong Chaonan (2008) Implementation of subway passenger's simulation system based on virtual reality. Comput Simul. J Syst Simul 2(4):258–262

6. Li Dewei (2007) Modeling and simulation of microscopic pedestrian flow in MTR Hubs [D]. Beijing Jiaotong University, Beijing
7. Liu Minzhu, Zhang Guobao (2004) A study of metro entry gate scheme based on Queuing system simulation. Urban Mass Transit 5:34–37
8. Kaakai F, Hayat S, El Moudni A (2007) A hybrid Petri nets-based simulation model for evaluating the design of railway transit stations. Simul Model Pract Theory 15:935–969
9. Lam WHK, Chung-Yu Cheung, Lam CF (1999) A study of crowding effects at the Hong Kong light rail transit stations. Transp Res Part A 33:401–415

# Chapter 7
# Comparative Study on the Transfer Efficiency of Triple-Line Transfer Station for Urban Rail Transit

**Chuanfeng Wan, Shaoman Liu, and Xiaobo Yan**

**Abstract** The paper defines different design conditions of triple-line transfer station type under simplified assumptions and presents the concept of transfer efficiency from the point view of urban rail transit network. Then, it builds simulation models of seven typical transfer patterns with Vissim, extracts the time of each transfer direction to compare the total transfer time by loading different passenger flows on different triple-line stations. Finally, taking into account other factors which have impact on station type selection, the paper provides guidance on lectotype design of triple-line transfer station.

**Keywords** Triple-line transfer station • Vissim simulation • Transfer efficiency • Lectotype design

## 7.1 Introduction

With the sustainable economic development and the accelerated pace of urbanization in China, the number of cities with urban rail transit constructed or planned is increasing rapidly. Rail transit in several big cities has entered the period of relatively mature network construction and operation. The urban rail transit network presents complex characteristics. As an important part of it, the transfer station plays an essential supporting role for the network transfer efficiency [1].

With the further construction of urban rail transit in our country, the number of multi-line transfer stations especially triple-line transfer station will increase [2].

C. Wan (✉) • S. Liu
School of Civil Engineering, Beijing Jiaotong University, 100044 Beijing, China
e-mail: chfwan@bjtu.edu.cn

X. Yan
China Railway Engineering Consultants Group Co., Ltd, 100020 Beijing, China

H. Xia and Y. Zhang (eds.), *The 2nd International Symposium on Rail Transit Comprehensive Development (ISRTCD) Proceedings*,
DOI 10.1007/978-3-642-37589-7_7, 

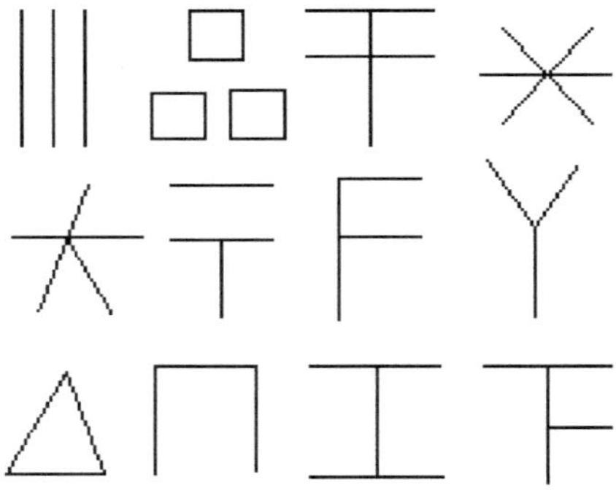

**Fig. 7.1** Triple-line transfer patterns

These stations attract a large number of passengers and the organization among transfer passengers is complex. Whether the lectotype design of triple-line transfer station is reasonable, or can meet the transfer requirements, will directly determine the efficiency of the entire network. However, the transfer station research home and abroad focuses more on two-line transfer station. There is no standard design guidance triple-line transfer station. Through modeling seven typical three-line transfer stations patterns and calculating transfer efficiency under the assumed same transfer passenger flows. The paper can provide the basis for lectotype design of triple-line transfer station.

### 7.1.1 Pattern Analysis of Triple-Line Transfer Station

For a single transfer station, the number of patterns is related to the number of lines which transfer station connects to. With the different mutual relationship of the two platforms, two-line transfer station has four different types: "十", "T", "L" and "—" [3]. Based on the four basic patterns, there will be 12 transfer types for triple-line transfer station. The specific combinations are shown in Fig. 7.1.

The transfer types of triple-line transfer station have a variety of combinations according to the difference of the line direction, the line position in space (ground, underground, elevated), and the layouts of the platforms. According to the number of line intersecting nodes, the triple-line transfer station has the following representative transfer type: "川", "品", "Y", "大", "H", "F", "△".

## 7.2 Method on Transfer Efficiency Analysis

### 7.2.1 Assumptions on Facilities

For the different transfer types, the layout and design of the facilities follows Code for Design of Metro. The maximum capacity of corresponding facilities is calculated with maximum passenger flows which is 14,400 people per hour.

Transfer passages are assigned with the width of 5 m without middle railings. Transfer stairs is 4 m wide. Stairs at the end of platform is 6 m wide. The escalator width is 1 m. All the platforms are compatible with 6 B-type cars with the width of 13 m island platform.

## *7.2.2 Assumptions on Simulation Model*

### 7.2.2.1 Basic Assumptions for Simulation Model

The model used in this study simulates the passengers' dynamic changes for triple-line transfer station based on the following basic assumptions [4]:

1. The platform, the station hall, stairs, escalators are simulated with different segments;
2. The passengers decide the path and, move with pre-assigned speed;
3. The model uses the time step of 10 simulation seconds/seconds;
4. The arrival of each train is independent and randomly;
5. The simulation will be finished when all passengers reach their expected transfer platform.

### 7.2.2.2 Assumption on passenger speed

Most passengers in subway hub have their own specific purposes especially in morning/afternoon peak hours commuters. Because of large passenger flow, there will be queuing or congestion for some facilities. So, the walking speed of passengers will greatly reduce. In this paper, the simulation speed for passenger flow is 1.1 m/s which is higher than peak hour as normal situations.

Passenger' speed on staircase and escalator is different from station hall and platform. The stair speed is 0.7 m/s and escalators speed is 0.65 m/s. When passengers reach the stairs or escalators, the walking speed changed to the pre-assigned speed.

### 7.2.2.3 Assumption on Transfer Facilities

Transfer facilities include stair, escalator, station hall and platform. Passengers' walking in the station hall and platform is similar to that in the segment. So the station hall and platform can be simulated as segment. The capacity of the stair and escalator is based on their width and numbers. So, it is possible to match simulation capacity with actual capacity by simulating stair and escalator as segment and taking speed deceleration measures. The platform layer, station hall, stair and escalator are all simulated with actual sizes.

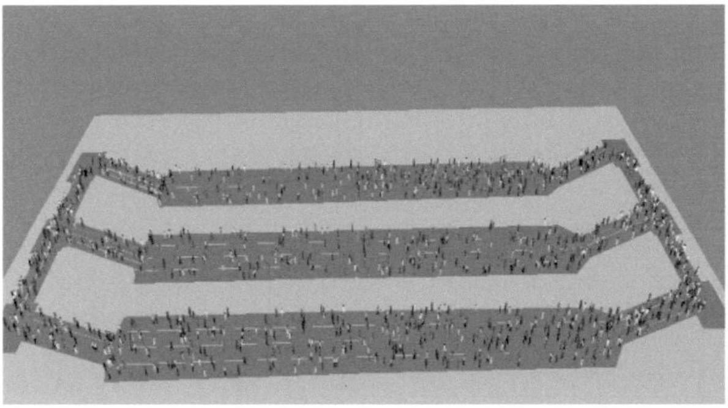

**Fig. 7.2** Simulation model for pattern of "川"

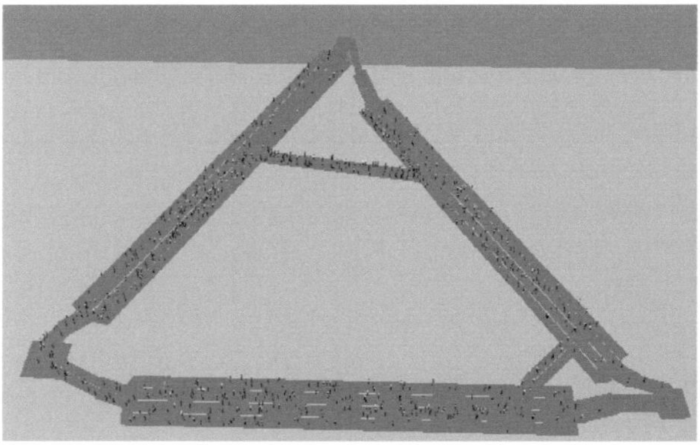

**Fig. 7.3** Simulation model for pattern of "Δ"

#### 7.2.2.4 Rule of Passenger Route Choice

The model assigns the same proportion of passenger route choice according to different transfer directions in the seven triple-line transfer station, and passenger walking route in the model is the same as the actual transfer routes.

### *7.2.3 Simulation and Calculation*

We Simplify passenger flow direction of triple-line transfer station into a triangle, each edge is a primary transfer direction, so each edge has eight transfer directions. The paper mainly simulates and calculates the transfer time of the three edges, and compares the differences of transfer patterns. The simulation process assumes that the transfer passenger is consistent which means the same passenger flows and consistent changes for different transfer direction.

Simulation models of seven typical three-line transfer stations are built as follows; the operational process is showed in the following Figs. 7.2, 7.3, 7.4, 7.5, 7.6, 7.7, and 7.8.

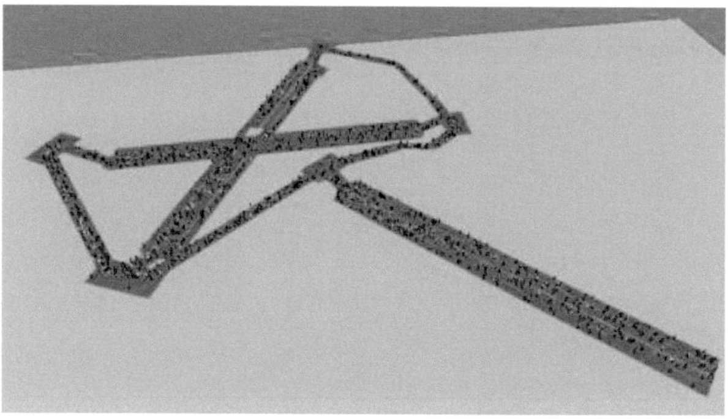

**Fig. 7.4** Simulation model for pattern of "大"

**Fig. 7.5** Simulation model for pattern of "Y"

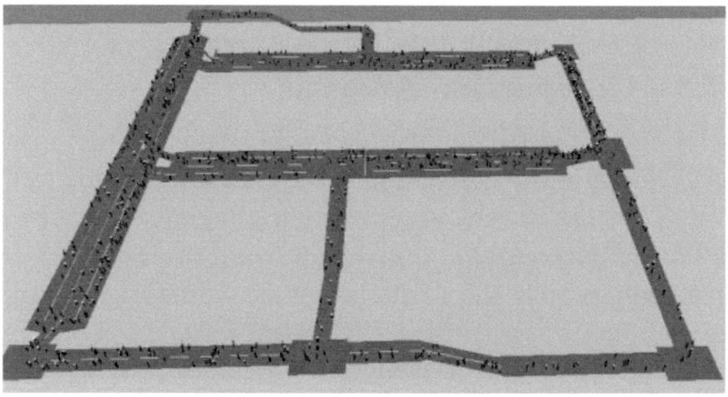

**Fig. 7.6** Simulation model for pattern of "F"

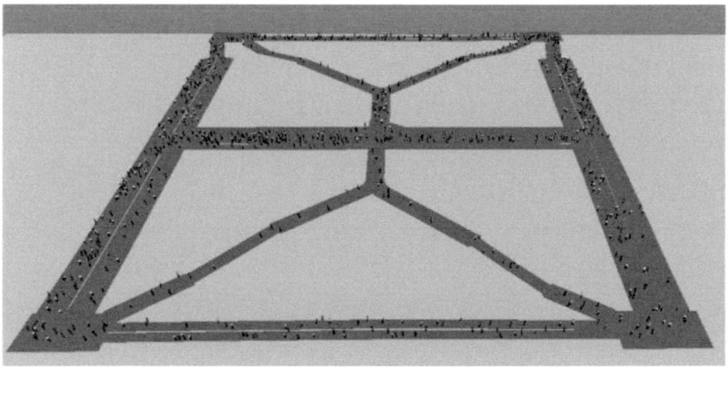

**Fig. 7.7** Simulation model for pattern of "H"

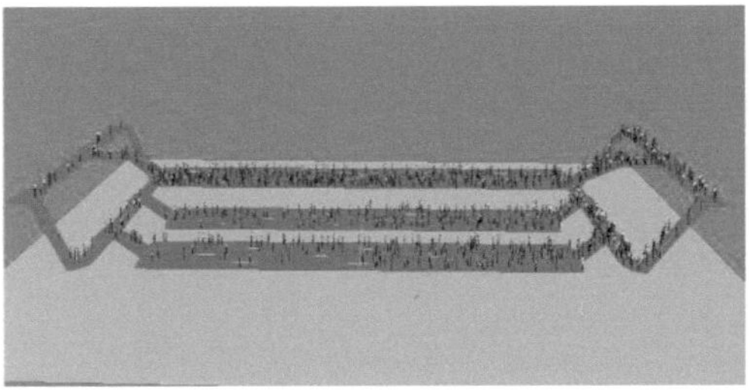

**Fig. 7.8** Simulation model for pattern of "品"

**Table 7.1** Transfer time

| Type/passenger | 500 | 1,000 | 1,500 | 2,000 | 2,500 | 3,000 | 3,500 | 4,000 | 4,500 | 5,000 |
|---|---|---|---|---|---|---|---|---|---|---|
| 川 | 350 | 701 | 1,057 | 1,418 | 1,788 | 2,163 | 2,546 | 2,941 | 3,336 | 3,740 |
| Δ | 460 | 921 | 1,387 | 1,858 | 2,336 | 2,831 | 3,329 | 3,841 | 4,251 | 4,877 |
| 大 | 622 | 1,245 | 1,872 | 2,506 | 3,149 | 3,802 | 4,463 | 5,128 | 5,807 | 6,502 |
| Y | 630 | 1,261 | 1,897 | 2,538 | 3,188 | 3,855 | 4,538 | 5,234 | 5,940 | 6,650 |
| F | 586 | 1,154 | 1,740 | 2,336 | 2,937 | 3,562 | 4,198 | 4,847 | 5,506 | 6,177 |
| H | 523 | 1,047 | 1,576 | 2,117 | 2,651 | 3,214 | 3,788 | 4,377 | 4,974 | 5,581 |
| 品 | 400 | 801 | 1,213 | 1,637 | 2,073 | 2,520 | 2,974 | 3,442 | 3,913 | 4,398 |

Passenger flows is loaded from 500 to 5,000 people per hour for 24 transfer directions in each model, and total transfer time is extracted shown in Table 7.1.

## 7.3 Comparison Analysis

As shown in the above chart, the transfer time of "川" is minimum, and the type of "Y" is maximum. According to the definition of transfer efficiency, when the transfer passenger is consistent, the efficiency is higher with the least transfer time, in which the sequencing of transfer efficiency is: "川" > "品" > "Δ" > "H" > "F" > "大" > "Y" (Fig. 7.9).

With the same passenger flows and consistent change of different transfer direction, the transfer time is mainly affected by the transfer distance. Pattern of "川" transfer station realizes six transfer directions while pattern of "品" realizes four directions. So, the efficiencies of these two types of transfer stations are higher than other patterns. The transfer distance for the pattern of "大" and "Y" is long, so the total transfer time is also long compared to others, and the transfer time of these two patterns is very close.

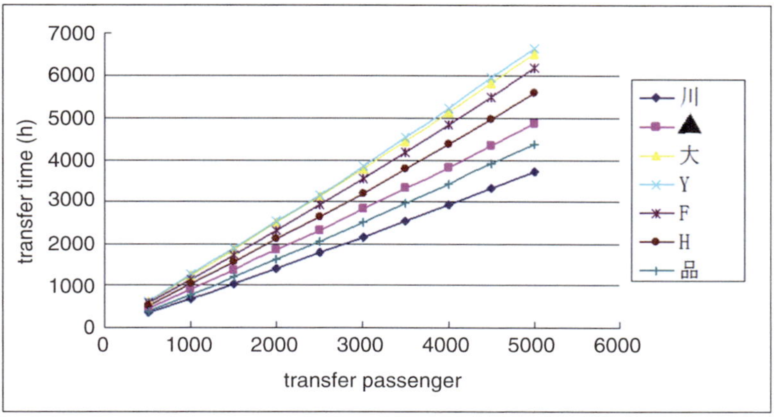


**Fig. 7.9** Comparison chart of results

## 7.4 Reasonable Pattern for Triple-Line Transfer Station

The triple-line transfer station is a huge project. It's hard to be finished at the same time. The more passenger flow direction in transfer stations, the more difficult to organize the passenger flow. The following factors should be considered when selecting the transfer station pattern:

### *7.4.1 Construction*

The construction cost directly affects the investment on subway because it accounts for a large part of the total cost. In general, the construction cost for underground station is higher than elevated station while the cost of ground station is minimum [5]. The construction cost of cross-platform interchange is higher than ordinary one, and the cost is also higher if not implemented at the same time. So, the pattern of-"川" transfer station is hard to be done if considering the feasibility of construction.

### *7.4.2 Project Reserved*

Combining the construction planning of the subway lines, it is less likely to construct three lines in the same time. Different type of transfer station has different reserved project, and the difficulty to reserve has a great difference. Considering the conditions of reservation node and transfer channel interface by stages, the more line intersections are, the more convenient the reserved project is. Therefore, the reserved works of type-" Δ" and " H" transfer station is convenient.

It's hard to construct three lines at the same time. Different transfer station pattern has different reserved project. Considering the conditions of reservation node and transfer passage interface for different phases, the more in pattern of "Δ" and "H" transfer station is more convenient for reserved works compared to others.

### *7.4.3 Transfer Passenger*

Transfer passenger volume is a very important part to select the transfer station pattern. Based on the changes of passenger flows, appropriate pattern of transfer station can be chosen to improve the transfer efficiency with the reduction of construction investment [6]. From the simulation, we can see that:

When passenger flow of each transfer direction are equal or have little difference between them, The cross-platform interchange for some direction should be selected as far as possible based on the line routes in order to achieve higher transfer efficiency. But, when they are not equal, if the transfer passenger flow is less on shorter transfer distance, the advantages of transfer fail to reflect obviously.

## 7.5 Conclusions

Based on the comparative analysis on transfer efficiency of triple-line station, the number of transfer nodes and the presence of the cross-platform interchange, the efficiency order is: The station with cross-platform has the highest efficiency; Usually, the transfer efficiency of triple-line station is based on the nodes. The efficiency with the intersection of three nodes is higher than that of two nodes which is higher than that of one node.

When planning and designing urban rail transit network, it's important to design line directions and transfer pattern appropriately. Triple-line transfer station should first ensure cross-platform interchange, and then try to realize the intersection of each two lines. Considering the aspects of construction costs, project reserved, the convenience for transfer passengers and the accessibility of the passenger flow organization and management, The order of transfer efficiency for triple-line station pattern is pattern of "Δ", "品", and "H".

## References

1. Wang Bo, Li Xiaoxia, An Shuanzhuang (2010) Analysis on transfer passenger flow characteristics and station design of urban rail transit[J]. Urban Rapid Rail Trans (2):55–58
2. Gao Feng, Wang Zhichen, Wang Lijuan (2008) Evaluation of transfer efficiency for urban rail transit [J]. Railway Transport and Economy 1:74–76
3. Zheng Pingping, Sun Haiwen (2011) Rail transit multi-line transfer design[J]. Friend Sci Amateurs (6):126–127
4. Zhao Yan (2008) Transfer efficiency analysis and countermeasures research of integrated urban passenger transport terminal [D]. Southwest Jiaotong University, Chengdu
5. Zhao Yugang (2011) On interchange of urban rail transit considering level of service [D]. Beijing Jiaotong University, Beijing
6. Gu Baonan, Ye Xiafei (2010) Urban rail transit engineering [M]. Huazhong University of Science and Technology Press, Wuhan

# Chapter 8
# Analysis on the Economic Effect of Urban Mass Transit Based on System Dynamics

**Yaodong Zhou**

**Abstract** The paper has an empirical research on urban economic effect of the rail transit based on system dynamics model. The result is on the 2007–2010 Beijing Transit Rail data, and concludes that (1) economic effects of transit rail is defined as the accessibility improvements caused by the concentration of population, and thus the formation of value of the land to enhance the prosperity of commercial services, and promote regional economic development; (2) empirical research from Beijing transit rail has proved that the increasing of passenger capacity is positive correlated with its economic effect; (3) In the short term, stimulation coefficient increased passenger traffic to improve the speed of its economic effects, In long term, the increasing of passenger capacity coefficient has little relationship with regional economic growth and the economic effect is even weakening.

**Keywords** Urban mass transit (UMT) • System dynamics • Influence coefficient

Because of the great capacity, quickness and convenience, Urban Mass Transit (UMT) has become an important long-distance vehicle in big cities. The construction of UMT has significant effect on regional economic growth. TOD (transit-oriented development) pattern will be dominant trend for urban development. Taking Beijing as the example, this chapter tries to analyze the economic effect of UMT with system dynamics model.

---

Y. Zhou (✉)
School of Economics and Management, Beijing Jiaotong University, BJTU, Beijing, China
e-mail: ydzhou@bjtu.edu.cn

H. Xia and Y. Zhang (eds.), *The 2nd International Symposium on Rail Transit Comprehensive Development (ISRTCD) Proceedings*,
DOI 10.1007/978-3-642-37589-7_8, 

## 8.1 Introduction

There are many papers on the economic effect of UMT. The available published documents mainly focus on the relationship between UMT and land value. Earlier research usually used comparison methods. Rice University Joint Center for Urban Mobility Research [1] studied the Lindenwold UMT in Philadelphia and found house prices rose 6.9 % to 4,500 dollars with UMT development. Han B-R [2] analyzed land and property price (1978–1991) along No. 2 and No. 4 Seoul subway. When two lines began running, commercial and residential property value increased a lot and commercial property price rose higher than residential property. Laakso [3] studied land value along the subway in Helsinki Finland and found land value within 1,000 m raised 6 %. Center for Land Policy Studies, UK [4] thought that London subway Jubilee had brought 1.3 billion since the subway began running.

On the study of specific price model, focused on the Rapid Transit system in California, Robert Cervero [5] found that commercial land value increased 120 % in urban center and 23 % in other area. M.-Y. Pior [6] analyzed the relationship between Tokyo Joban Shinkansen and land value with GIS technology. With the investment of 700 million JPY, Tokyo Joban Shinkansen had run 8 years and brought 20 trillion JPY economic returns. Boucq [7] pointed that T2 tramway influenced the surrounding house price significantly (housing price raised 5 %). Deloukas [8] reported that Athens No.2 subway which run for Athens Olympic Games 2004 played key role in surrounding house price rising.

More domestic economists do research on UMT with the development of China UMT. There're many papers on the relationship between UMT and land value. On the study of influence area, Zheng Jiefen [9] said the significant influence area of Shenzhen Metro is within a radius of 400–600 m. The average increase is 23.03 % within 400 m radius and 16.95 % within 600 m radius. Wang xia [10] said the influence area of Beijing No. 13 subway is within 1km. Linxiaoyan [11] analyzed residential price along the Metro line with land value function model. According to the research, the distance between Metro station and city center is negative related with residential price fluctuation. Liu Qinyuan [12] analyzed Shanghai No. 4 Metro line and found the residential price influence area is within 1,000 m. On the study of influence factors, Gao Xiaohui [13] selected 500 sample data and found UMT had different influence on surrounding residential price in different area. The influence factors include geographic location, commercial concentration, infrastructure construction, etc. Some research focus on other field, such as time effect, environment effect, traffic effect, economic effect and so on [14, 15].

Nowadays, present research on UMT economic effect pays more attention on the relationship between UMT and land increment. The past studies are mainly from micro aspect. This paper will build system dynamics model to analyze UMT economic effect in macroscopic view.

## 8.2 Economic Effect of UMT

The main function of UMT includes: (1) UMT relief urban tense transportation and complete public transportation system. (2) People spends less time on traffic and individual labor productivity raises.

By improve the accessibility, UMT shorten the distance between customers and producers. UMT also has great diffusion and assemble effect on population migration. The economic effect of UMT has two aspects:

1. Inductive Effect. With the construction and operation of UMT, plenty of residences, shopping malls, hospitals and school set up surround the Metro line. These economic activities around Metro line optimize industrial structure and provide more jobs for thousands of people. On the other hand, people gathering stimulate traffic demand and promotes UMT construction.
2. Corridor Effect. By improve the accessibility, business return along the Metro line rises. UMT improves commercial environment and attracts more business opportunities. Capital, information and technology resources are gathered and explored. Because of UMT development, regional economies grow quickly.

Therefore, the economic effect of UMT based on the improvement of accessibility. The development of UMT increases passenger volume and enhances surrounding land value. Regional investment is improved and service industry grows rapidly. UMT promotes commercial prosperity and economic growth (Fig. 8.1).

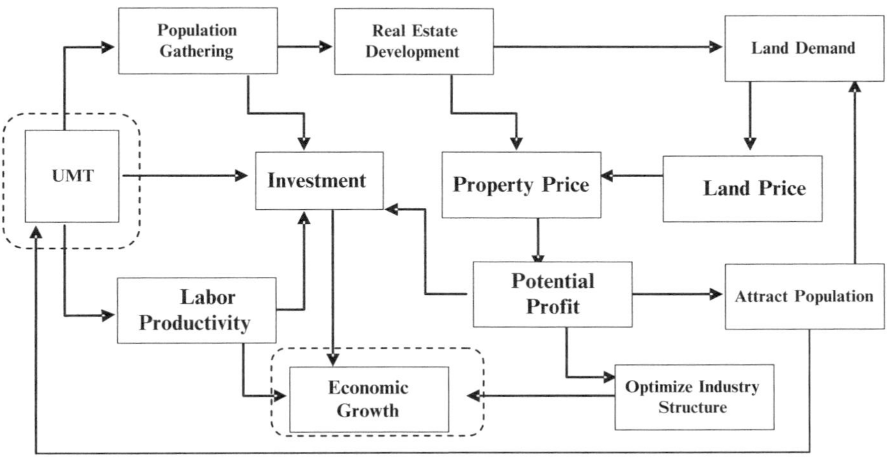


**Fig. 8.1** Interaction between UMT and economic factors

## 8.3 Analytical Methods

The paper has a positive research which rail transit has an impaction on urban economic with system dynamics method. The method of system dynamics is considered as a kind of tools of the use of computer technology to the system simulation, in order to deal with the complex, nonlinear, and system problems with the delayed signals through a structured approach and analysis of information feedback. The Professor, Jay. W. Froroster, thought system dynamics model should be made up of five aspects: system boundaries, dynamic hypothesis, structural equation, simulation and testing, parameter adjustment and policy application [16]. The advantage of the system dynamics method to study the economic effects of rail transit is: (1) to solve complexity problems. Because of complexity urban economic effects, it is difficult to set up univariate regression equation to explain relationship amongst the economic effect variables, but system dynamics are able to build more complex model of economic effects through the expression of causality relationship; (2) to solve progressive relationship among variables. Meanwhile, there are kinds of mutual transmission between variables in UMT economic effect. It is also difficult to express progressive relationship between variables through multi-factors regression equation, but method of system dynamics is conducive to combing the logical progressive relationship between variables.

Section 8.3.1 defines the boundaries of the system and set up a series of indicators UMT economic effects; Sect. 8.3.2 builds system flow diagram and structural equation to express economic effect relationship between UMT and urban economic; Sect. 8.3.3 tests results through simulation; and Sect. 8.3.4 is parameter adjustment and policy application.

### *8.3.1 Indicators Selection*

This chapter will use length of Metro line to describe UMT condition and use passenger volume as the main indicator to connect UMT with regional economy (Tables 8.1 and 8.2).

### *8.3.2 Flow Graph and Model Equation*

In system dynamics model, flow graph can show the causation relationship among influence factors. UMT influence indicators include operating length and passenger volume. The economic indicators mainly include population gathering, the tertiary industry investment, regional GDP and social fixed assets investment. UMT influences regional economic growth, in return, these economic indicators also will

**Table 8.1** Indicator illustration and calculation

| Evaluating indicators | Statistical indicators | Indicator illustration and calculation |
|---|---|---|
| UMT development | Operating length of UMT and passenger volume | The increase of operating length and passenger volume will reflect UMT investment and construction level |
| Regional economy | GDP | GDP could reflect the overall level of regional economic development |
| Social fixed assets investment | Social fixed assets investment | Social fixed assets investment could reflect the level of regional investment |
| The tertiary industry development | The tertiary industry GDP | UMT mainly stimulates the tertiary industry to affect regional economic growth |
| Regional population gathering | Residential population | UMT promotes regional population gathering and residential population is one of the main indicators to reflect regional population |
| Regional land value | Land transaction price index | Land transaction price index could reflect the fluctuation of regional land value |
| Real estate investment | Real estate investment | Real estate investment could reflect the real estate development level |
| Service industry investment | Service industry investment | Service industry investment could reflect the development level of tertiary industry except the real estate industry |

Data of indicators above is mainly from Beijing Statistical Yearbook (2008–2010), part of data is from Urban Traffic Yearbook and websites

**Table 8.2** Part of original data

| Year | | 2007 | 2008 | 2009 | 2010 |
|---|---|---|---|---|---|
| Operating length of UMT (km) | | 142 | 198 | 228 | 336 |
| Passenger volume | Hundred million persons | 6.55 | 12.17 | 14.33 | 18.49 |
| Residential population | Ten thousand | 1,633 | 1,695 | 1,755 | 1,961 |
| Land transaction price index | | 109.4 | 111.6 | 114 | 116.5 |
| Real estate investment | Hundred million RMB | 1,995.8 | 1,908.7 | 2,337.7 | 2,901.1 |
| Service industry investment | Hundred million RMB | 1,470 | 1,525.7 | 2,051.8 | 2,021.2 |
| The tertiary industry GDP | Hundred million RMB | 7,236.1 | 8,375.8 | 9,179.2 | 10,330.5 |
| GDP | c RMB | 9,846.8 | 11,115 | 12,153 | 13,777.9 |
| Social fixed assets investment | Hundred million RMB | 3,966.6 | 3,848.5 | 4,858.4 | 5,493.5 |

influence UMT investment and then have effect on operating length of UMT. The flow graph is shown in Fig. 8.2.

From the flow graph, we can see that the economic effect of UMT mainly work through the increasing of passenger volume. Passenger volume increase originates from operating length extending and length expanding needs more investment. Recent years, Beijing UMT investment keeps growing quickly. According to

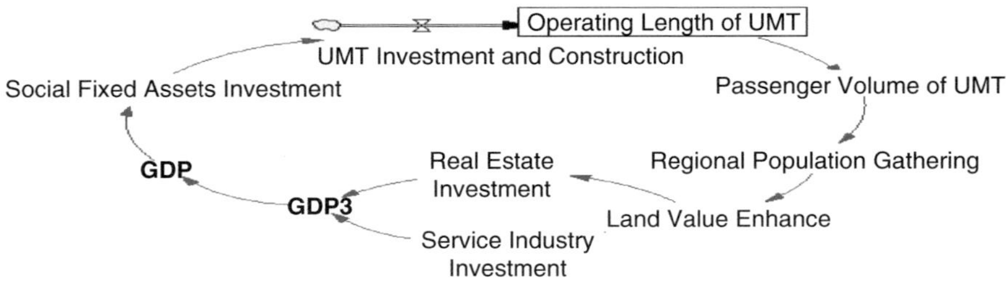


**Fig. 8.2** Flow graph of UMT effect on urban spatial distribution

Beijing UMT planning, the total investment will be 500 billion yuan and operating length will reach 561 km till 2015. On that day, the daily passenger volume will be eight million. After 2015, the investment will decrease and the increasing of operating length will slow down. The path of UMT investment and operating length is nonlinear, therefore, system dynamics model will use Logit function. The equations are as follows:

1. UMT Investment = LN (Social Fixed Assets Investment) × Influence Coefficient (Social Fixed Assets Investment – UMT investment)
2. UMT Operating Length = INTEG (UMT investment, Initial Operating Length)
3. UMT Passenger Volume = LN (UMT Operating Length) × Influence Coefficient (UMT Operating Length – UMT Passenger Volume)
4. Regional Population Gathering = UMT Passenger Volume × Influence Coefficient (UMT Passenger Volume – Regional Population Gathering)
5. Regional Land Value = Regional Population Gathering × Influence Coefficient (Regional Population Gathering – Regional Land Value)
6. Real Estate Investment = Regional Land Value × Influence Coefficient (Regional Land Value – Real Estate Investment)
7. Service Industry Investment = Regional Land Value × Influence Coefficient (Regional Land Value – Service Industry Investment)
8. GDP3 = Real Estate Investment × Influence Coefficient (Real Estate Investment – GDP3) + Service Industry Investment × Influence Coefficient (Service Industry Investment – GDP3)
9. GDP = GDP3 × Influence Coefficient (GDP3 – GDP)
10. Social Fixed Assets Investment = GDP × Influence Coefficient (GDP – Social Fixed Assets Investment)

According to the equations above and Beijing statistics from 2007 to 2010, we do regression analysis and get the indicators' influence coefficients as follows Fig. 8.3. From Fig. 8.3, it is depicted that mutually increasing relationship exists between each variable, because of all coefficients positively. From in the degree of effect influence, the amount of coefficients of land value-real estate investment (134) and land value-service industry investment (90) get to the highest level in all coefficients. Secondly highest level is coefficient of passenger volume to population gathering (26). Others is all small.

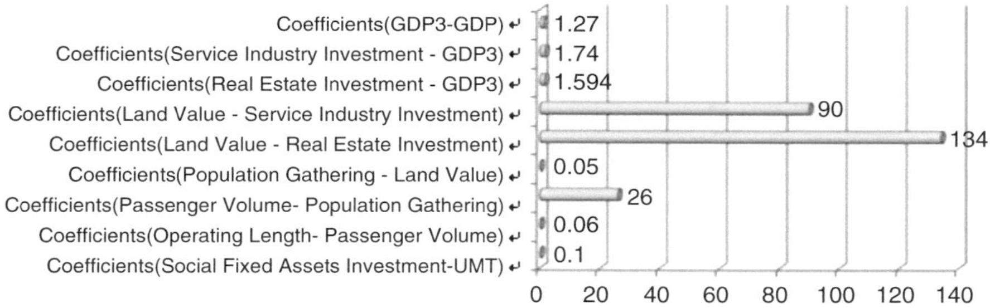


**Fig. 8.3** UMT economic effect influence coefficients

**Table 8.3** Comparison between simulation and empirical data of residential population

| Year | Simulation results | Empirical data | Error | Error degree (%) |
|---|---|---|---|---|
| 2007 | 1,644.84 | 1,633 | 11.84 | 0.73 |
| 2008 | 1,730.15 | 1,695 | 35.15 | 2.07 |
| 2009 | 1,799.47 | 1,755 | 44.47 | 2.53 |
| 2010 | 1,865.57 | 1,961 | 95.43 | 4.87 |

### *8.3.3 Model Testing and Simulation*

To unify the statistic standard, we compare the empirical data with simulation results to test the credibility. This chapter chooses residential population as example, the error results are as follows (Table 8.3):

From the table we can see that 4 years' error degree between simulation results and empirical data are all under 5 %. Usually error degree under 5 % is acceptable.

We simulate Beijing UMT situation from 2007 to 2025 with Vensim software. The simulation diagram is as follows (Fig. 8.4):

The simulation shows that UMT expanding could promote regional economic development. With the investment and construction of UMT, passenger volume increases every year. More people move to the area along Metro line and regional land value enhances a lot. At the same time, rising land price stimulates the increasing of real estate and service industry investment. The tertiary industry fixed assets investment grows fast and makes great contribution to regional economic development. In return, regional economic development increases social fixed assets investment and UMT investment. Based on abundant capital support, more metro lines are explored and passenger volume keeps rising. UMT development interacts with economic factors which forms a virtuous cycle.

### *8.3.4 Parameter Adjustment and Policy Application*

Sensitivity analysis can forecast indicators fluctuant trends by adjusting the influence coefficients. According to sensitivity analysis, all economic indicators are

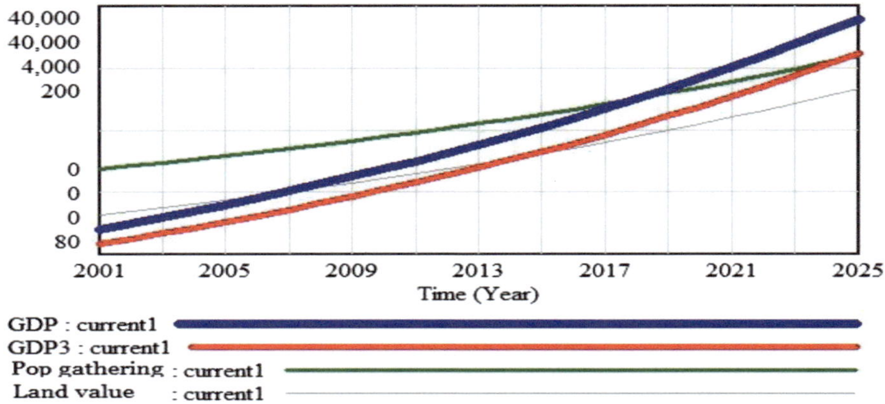


**Fig. 8.4** Simulation diagram of system dynamics analysis

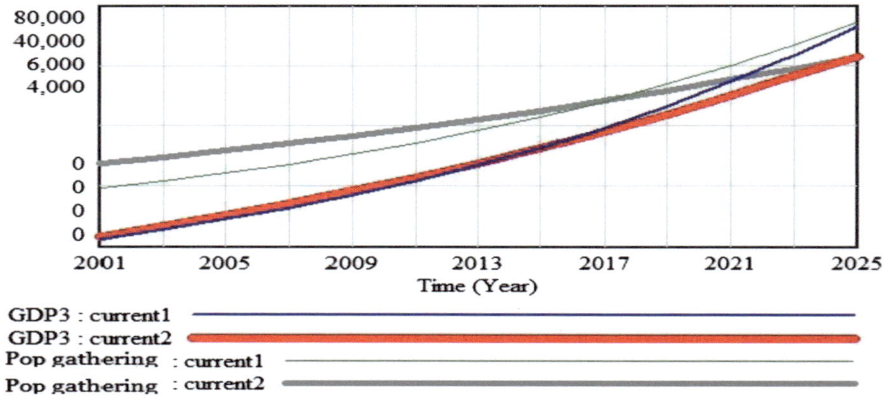


**Fig. 8.5** The sensitivity of passenger volume influence coefficient to GDP3 and population gathering

positive relevant with UMT passenger volume. By increasing influence coefficients, GDP grows faster.

We use Automatically simulate on change function of Vensim software to run synthesim. By adjusting passenger volume influence coefficient to population gathering, we do sensitivity analysis of GDP3 and GDP. In Figs. 8.5 and 8.6, passenger volume influence coefficient of Current 2 is higher than Current 1. From the sensitivity analysis we can see: (1) By increasing the passenger volume influence coefficient, population assemble effect is more significant and GDP grows faster. (2) Around 2020 in Current 2, population assemble effect begins weaken. Correspondingly, growth of GDP3 and GDP in Current 2 is lower than Current

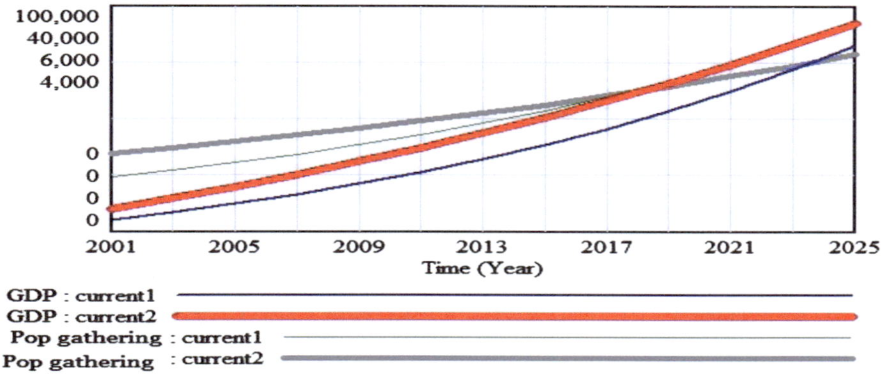


**Fig. 8.6** The sensitivity of passenger volume influence coefficient to GDP and population gathering

1 after 2020. This demonstrates that the stimulation of UMT on economic growth is slowdown. Likewise, we do sensitivity analysis to other influence coefficients and get the same conclusion.

## 8.4 Conclusion

By improve the accessibility, the development of UMT promotes population expand and enhances land value. The expanding of UMT brings business and service industry boom. According to the analysis we can conclude that: (1) Beijing UMT passenger volume is positive relevant with economic effect. (2) The influence coefficient has critical point. In short term, increasing the influence coefficient could enhance the economic effect significantly. In long term, increasing the influence coefficient has little relationship with regional economic growth. In existing technology condition, influence coefficient of UMT has notable positive correlation with regional economic development before 2020. However, economic sensitivity to UMT is weakening.

**Acknowledgement** This chapter get support from Beijing Jiaotong Scientific Research Fund: The Research of UMT Impact on Beijing Spatial Distribution (KBJB10012536) and Beijing Social Science Plan Project: Research on Institution Problems and Policy of Comprehensive management of Traffic Congestion in Beijing (11JGB034).

## References

1. Joint Center for Urban Mobility Research, Rice University (1987) Assessment of changes in property values in transit areas [R]. Surrey: ERA, pp 14–15
2. Han B-R (1991) Neighborhood land value changes from subway construction: case study generalized least squares [J]. Dankook Univ Reg Stud 11:125–146
3. Laakso S (1992) Public transportation investment and residential property values in Helsinki [J]. Scand Hous Plan Res 9:217–229
4. Riley D (2001) Taken for a ride: trains, taxpayers, and the treasury [J]. Center for Land Policy Studies, London
5. Cervero R, Landis J (1997) Twenty years of the bay area rapid transit system: land use and development impacts [J]. Transp Res Part A Policy Pract 31(4):309–333
6. Pior M-Y, Shimizu E (2001) GIS-aided evaluation system for infrastructure improvements: focusing on simple Hedonic and Rosen's two-step approaches [J]. Comput Environ Urban Syst 25:223–246
7. Boucq E (2007) The effects accessibility gains on residential property value in urban areas: the example of the T2 Tramway in the Hauts-de-Seine Department [R]. In: Proceedings of the 2007 AET conference, Boston
8. Deloukas A (2008) Capitalisation of urban rail investments in land value: the case of Athens [R]. In: Proceedings of the 2008 AET conference, Leeuwarden
9. Zheng Jie-feng, Liu hong-yu (2004) The impact of URRT on house prices in Shenzhen [J]. J China Railw Soc 27(5):11–18 (in Chinese)
10. Wang Xia, Zhu Dao-lin, Zhang Ming-ming (2004) Analysis of the influence of rail transit on the distribution layout of real estate prices – a case study of Beijing light rail line NO. 13 [J]. Urban Prob 6:30–42 (in Chinese)
11. Chen You-xiao, Lin Xiao-yan, Liu Yun-hui (2005) Impacts of rail transport construction on land value, valuation model and empirical study – case study on Beijing Light Rail [J]. J Beijing Jiaotong Univ (Soc Sci Ed) 4(3):7–13 (in Chinese)
12. Liu Qing-huan, Ke ren (2007) Analysis of the influence of rail transit on house prices in Shanghai Light Rail line NO. 4 [J]. Contemp Econ 12:112–114 (in Chinese)
13. Gao Xiao-hui, Liu Fang (2011) Impact on URRT on house prices – a case study of Shanghai [J]. Urban Probl 12:41–46 (in Chinese)
14. Xu Yi (2004) The effect of rail transport on urban environment and economy [J]. Urban Public Transp 4:19–21 (in Chinese)
15. Zou Wei, Liu Shu Guang (2008) Study between unit investment and financing models and profit model based on the sustainable development [J]. J Beijing Inst Financ Commerce Manage 6(2):28–31 (in Chinese)
16. Forrester JW (1968) Principles of systems[M], 2nd ed. Pegasus Communications, Waltham

# Chapter 9
# The Arrival Passenger Flow Short-Term Forecasting of Urban Rail Transit Based on the Fractal Theory

**Liangliang Zhang, Yuanhua Jia, Xihui Yin, and Zhong-hai Niu**

**Abstract** According to fractal characteristics of the arrival passenger flow volume of the urban rail transit station, the paper established the arrival passenger flow volume short-term forecasting model based on the fractal theory. Then, the paper took the *AHQB* station of MRT Line Four in Beijing as an example and forecasted the short-term arrival passenger flow volume. Lastly, the paper gave a comparison between the predicted value and the actual value, and found that the maximum error was less than 7 %. It indicated that the model established could be used to forecast the short-term arrival passenger flow volume of the urban rail transit, and had better adaptability.

**Keywords** Urban rail transit • Arrival passenger flow volume • Fractal theory • Short-term forecasting

## 9.1 Introduction

Urban rail transit is the important part of the urban comprehensive traffic and transportation, and plays a great role in guiding and improving construction location of city. In recent years, urban rail transit has developed especially rapidly in China. In some big cities, the urban rail transit network has begun to form rail transit network, for example Beijing, Shanghai, Guangzhou and so on. Other cities are also actively prepared for the construction of urban rail transit. According to an urban rail transit planning, there are about 40 cities that have urban rail transit by 2020, in China, and the total mileage of the rail transit will be 7,000 km.

Not only planning and construction, but also the department organized the passenger flow when it had put into operation, did urban rail transit passenger

L. Zhang • Y. Jia (✉) • X. Yin • Z-h. Niu
Institute of Systems Engineering and Control, School of Traffic and Transportation, Beijing Jiaotong University, Beijing, China
e-mail: 10114212@bjtu.edu.cn

H. Xia and Y. Zhang (eds.), *The 2nd International Symposium on Rail Transit Comprehensive Development (ISRTCD) Proceedings*,
DOI 10.1007/978-3-642-37589-7_9, © Springer-Verlag Berlin Heidelberg 2014

flow volume forecasting be very significant. At present, there were many researches on urban rail transit passenger flow volume forecasting in domestic and foreign. They used Four-stage Model [1–3] and Disaggregate Model [4, 5] for forecasting urban rail passenger flow volume, but they mainly focused on the long-term forecasting. At the same time, domestic and foreign researcher used linear neural network [6–8] and other methods which used a historical data of some day to forecast passenger flow volume of other day. According to analyzing the history of their researches and practices, the paper found that current researches on urban passenger flow volume forecasting were more macro, and could not reflect the unbalanced passenger flow, especially the peak hours and off-peak hours. Therefore, in order to achieve short-term arrival urban rail station passenger flow volume forecasting research, the paper established the short-term arrival passenger flow volume forecasting model based to the fractal theory, and hoped the method and the forecasted result could be used for passenger flow organization and optical allocation facilities.

## 9.2 Fractal Theory

The fractal theory was founded by B.B. Mandelbrot, and could effectively describe a variety of complex graphics in nature and scientific research [9]. Since it was founded, the theory had been widely used in mathematics, physics, and so on. In recent years, it was used for many other fields (for example, the linguistics, economics, and social science), and became modern and an important branch of nonlinear science. Along with in-depth research and applications expanded in more and more fields, it contained several meanings [10].

1. Fractal could be geometry and also could be mathematical models that constructed by using the functions and information.
2. Fractal had self-similarity of three aspects including the morphology, function and information, also could have self-similarity of one aspect.
3. Fractal set has some form of self-similarity, and may be approximate self-similar or statistical self-similar.
4. Self-similarity of fractal had some variation in the level, and fractal has infinitely nested hierarchy in mathematics.
5. Fractal could contain complex detail of all elements in different scale, and also could give a quantitative description about fractal geometry of the complex nature.

## 9.3 Model Formulas

### 9.3.1 *Fractal Model*

The fractal dimension was a significant parameter that could be used to describe quantitatively characterization of fractal. Based on the definition of fractal theory, the fractal distribution could be defined a formula of power-exponent [11].

$$N = \frac{c}{r^{D}} \tag{9.1}$$

Where N which related r variable was a variable, r variable was characteristic length (time, length, etc), D is fractal dimension, and c was a constant.

Then the paper used natural logarithm (base e) to change the formula (9.1), and obtained the formula (9.2).

$$\ln N = \ln c - D\ln r \tag{9.2}$$

In common, D which was fractal dimension was a constant, so that the formula (9.2) was an approximately straight line in the logarithmic coordinates. Therefore, according to two points ($\mathbf{N_i},\mathbf{r_i}$) and ($\mathbf{N_j},\mathbf{r_j}$) of the approximately straight line, the paper could calculate D and c.

$$D = \frac{\ln\left({}^{N_i}/_{N_j}\right)}{\ln\left({}^{r_j}/_{r_i}\right)} \tag{9.3}$$

However, the raw data could only get some discrete points in practical application, and the function was unknown. In this case, the paper needed to use the transformed method, so that the data could be associated with a fractal distribution model when they were done a series of transformation. In prior researches, a commonly transformed method is CUSUM. The paper gave the detailed steps in following content.

(1) The paper layout the point ($N_i$,$r_i$), ($i = 1,2,3,\ldots, n.$) in double logarithmic coordinate. The raw data may be some discrete points in common. The paper must use transformed method if the raw data was discrete points.
(2) The paper established various orders CUSUM that was $\left\{S_i^{(n)'}\right\}$: first order was $S_i' = N_1 + N_2 + \cdots + N_i$, i = 1, 2, 3,.., n; second order was $S_i'' = S_1' + S_2' + \cdots + S_i'$;and n order $S_i^{(n)'} = S_1^{(n-1)'} + S_2^{(n-1)'} + \cdots + S_i^{(n-1)'}$, i = 1,2,3,.., n.
(3) On the basic of step (2), the paper established fractal model, calculated various orders CUSUM's fractal dimension, and determined the final result of D and c.

(4) Finally, the paper forecasted the result, $\mathbf{S}_{\mathbf{i+1}}^{(\mathbf{n})'}$. According to the formula $\left(S_{i+1}^{(n-1)'} = S_{i+1}^{(n)'} - S_{i}^{(n)'}\right)$, the paper obtained the $N_i$.

### 9.3.2 *R/S Analysis*

Nowadays, R/S analysis was used to analyze whether a time series has fractal property or not. In common, the progress of the R/S analysis contained four steps.

1. The paper assumed the existence of time series $x = x_1, x_2, \ldots, x_n$;
2. The paper calculated the mean value E(x) and standard deviation S(n);
3. The paper calculated accumulation's deviation $X(i,n) = \sum x_n - E(x)_n$;
4. The paper calculated ranges $R(n) = \max X(i,n) - \min X(i,n)$.

Hurst index was determined by formula $R(n)/S(n) = c * n^H$. According to Brownian movement, the fractal dimension could be a formula $D = 2 - H$. Hurst index could be calculated from R/S analysis, and judging the trend of development of the time series. When the H was 0.5, the time series was a common Brownian movement; the time series also had fractal dimension when the H was not 0.5.

## 9.4 Computational Experiments

The paper took the *AHQB* station of MRT Line Four in Beijing as an example. Due to the arrival passenger flow volume during peak hours was more significant that the station organize the passenger flow, the paper took the arrival passenger flow volume from 7:00 to 8:30 as object of research, and divided the one and half hours into 18 periods of time in cycle lasting 5 min.

For the convenience of calculation, the paper defined 7:00–7:05 as 1 period of time (called $r_1$), and the arrival passenger flow volume was $N_1$; the paper define 7:06–7:10 as 2 period of time ($r_2$), and the arrival passenger flow volume was $N_2$; and so on, the 8:26–8:30 as 18 period of time (call $r_{18}$), and the arrival passenger flow volume was $N_{18}$. Based on the definitions above all, the paper draw line graph of the actual passenger flow value, in Fig. 9.1.

According to the process of R/S analysis, the paper got the result of the Hurst index. In order to reduce error, the paper chose the some points. Finally, the Hurst index was 0.031, and the paper verified the arrival passenger flow of urban rail transit had fractal property. So the paper could use fractal model to forecast the arrival passenger flow.

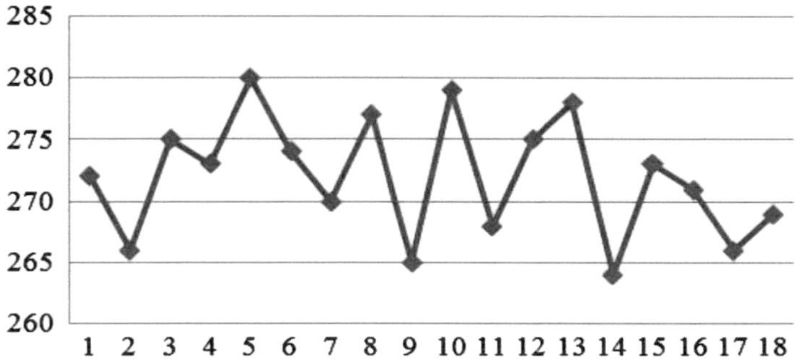


**Fig. 9.1** The arrival passenger flow volume of *AHQB* Station 1–18 periods of time (x-axis was $r_i$; y-axis was $N_i$)

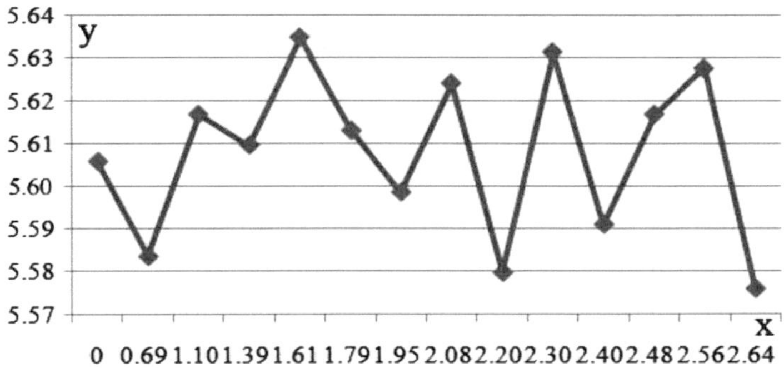


**Fig. 9.2** The point ($N_i$, $r_i$) were layout in double logarithmic coordinate (x-axis was $\ln(r_i)$; y-axis was $\ln(N_i)$)

In the forecasted process, the paper used the first 14 period of time (i.e. 7:00–8:10) and lay out the point ($N_i$,$r_i$), ($i = 1, 2, 3,\ldots, n.$) in double logarithmic coordinate. Then, the paper obtained the Fig. 9.2.

From the Fig. 9.2, we could conclude that the raw data was discrete points. Therefore, the paper should use transformed method to obtain the known function. According to the formula (9.3), the paper calculated the results of the various orders CUSUM, and their fractal dimension. The final results of fractal dimension were shown in Table 9.1.

In the Table 9.1, DS represented the calculated results that used initial point ($N_i$,$r_i$) and point ($N_{i+1}$,$r_{i+1}$) according to formula (9.3), $DS'$ represented the results that used first order CUSUM method, $DS''$ represented the results that used second order CUSUM method; $DS'''$ represented the results that used third order CUSUM method; and $DS''''$ represented the results that used firth order CUSUM method.

From the Table 9.1, we could conclude that the results of the fractal dimensions were not only positive but also negative if calculation used the initial values.

**Table 9.1** the results of fractal dimension which used the first 14 periods of time

| | DS | DS′ | DS″ | DS‴ | DS⁗ |
|---|---|---|---|---|---|
| 1 | 0.0322 | −0.9840 | −1.5743 | −1.9920 | −2.3155 |
| 2 | −0.0821 | −1.0183 | −1.7141 | −2.2599 | −2.7077 |
| 3 | 0.0254 | −1.0064 | −1.7808 | −2.4120 | −2.9460 |
| 4 | −0.1135 | −1.0280 | −1.8297 | −2.5147 | −3.1101 |
| 5 | 0.1188 | −1.0027 | −1.8551 | −2.5857 | −3.2294 |
| 6 | 0.0954 | −0.9887 | −1.8705 | −2.6369 | −3.3197 |
| 7 | −0.1917 | −1.0142 | −1.8885 | −2.6776 | −3.3909 |
| 8 | 0.3760 | −0.9711 | −1.8938 | −2.7082 | −3.4480 |
| 9 | −0.4886 | −1.0228 | −1.9082 | −2.7347 | −3.4952 |
| 10 | 0.4220 | −0.9822 | −1.9129 | −2.7560 | −3.5348 |
| 11 | −0.2963 | −1.0083 | −1.9212 | −2.7744 | −3.5686 |
| 12 | −0.1356 | −1.0182 | −1.9295 | −2.7908 | −3.5979 |
| 13 | 0.6973 | −0.9674 | 1.9298 | −2.8040 | −3.6233 |

**Table 9.2** The forecasting value compared with actual value and the relative error

| | Actual value | Forecasting value | Relative error (%) |
|---|---|---|---|
| 15 | 273 | 260 | −4.76 |
| 16 | 271 | 259 | −4.41 |
| 17 | 266 | 272 | 2.26 |
| 18 | 269 | 252 | −6.32 |

However, all of the fractal dimensions were negative when the paper used the various CUSUM. By contrast, the results of the second order CUSUM were better than others', and the last two results' deviation was 0.0003.

So, the paper chose the arrival passenger flow volume of 13th and 14th period of time to calculate the fractal dimension D.

$$D = -1.9298.$$

Then, according to formula (9.2), the paper calculated the result of c.

$$c = 175.8387.$$

Lastly, according to formula (9.1), the paper calculated the results of $S''_{15}$, $S''_{16}$, $S''_{17}$, and $S''_{18}$. On the basis of result of $S''_{15}$, $S''_{16}$, $S''_{17}$, and $S''_{18}$, the paper used $S'_{i+1} = S''_{i+1} - S''_i$ to calculate the arrival passenger flow volume of the 15th, 16th, 17th and 18th period of time. The results were shown in Table 9.2. The paper compared the actual value of the 15th, 16th, 17th and 18th period of time with the forecasting value of the 15th, 16th, 17th and 18th period of time, and found that the maximum relative error was −6.32 %.

## 9.5 Conclusions

The arrival passenger flow volume, especially the arrival passenger flow volume during peak hours, was the base of urban rail transit station which organized passenger flow and optical allocation facilities. Therefore, the paper established the short-term arrival passenger flow volume forecasting model, took the AHQB station of the MRT Line Four in Beijing as an example. Lastly, the paper gave a comparison between the predicted value and the actual value, and found that the maximum error was less than 7 %. It indicated that the established model could be used for short-term arrival passenger flow volume forecasting However, due to the fractal dimension D and the constant c were approximate, and the relative error was increased if the expansion of the forecasting increased.

**Acknowledgments** Special thanks to Yuanhua Jia, my teacher, and Xihui Yin, Zhong-hai Niu, my classmates, for all information and advises.

## References

1. Tan Ben-xuan (2009) Research on urban rail transit passenger flow forecasting technology [D]. Wuhan University Technology, Wuhan
2. Li Qing-hua (2011) The comparison of urban transit demand forecasting method [J]. Technol Econ Areas Commun 1(63):54–57
3. Lu S, Guo X-c (2001) Method of the URT's passenger flow forecast [J]. Syst Eng Theory Pract 21(8):106–110
4. Yan Ji, Shao Chun-fu (2001) Passenger flow forecast of urban rail line based on multi-logit method [J]. J Xi'an Highway Univ 3:68–72
5. He Ming, Guo Xiu-cheng (2010) Forecasting rail transit split with disaggregated MNL model [J]. J Transp Syst Eng Inf Technol 10(2):136–142
6. Ran Y (2010) Study on passenger flow forecasting and operation scheduling of urban rail transit [D]. Beijing Jiaotong University, Beijing
7. Wang Li-na (2011) Traffic forecasting and scheduling of the urban passenger rail network based on historical data [J]. Beijing Jiaotong University, Beijing
8. Mao Jing (2011) Urban rail traffic passenger flow short-term forecasting method and empirical research [D]. Beijing Jiaotong University, Beijing
9. Mandelbrot BB (1997) Fractals: form, chance and dimension [M]. W H Freeman, San Francisco
10. Lin Hong-yi (1992) Fractal theory-singularity discussed [M]. Beijing Institute of Technology Press, Beijing
11. Turcotte DL (1992) Fractals and chaos in geology and geophysics [M]. Cambridge University Press, Cambridge

# Chapter 10
# Analysis on Influence of Railway Station on Planning and Design of Surrounding Region

**Linfei Han, Xiawen Wei, and Lan Chen**

**Abstract** As the country with largest population in the world, China's urbanization is developing rapidly and large populations migrated to central cities, forming excessive passenger flows during specific periods (e.g. spring festival, summer holiday and golden week). During the leaping development of railway and rapid development of contemporary China's urban construction, various urban material environmental, urban economical and urban social problems concerning traffic efficiency, city images, commercial development, estate development, house removal and public security were caused in surrounding areas of urban railway stations due to the excessive passenger flows caused by sharp increase of inter-city passenger flows and increasingly highlighted pressure on stations and their surrounding areas. For adapting to the passenger transportation demand in high-speed railway times and optimizing urban living environment, it is necessary to study how to increase use efficiency of urban space to the maximum extent and perfect urban functions when achieving the collective and distributive function of railway passenger transport. The urban planning and reasonability as well as high efficiency of urban design near the passenger stations not only can directly affect the prosperity and development of sub-centers, but also are related with the revival of old city center.

**Keywords** Railway station • Surrounding region • Planning

L. Han • X. Wei (✉) • L. Chen
School of Architecture and Design, Beijing Jiaotong University, 100044 Beijing, China
e-mail: w-xw@live.cn

H. Xia and Y. Zhang (eds.), *The 2nd International Symposium on Rail Transit Comprehensive Development (ISRTCD) Proceedings*,
DOI 10.1007/978-3-642-37589-7_10, 

## 10.1 Introduction of Urban Planning and Design Surrounding Railway Station

The railways connect the cities together. Being the place for entering into and leaving the city, railway stations join future careful creation of cities and regions and share the same destiny with the city and the region. Currently, urban areas surrounding railway stations in China is facing with different land usage natures, different construction qualities, and poor environmental situation, easily causing misconducts. Consequently, it is necessary to have good overall urban planning and design.

### *10.1.1 Development Background of Railway Station Construction in China*

As the door of cities, urban railway station is the congruent point of traffic, business, culture, and life. For cities, fast urbanization demands mutual adaptation of urban traffic and urban development scale. Namely, the fast traffic should be realized based on lowest energy and area. For external development of the city, the railway station drives the development of surrounding regions and the city as the key component of urban traffic system and intercity bridge [1]. Newly constructed railway station in China is a system with complete function and space environment rather than a single construction as before. It not only has independent metabolic ability, but also is separated from the city as an intercity communication node and a part of urban function.

### *10.1.2 Analysis on Current Situation Problems of Areas Surrounding Railway Station in Chinese City Center*

China's Middle and Long-term Railway Network Planning (2004–2020) and railway great-leap-forward development planning attach high attention on the passenger railway setting plan, an effective method to solve the current insufficient transportation capacity and low transportation quality [2]. With the accelerated development of China's railway, urban railway station plays an increasing important role in the city.

1. Urban material environmental problem existed in the development and construction

   Urban traffic system is the foundation for urban vitality and development. The efficiency of traffic system determines the correlation among various urban components and is the great motivation of urban development. However, passenger flows in railway stations increase sharply with the progress of China's, railway transportation and economic development, resulting in traffic jam and low efficiency of railway station and affecting the urban vitality directly.

2. Urban economic problem existed in the construction and development

   Traditional station layout disagrees with the urban land value. Under the rapid urbanization development in China, both newly constructed and reconstructed railway stations locate in the city center or planned development zone. However, most railway stations in China choose plane layout with little association with surrounding urban construction, reducing the land usage strength. In addition, the excessive plane traffic area restricts the great value of urban land, wastes the precious urban construction land resource, and reduces the internal urban development motivation significantly.

3. Urban social problem existed in the development and construction

   Residents in city center suffer poor living conditions due to various problems like excessive crowd, serious urban environmental pollution, and poor traffic condition caused by over concentrated populations. Moreover, there are insufficient facilities for business service and education in border community, resulting in various phenomena.

### *10.1.3 Era Demand for Urban Planning and Urban Design in Areas Surrounding Railway Station in Current Chinese City Center*

Railway stations in China often maintain some characteristics from modern industrial society. For example, the lands are mainly used for industrial transportation and storage. In some cities, the rearrangement of industrial storage produced numerous idle lands surrounding the railway station [3].

1. There's no bridge connecting the railway station and the city

   Although a batch of new railway stations constructed from late 1980s and early 1990s have been tried to unite the city separated by railways together by station-crossing-railway, it received no expected effect because of various reasons, including technology, fund, and management.

2. With single function, the railway station fails to take the full advantage of traffic and information mode and fails to become the aggregation of urban functions.

   Generally, it is impossible to realize vertical traffic transfer (including subway, bus, long-distance passenger transportation, and sedan car) between the intercity railway transportation and urban traffic system and one-stop transfer.

3. There is no public space to attract urban life and leisure passenger flow and no connection with urban pedestrian system.

   Generally, railway stations are crowded with vehicles and passengers but equipped no greenbelt and leisure facilities. Moreover, there is no exit to urban pedestrians.

## 10.2 The Influence on Urban District Design and Planning from the Construction of Urban Railway Station Construction

Currently, the construction background and mode of railway station in China change significantly with the development of China's railway and promotion of economic level. The importance of railway station in urban planning has been highlighted with the promotion and popularization of intercity high-speed trains. However, the district construction surrounding railway stations is facing with various problems due to historical and realistic problems as well as its role in urban functions neglected by the traditional railway station design theory, which affect the urban development.

### *10.2.1 Influence of Railway Station Construction on Urban Traffic System*

The progress of Chinese railway transportation and economic development has lead to great increase in passenger flow of railway stations, resulting in traffic jam and low efficiency in the railway station, which has directly influenced the urban vitality. These influences include:

1. The arrangement mode of railway station does not match the development and differentiation of modern urban traffic.

   Currently, most domestic railway stations apply plane arrangement mode, in which the station square is used to connect the railway passenger transportation and urban traffic. Numerous passengers and vehicles into and out of the railway station and urban public transit go through the station square. In addition, the station square also functions as parking lot and landscape area. Numerous passenger flow and diversified transportation mode lead to the following phenomenon: over-huge of the square, serious disturbance in function division and various lines, inconvenient transfer, and traffic jam. For example, in Taiyuan Railway Station (Fig. 10.1), the square, station building, and the station yard are arranged on the plane in order, resulting in three-stop structure, which is classical plane arrangement mode. The urban traffic and the function are of railway station are on the same plane (Fig. 10.2). Various traffic transfer stops surround the aggregation plaza, so the traffic of the railway station is seriously disturbed by the urban traffic, passenger flow and vehicle flow. Meanwhile, the land usage efficiency is low and various functional lands have conflicts. Especially, the railway station is located in city center of Taiyuan, close to the business district, Liuxiang. The serious traffic jam in Yingze Avenue (Arterial Street) and the square greatly influences the urban structure and urban traffic.

Fig. 10.1 Station square in front of Taiyuan Railway Station

Fig. 10.2 Bus station in front of Taiyuan Railway Station

2. Railway station construction separates the overall urban traffic

   Due to the railway station construction and penetration of railway, the city is divided into two parts or more parts. It is inconvenient to connect all parts in space and traffic. Especially, the district far from the city center develops slowly with no vitality. Some districts are almost half-closed. The streets are narrow. Road network transformation has been terminated for a long term. Constructions in the streets are old. The life service facilities and municipal facilities are out of date. Moreover, numerous urban lands are idle. All of these greatly blocks the overall urban development.

3. Backward non-motor traffic system

   The backward non-motor traffic system greatly blocks the overall vitality and development of the station district can be summarized as two aspects: firstly, for the district with huge motor flow, the continuity of walking system commonly relies on the passenger foot-bridge and underground passage. However, due to lack of uniform planning and design, these facilities are unconnected with surrounding constructions, resulting in seldom usage. As a result, people go together with the vehicles, influencing the traffic seriously. Secondly, bicycle is still an important commuting vehicle in China, however, most railway stations fail to consider the parking for bicycle, so numerous bicycles are parked in front

**Fig. 10.3** Chaotic phenomenon in the Commerce & Trade Market surrounding the Zhengzhou Railway Station

of railway station chaotically, which not only causes negative influence on urban landscape, but also occupies traffic road, square, and parking lot of motor vehicle, leading to greater decrease of traffic efficiency.

### *10.2.2 The Influence of Railway Station Construction on New Urban Functions*

As important construction in the urban area, railway station is inevitably to undertake more urban functions to develop harmoniously with the urban area. However, current railway station construction has increasing influence on the urban functions.

1. The conflict between traditional railway station arrangement and urban land value

   Most railway stations in China are designed as plane structure, with no comprehensive coordination with surrounding urban constructions, which reduces the land usage rate.
2. The conflict between the overall environment of railway station and urban development

   China has ignored the urban functions of the railway station for a long time, resulting in chaotic traffic, poor environment, greenbelt inadequacy and leisure facilities shortage, and inability to attract non-travel passenger flow around railway station (Fig. 10.3).
3. The railway station construction is unable to adapt to the urban development and integrate into the surrounding areas. Moreover, the railway station even becomes an island separated from the city, which blocks the upgrading of surrounding land usage in urban structure [4].

### 10.2.3 Influence of Railway Station on Urban Spatial Form and Environment

As the representative of urban image and important urban center, the urban environment of this district directly will heavily determines the development of commerce industry and service industry. However, many railway stations in China seriously affect the planning and design of entire urban spatial form. The negative influence on urban environment of the railway station construction is also a great problem.

1. Many railway station designs in China focus on lighting urban image and pursue novelty and difference. Some railway stations designs even apply the pattern mechanically to cater to so-called political influence. As a result, the construction has insufficient local cultural characteristics, which is different from urban form elements such as cultural content, landscape texture, city style, and characteristic in surrounding districts.
2. Traditional railway station construction greatly divides the city, which leads to a terminal in urban space. The station building and the station square become the terminal of the city. The continuity of the city is blocked by the external elevation of the station building. Especially, when the city is enlarged gradually, current railway station in urban border becomes the city center, resulting in serious damage to urban texture and space.
3. Due to the development of the railway station and surrounding areas, traffic flow increases greatly. Moreover, many sections have to be the corners with no passengers, which worsen the urban environment of the sections. In addition, it also influences the commercial service and development of other urban functions in the surrounding areas.

### 10.2.4 The Role Transfer of Railway Station and Corresponding Influence on the City

Constantly developing new conditions raise new requirements for the railway station construction and cause certain role change of railway station. New requirements on traditional passenger transportation should be proposed. In addition, more interactions with urban functions and structures should be created.

1. when the railway stations develop into comprehensive urban traffic mode, the overall traffic system of corresponding urban area must combine various traffic modes. As a result, the district should satisfy various demands of railway traffic and undertake the urban traffic connection functions as a part of urban traffic system to guarantee the smooth urban commute. Therefore, for this objective, the urban node with various complex traffic functions must integrate various traffic modes organically with reasonable economic and effective solutions.

2. The railway station construction brings in more diversified derived functions, including commercial service and cultural information displaying, which effectively meets different demands of urban population. In this way, the ideal urban living environment for people can be created [5]. Due to this change, railway station design should be promoted to the overall urban scope, instead of focusing on the construct itself.
3. Higher requirements on volume and modeling of railway station comes out based on stronger comprehensive national power and increasing appreciation level of people. Compared with traditional railway stations, new railway station are required to have rich sense of era, regional characteristic, sense of location and spatial experience by travelers who also pay more attention to details of railway station and human care of facilities [6].

## 10.3 An Analysis on the Urban Design Strategy in the Surrounding Areas of the Railway Station

Due to the restriction of current management system in our country, insufficient understanding of construction of the station area's impact on the city by academia, passenger station gets awkward position in the city presenting some contradictions. Of course, there are many reasons causing these problems. Fundamentally, the acceleration of speed of economic growth in our modern society and urbanization leads to the close association of elements constituting urban morphology. As our country accelerates its development in all directions, the large and medium-sized railway station will be definitely faced with more sharpened contradictions, making the construction of the railway station no longer independent from overall the development of the overall urban morphology. Therefore, we should start with the urgent problem in this area, focus on the urban morphology and rely on the adaptive design methodology so as to propose the strategies of the comprehensive construction for our railways station and surrounding areas.

1. The design of urban transport system should plan site selection as whole and combine the city to take the transport system in surround areas of railway station into overall considerations.

Currently, the railway station is transformed from the front door of the city to the heart of the city, enjoying a more significant status in urban transport system. The newly-constructed metro and light rail in large city, increasing private car, increasing bus and tax make the transport system in this area unprecedentedly complicated.

Due to the relationship between the site selection of railway station and urban transportation, research community of traditional theory views the conclusion in Fundamentals of City Planning (the 3rd edition) and Urban Road and Transport Planning as the most representative viewpoint, that is the distance of 2–3 km from station area to urban center is appropriate. Then we can get reference data based on the transfer factor, transportation costs and psychological factors of ordinary crowd of tourists and city. Combining the urban planning and construction, we can finally the relationship between the site selection and the city.

As the significant distribution point of crowd in the city, the area of railway station is significantly guided by the concept "Generalized reachability". Therefore, we should make the transport system in station area fully bend into the whole transportation network in the city, pay more attention on the reachability in different lots and functional areas and plan the transportation under the background of urban planning of commercial service area and future development. Full combination between the operation of different transportation systems in railways station and urban elements should be kept to make the passengers freely access to different areas.

2. An analysis on the urban design strategy on the function and environment in the vicinity of the railway station

According to the current problem and development direction, the overall regional planning and design should adopt the three-dimensional composite structure. In addition, for the regional plan and design, the demand for diversity of different users should be fully taken into consideration to the function settings and organizations, which is core embodiment of people-oriented design.

Due to the expanding urban economic development, most of our railway stations planed in the old days have gone through the tremendous change in regional location. Since most of the lots lie in the urban center, the railway station and railway lines cut the urban space and overall shape apart and damage of surrounding environment is fully presented. In order to solve this, it is necessary to boldly change the design philosophy of railway lines in station area and lines in the urban center. We have already had the preliminary exploration on the theories and practices in some western developed countries, that is to say, for the railway lines in the urban center, the overall elevated and buried methods should be adopted to minimize the negative effects on cities.

## 10.4 Conclusion

1. The uniform planning of railway station is critical for the urban development.

   Judging from the internal development of city, as the urban portal, railway station should combine the urban traffic with business, culture and life, and realize efficient light traffic based on energy-and-land saving. For the external development of the city, the railway station shall drive the intercity development and become an indispensible integrated part.
2. Railway station completes the transition between external urban traffic and internal urban traffic.

   With the launch of intercity train and popularization of high speed train, the railway station should connect the traffic of inner city and intercity efficiently. Previous railway station usually ineffectively dredges the passenger flow, vehicle flow, entrance, and exit through plane arrangement. However, vertical traffic facilities rise in recent years have certain defects due to design order in later supplementary stage, including poor location, inconvenient usage, insolated bridge, winding gallery, hidden parking lot, and narrow entrance and exit. All of these directly lead to poor regional traffic environment.

3. Appropriate commercial development surrounding the railway station

   As the urban portal, railway station must reconstruct and upgrade its regional functions and public environment with the social development and change of traffic load, which is the inevitable result of urban update. However, some cities also carry out blind commercial construction with the name of updating the railway station, which not only damages the original street atmosphere of railway station area, but also leads to great governmental burden and decline of urban space based on large-scale comprehensive development.

## References

1. Han Lin Fei, Yao Zhuo (2005) Railway construction and urban development [J]. Archit Works 10:22–27
2. Zhao He Sheng (2000) Urban development and urban regeneration. Southeast University Press, Nanjing
3. Kenneth Powell, Yu Wang (Trans.) (2002) Evolution of the city-urban architecture at the beginning of 21th century. China Building Industry Press, Beijing
4. Zhang Bei Bei (2009) The adaptive urban design study in the areas of large-scaled railway station in China. Master's degree thesis in Southwest Jiaotong University, Nanjing
5. Liu Jie (2004) Integration of urban form[M]. Southeast University Press, Nanjing, p 106
6. Sorensen A (2000) Land readjustment and metropolitan growth: an examination of suburban land development and urban sprawl in the Tokyo metropolitan area. Progress Plan 53:217–330

# Chapter 11
# The Fallacy and Planning Strategies on Land-Use of Rail Transit Site Area in China

**Zhuoran Shan and Yaping Huang**

**Abstract** Research the development of rail transit site area in China's major cities. There are three major characteristics in development and construction of rail transit site area, including efficient-tending of land-use, increasingly complex of city function, the laws of economics obeying. Four developing fallacy mechanism are at the prevalence which can be summarized as convergence of land development methods, the serious space closure, and traffic transfer's complex and single and the marginalization of public facilities. Research recommendations that five active planning guide strategies will maximize help cities to solve the pervasive problem of rail transit site region, including the land use go-mode strategy, ecological open space reserving policy, low-carbon transportation transferring-access strategy, the closing strategy of public service facilities.

**Keywords** Rail transit • Land use • Fallacy mechanism • Planning strategies

## 11.1 Characteristics of the Development and Construction of Rail Transit Site Areas in China's Major Cities

### *11.1.1 Land-Use Tends to Be Efficient*

First, local governments and planners are gradually realized that the efficiency of urban land use is worth to improve by the influence of rail transit which has also become the characteristics of the development and construction. The prominent reflection is triple jump phenomenon of strength, density and height. Among them, the most affected is building height and building strength. The local government

Z. Shan (✉) • Y. Huang
Architecture and Urban Planning Institute, Huazhong University of Science and Technology, 430074 Wuhan, China
e-mail: 371760860@qq.com

H. Xia and Y. Zhang (eds.), *The 2nd International Symposium on Rail Transit Comprehensive Development (ISRTCD) Proceedings*,
DOI 10.1007/978-3-642-37589-7_11, 

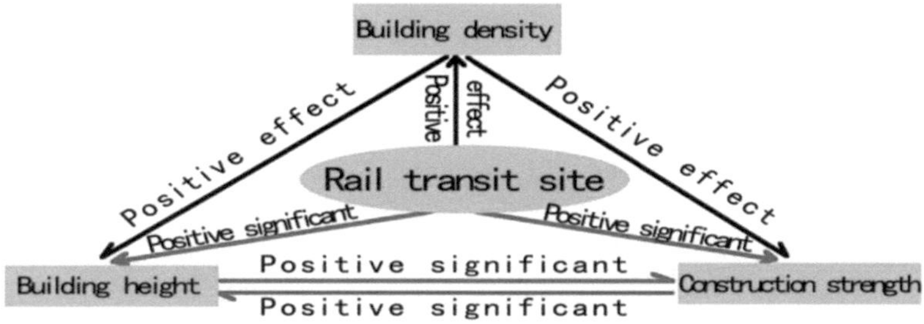


**Fig. 11.1** "One point three degree" mechanism

seems to think that rail transit will become a fertilizer, the activation of the land and makes it continue to play performance of "one point, three degrees" (Fig. 11.1).

### *11.1.2 Urban Functions Are Increasingly Complex*

Secondly, with the efficient of land-use, surrounding areas become increasingly complex. The positive sense is to break the drawbacks brought by rigid functional partition and to shape urban functions integrated regional with composite functional organizations. And the negative sense is at its organization unscientific and piled, which often format the mode of the business office, business hotels, high-rise residential and creative industries.

### *11.1.3 Development Subjects to the Laws of Economics*

Moreover, rail transit site area development tends to obey the rules of market economy and rent bidding curve (Fig. 11.2). It's difficult to use competition relationship to explain, and also do not belong to the basic means of macroeconomic regulation and control.

## 11.2 The Fallacy of Land Use on Rail Transit Site Area in Major Cities

### *11.2.1 Land Use Pattern Convergence*

Development model of land in the vicinity can be summarized as "function collage, strength first, chaotic composite". Its consequences is not only the city-one-side,

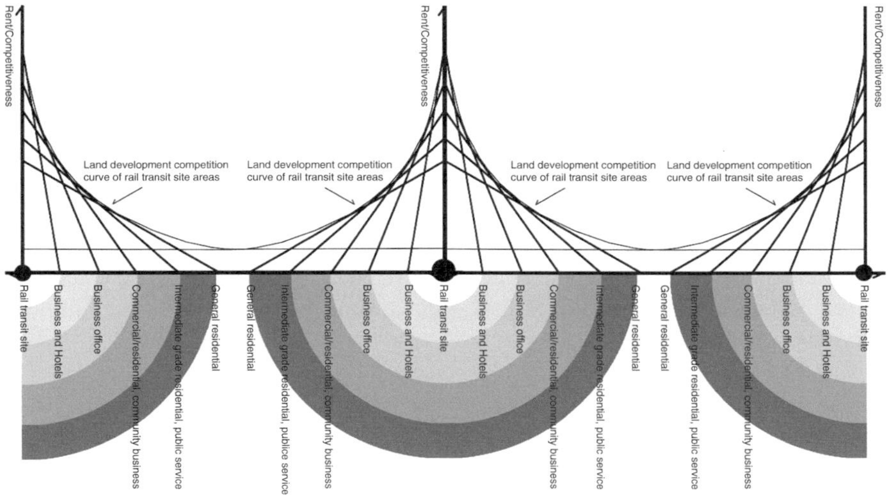


**Fig. 11.2** The competitive curve and land use mode of rail transit site surrounding land development

will also be derived from cultural loss, long-term low-level equalization of living standards, negative attitudes and hedonistic consumerism leading, innovation capacity decreased, the quality of living environment continued to decline and the increasing degree of confusion.

### *11.2.2 Serious Space-Close Phenomenon Leading Open-Space Displacement*

Foreseeable problem is the space closed. In China, this problem reflects in the ecological green space's disappear or replacement. Nature of the replacement is the "rediscovering" on value of the land, but the result is a lower quality of human settlements environment. Closed urban space will bring many fundamental issues, such as urban ventilation obstruction, incomplete regional ecosystems, crowded city repression and space.

### *11.2.3 Traffic Transfer's Complex and Single*

By the influence of construction sequence, most Chinese rail transit is not so easy to transfer, harder for connection level. Most of them are still unable to form the ideal mode of multiple transportation modes. And complex place lies in citizens' usability to use the shortest time through the transfer and spend much time waiting for another.

### *11.2.4 Marginalized Public Facilities*

Public facilities are often in marginalized situation. Although it is the inevitable under the market economy and free competition, but it is also the consequences due to acquiescence of government failure which means the reducing reach ability and service capabilities.

## 11.3 Planning Guide Strategy of Rail Transit Site Areas in Large Cities

### *11.3.1 The Strategy of Modeling Against of Land-Use*

#### 11.3.1.1 Shift from the Multi-legged Race to Focus Driven

By strengthening the degree of coordination of land use along rail transit lines, it coordinated land function, development intensity; residence employment population, which is the city performance basis [1]. The phenomenon of high-rise commercial office, high-end commercial and upscale residential "three high" has become typical. To this end, the development and construction must priority be to determine the focus driven concept. The focus driven development model under the target of Healthy Cities and Smart City will benefit the common upgrade of urban space protection and promoting [2].

#### 11.3.1.2 Transformation from the Site Relying to the Communities Relying

Clear traffic-oriented plays a positive role in promoting urban development [3]. This center will surely be the information flowing gathering. Urban entity functional elements need to reboot information collection in a new starting point. Whether it is a function of culture and leisure, eco-tourism function, the high-tech industry or government service functions, it should "abandon center, with a view to the edge of self-reliance" (Fig. 11.3).

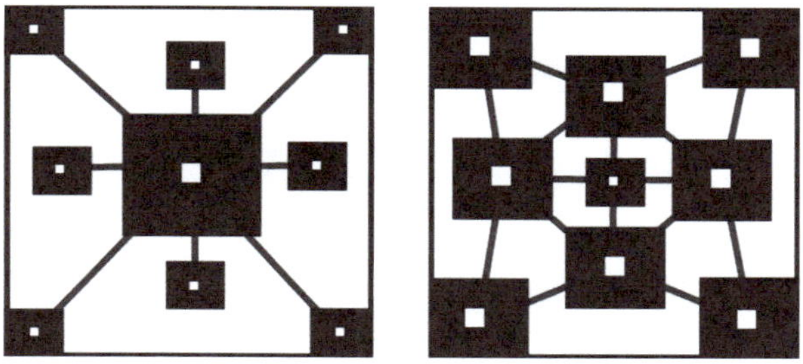

**Fig. 11.3** The transform from center leading to edge independence

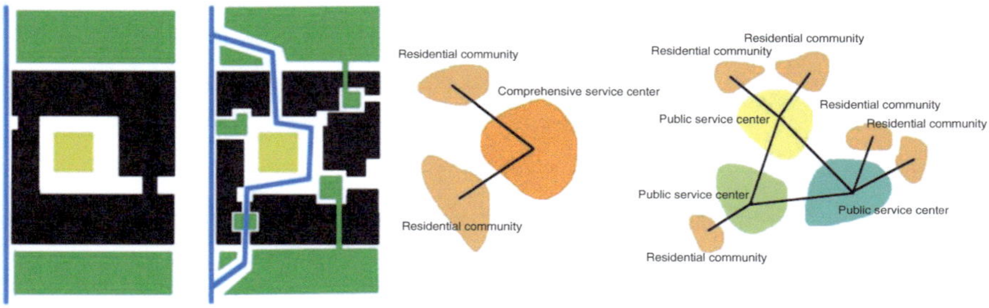


**Fig. 11.4** Improving open space level and promoting to multi-center combination – multi-strength

## *11.3.2 The Strategy of Reserving for Ecological Open Space*

### 11.3.2.1 Improve the Level of Open Space

By the affect of land economic factors, the development intensity leads to the low level of open space in rail transit site areas. What is serious that the closed spatial pattern formatted around the site which is very detrimental to the sustainable shaping of urban space, especially for highly populated areas and frequent flowing areas because the closing means the blocking of information. Therefore, although the level of broad spaces is not completely in line with the model of land economic benefits under the driver of rail transportation, but it is significant for the spatial structure. Because of the structure of highly dense itself represents the tense, complex, disorderly and congestion which are most not the wanted of urban construction in the future (Fig. 11.4).

### 11.3.2.2 Improve the Level of Network and Hierarchical of Open Space

Planning for open space must have the network level, including the transparent and connectivity of surrounding buildings floor space, ground greening, vegetation cover and building green roof height differences of processing, including water and wetland system, drainage system, commercial street, City Square Park and the organic combination.

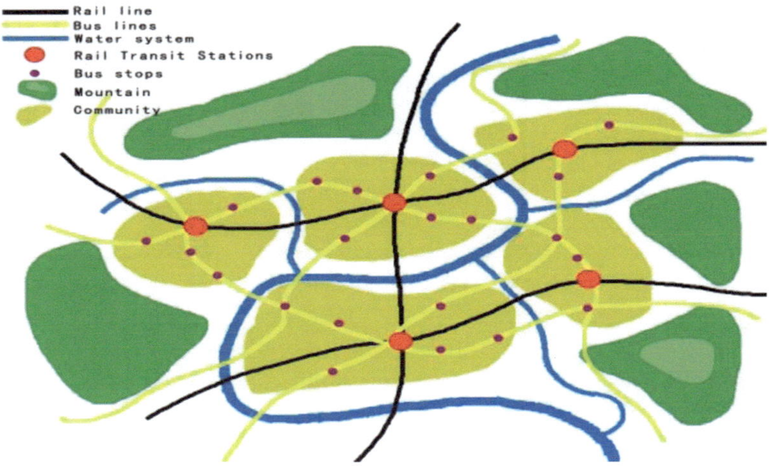


**Fig. 11.5** Schematic diagram of park-ride and community divided way based on reach ability regional

## *11.3.3 Low-Carbon Transportation Transfer and Connection Strategy*

### 11.3.3.1 Public Car Transfer and Bicycle Transfer

Transportation transfer is an important driving force for promoting the development of rail transportation better. People hope to transfer from crowded rail transit as a loose free, time-controlled manner. Bus and bike transfer can provide a possible for residents. In addition, study suggests that private car parking building, City Air Terminal, train station and all levels of passenger station should be combined with the rail transit site organic.

### 11.3.3.2 The Division and Exploration of the Reach-Ability Region and Community

Research thinks, reachable region can determine synthetically through the transfer station, transfer time and trip distance. From the rail station, 5 min after the waiting time, you can take 2 station roads within distance off, and most cost time-consuming 10 min to get home. According to the time dimension, it is the maximum consumption time of 20 min. But if according to distance dimension consideration, it is the formation of the site as the center of the circle, with about 1,000 m to the radius of a specific region: an area of about 3 km$^2$. Research suggests, reachable region as the unit of 3 km$^2$ in large cities is roughly reasonable planning object (Fig. 11.5).

### *11.3.4 The Strategy of Public Service Facilities Approaching*

Cultural and medical facilities in spatial distribution should be close to the site. Educational facilities may be deep within community, be close to bike and walk system, within walking distance is important. Therefore, scope of its services may be much greater than community-wide, complete design of short distance walking system is critical.

## 11.4 Conclusions and Outlook

Tend to be efficient, complex urban functions, developed obey economic laws are major characteristics of regional construction of the mass transit stations. Four fallacy mechanisms include convergence of land development methods, serious space closed, complex and single traffic transfer, marginalized public facilities. Positive guidance strategies include land-use go-mode strategy, ecological open-space reserve policy, low-carbon transportation transfer access strategy and public-service closing strategy.

**Acknowledgment** NSFC:51178200《The research of path, patterns and spatial organization on new urbanization of county in the central of China – Take Hubei Province as an example》.

## References

1. WenQi Yao, YuYan Zheng (2011) Measures to improve urban rail transit efficient. Urban Plan Forum 5:73–79
2. ZhuoRan Shan, XianChun Zhang, YaPing Huang (2011) Duet attributes of healthy city system: indemnificatory and propel. Planners 4:14–18
3. HaiXiao Pan (2012) China urban transport top level architecture design. Constr Sci Technol 14:17–22

# Chapter 12
# Short Time Forecasting of Rail Transit Passenger Volume

**Haifeng Song, Tao Tang, Chenling Li, and Yi Ding**

**Abstract** The theory and application of railway passenger traffic volume (RPTV) forecasting is one of the focuses in the research areas of traffic transportation. In China, the passenger traffic system is composed of railways, highways, waterways and civil airlines. However, the railways have been played a main role in the national traffic system and the RPTV has attracted more and more attention because of its vital effect in economic evaluation of railway project, national resources allocation, and readjustment of investment structure within railway enterprises.

**Keywords** Railway • Passenger traffic volume • Forecasting • Short time

## 12.1 Introduction

Rail transit has been an important transportation method nowadays, and occupied a more and more important position in people's daily life. In the mainland of China, urban rail transit involves light rail, subway, urban railway and so on. While in national railway, the quick development of the high speed railway makes rail transit achieved an unprecedented height. Compared with road traffic and civil aviation, it has many advantages such as large traffic volume, less pollution, no level crossing with the urban road, prompt, security and comfort [1]. All the countries in the world recognize the main way to solve city traffic problems is paying more attention to the development of railway. While China has high density of urban population and limited resources of road space, it determines the ways that regard the urban mass transit system as the priority is the inevitable choice.

H. Song (✉) • T. Tang • C. Li • Y. Ding
State Key Laboratory of Rail Traffic Control and Safety, Beijing Jiaotong University, No. 3 Shang Yuan Cun, Hai Dian District, Beijing, China
e-mail: songhaifeng2011@gmail.com; ttang@bjtu.edu.cn; 12111047@bjtu.edu.cn; 1125061@bjtu.edu.cn

H. Xia and Y. Zhang (eds.), *The 2nd International Symposium on Rail Transit Comprehensive Development (ISRTCD) Proceedings*,
DOI 10.1007/978-3-642-37589-7_12, 

Design and operation of rail traffic involve many aspects, including route selection, station build, electrification design, train operation diagram, and so on, the works' entire objective is to complete transportation. The subject of transportation is locomotive, and the object is people and goods, when we put the subject and object together, the problem we want to solve is the passenger flow. From the route selection to the operation, the passenger flow forecasting and distribution play a major role.

## 12.2 The Important of Railway Passenger Traffic Volume

With the reforming of China's railway passenger market, especially in recent years, China's highway, civil aviation has been greatly developed through the open management, while railway transportation has been challenged powerfully, the competition between different modes of transportation will become increasingly intense. Highway, civil aviation, pipeline transportation and other transportations, are intended to be further extended to the advantaged area of traditional railway, and the quality, efficiency of transportation will be the focus of competition in the future.

Passenger flow forecasting methods include long time forecasting and short time forecasting. Short time forecasting studies the pattern of data over time on huge amount of history record of passenger transportation. The main function is to provide a guidance for the current running lines, according to the results of short-term passenger forecasts to modify the operation diagram and solve the peak passenger flow in holidays, as well as to reduce the usually unnecessary Train running, energy saving, so that make the railway transportation efficient and environmentally.

## 12.3 Short Time Forecasting for the Train Passenger

As we all know, every year during the holiday railway stations are facing a peak passenger flow, and in the city daily commutes are also a great challenge to the subway, so after the building of rail transport was finished, how to make it meet the passengers travel demand is also a big mission.

Short time forecasting for the passenger flow makes a great contribution to the operation of railways, it helps the railway department make a suitable train operation diagram.

Recently various prediction techniques have been proposed in railway passenger volume forecasting, some of them are outlined here.

Support vector machine (SVM) is a new machine learning method based on the statistical learning theory [1, 2], which solves the problem of over-fitting, local optimal solution and low convergence rate existed in ANN and has excellent generalization ability in the situation of small sample [3–7]. However, the select of appropriate SVM parameters is difficult. Whereas GA has strong global search capability, support vector machine optimized by genetic algorithm (GA-SVM) is proposed to forecast the railway passenger volume [4–5], among which GA is used to determine training parameters of support vector machine.

### 12.3.1 SVM Model

Given a set of training samples $\{x_i, y_i\}$, where $x_i$, $y_i$ represent the input vector and the corresponding output respectively, the function of support vector machine can be approximated as follows [2, 8–9]:

$$f(x,w) = w^T\phi(x) + b = (w, \phi(x)) + b$$

Where $W$ denotes the weight vector and $b$ denotes the bias term, $\phi(x)$ denotes the high-dimensional feature space.

Define loss function:

$$L(y, f(x)) = (y - f(x))^2 = e^2$$
$$s.t. y(x) = w^T\phi(x) + b + e_i (i = 1, 2, \ldots, n)$$

Two positive slack variables $\xi$ and $\xi^*$ are introduced to represent the distance from actual values to the corresponding boundary values of the $\varepsilon - tube$ [10]. Then, the regression problem is transformed into the following constrained form:

$$Min\left[\frac{1}{2}w^T w + C\sum_{i=1}^{n}(\xi_i + \xi_i^*)\right]$$

Subject to

$$\begin{cases} y_i - w \cdot \mu(x) - b \le \varepsilon + \xi_i \\ w \cdot \mu(x) + b - y_i \le \varepsilon + \xi_i^* \\ \xi_i, \xi_i^* \ge 0 \end{cases}$$

where $C$ represents the regularization parameter.

This constrained optimization problem is solved using the following Lagrange form [2]:

$$\left[\sum_{i=1}^{n} y_i(\alpha_i - \alpha_i^*) - \varepsilon\sum_{i=1}^{n}(\alpha_i + \alpha_i^*) - \frac{1}{2}\sum_{i,j=1}^{n}(\alpha_i - \alpha_i^*)\left(\alpha_j - \alpha_j^*\right)k(x_i, x_j)\right]$$

Subject to

$$\begin{cases} \sum_{i=1}^{n}(\alpha_i - \alpha_i^*) = 0 \\ 0 \le \alpha_i, \alpha_i^* \le C \end{cases}$$

where $\alpha_i$, $\alpha_i^*$ are Lagrange multipliers. $k(x_i,x_j) = \phi(x_i)\ \phi(x_j)$ is called the kernel function. Finally, support vector regression function can be written as follows:

$$f(x) = \sum_{i=1}^{n}(\alpha_i - \alpha_i^*)k(x_i, x) + b$$

Least squares support vector machine is an improved support vector machine, which can use equality constraints for the error instead of inequality constraints. The regression function of least squares support vector machine can be approximated by solving the optimization problem [9–10]:

$$Min\left[\frac{1}{2}\left(w^T w + C\sum_{i=1}^{n}\varepsilon_i^2\right)\right]$$

Subject to

$$y(x) = w^T\phi(x) + b + e_i(i = 1, 2, \ldots, n)$$

where $e_i$ represents the error.

Then, obtain the dual Lagrange form by using the Lagrange multipliers:

$$L(w, b, e, \alpha) = \frac{1}{2}w^T w + \frac{1}{2}C\sum_{i=1}^{n} e_i^2 - \sum_{i=1}^{n}\alpha_i\left[y_i - w^T\phi(x) - b - e_i\right]$$

where $\alpha_i$ are the Lagrange multipliers.

The partial derivative of above the dual Lagrange formula is given as follows:

$$\begin{cases} \dfrac{\partial L}{\partial w} = 0 \rightarrow w = \sum_{i=1}^{n} \alpha_i \phi(x_i) \\ \dfrac{\partial L}{\partial e_i} = 0 \rightarrow \alpha_k = C \cdot e_i \\ \dfrac{\partial L}{\partial b} = 0 \rightarrow \sum_{i=1}^{n} \alpha_k = 0 \\ \dfrac{\partial L}{\partial \alpha_i} = 0 \rightarrow y_i = w^T \phi(x_i) + b + e_i \end{cases} \quad (i = 1, 2, 3, \ldots, n)$$

Finally, regression function of least squares support vector machine can be written as follows:

$$y(x) = \sum_{i=1}^{n} \alpha_i k(x, x_i) + b$$

SVM constructed by radial basis function $K(x_i, x_j) = \exp\left(-\frac{\|x_i - x_j\|}{2\sigma^2}\right)$ has excellent nonlinear forecasting performance. Thus, in this work, RBF is used in the SVM, where $\sigma$ is width of Radial Basis Function.

## 12.3.2 *Parameters Optimization of SVM Based on GA*

The training parameters $C$, $\sigma$ and $\varepsilon$ greatly affect the forecasting performance of SVM. However, the appropriate select of these SVM parameters is very difficult. GA has strong global search capability, which can get optimal solution in short time. So GA is used to search for better combinations of the parameters in SVM [9–10]. After a series of iterative computations, GA can obtain the optimal solution. As shown in Fig. 12.1, the process of optimizing the SVM parameters with genetic algorithm is presented [1–2], which is described as followings.

### 12.3.2.1 The Principle and Structure of GA

GA is a global optimization technique derived from the principle of nature selection and evolutionary computing or technique, GA has been theoretically and empirically proven to be robust search technique [4]. Each possible point in the search space of the problem is encoded into a representation suitable for applying GA [11]. GA transforms a population of individual solutions, each associated with a fitness value into a new generation of the population, using the Darwinian principle

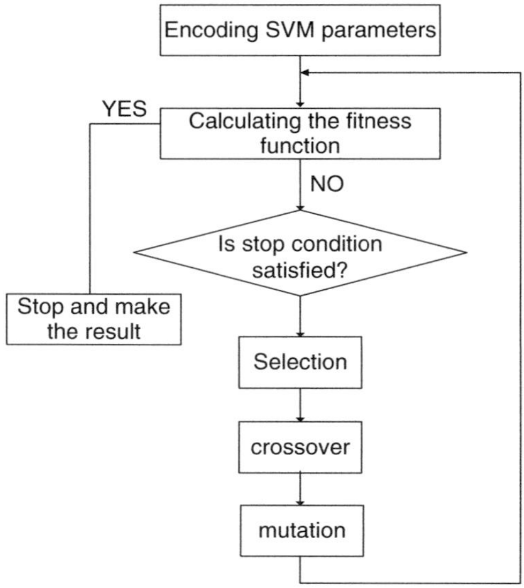


**Fig. 12.1** SVM parameters

of the survival of the fittest. By applying genetic operators such as crossover and mutation, GA produces better approximations to the solutions. At each iteration, new generation of approximations is created by the process of selection and reproduction. The next generation of population is a generation population after selection, crossover and mutation after the formation of the, the next generation of the population than the last generation has a strong suitability population and so on and so on, until the termination conditions meet [12]. The structure of GA can be shown as follows:

#### 12.3.2.2 Use Genetic Algorithm to Optimize $\sigma$ and $C$

Encoding SVM parameters and initial population

Encoding and generating initial population, and use floating-coding generic algorithm. In general the population size is about 10–160, in here we choose 50 [2, 4].

Fitness definition

Taking $\sigma^2$ and $C$ as the combination of two variables to figure out the fitness function of the GA. Considering the least squares support vector machines (LS-SVM) has the smallest error, and the fitness function is to realize the maximization [2], so the fitness function can be written as follows:

$$f(x) = \frac{1}{R(\sigma^2, C)}$$

$$R\left(\sigma^2, C\right) = \sum_{i=1}^{n} (y_i - f(x_i))^2$$

Where $R(\sigma^2,C)$ is square sum of error (SSE), which changing with the parameters of SVM $\sigma^2$ and $C$. $(x_i,y_i)$ is validation sample set, $f(x_i)$ is outputting of $(x_i,y_i)$, $n$ is the size of $(x_i,y_i)$.

Choosing rules

Populations with better fitness should be chose, according to Darwin's evolution theory the best one should survive and create new offspring. In order to make the best one have more probability to be selected, random competition method is used. There are many methods available for selecting the better one such as roulette wheel selection, steady state selection, tournament selection, elitism selection, etc. In this paper, the roulette wheel selection method is used.

Crossover and mutation

Real coded is used in the gene encoding, in order to generate new parameter after crossover, the arithmetic cross technique among different groups was used.

Is stop condition satisfied?

If the stop condition is satisfied turn to step f, else turn to step b. There are two kinds of stop condition are defined: threshold end means when the optimal solution set has reached the pre-set threshold, the progress stop; iterative steps end means when the iterative steps have reached the pre-set volume, the progress stop.

Termination algorithm

Outputting the optimized results $\sigma^2$ and $C$.

### 12.3.3 Modeling and Procedures

1. Get the original data of passenger flow for short-time forecasting.
2. The training sample sets can be established by the formula [12]:

$$X = \begin{bmatrix} x_1 & x_2 & \cdots & x_m \\ x_2 & x_3 & \cdots & x_{m+1} \\ \vdots & \vdots & \vdots & \vdots \\ x_{n-m} & x_{n-m-1} & \cdots & x_{n-1} \end{bmatrix}, \quad Y = \begin{bmatrix} x_{m+1} \\ x_{m+1} \\ \vdots \\ x_n \end{bmatrix}$$

   where $X$ denotes the input vector and $Y$ denotes the corresponding output, $R^m \to R$, $m$ is embedding dimension.
3. Utilizing the passenger flow, the training sample sets and forecasting sample sets' input and output is determined through analyzing the embedding dimension $m$.
4. RBF kernel function is used.

5. Establish LS-SVM regression function:

$$f(W_j) = \sum_{n}^{n-m} \alpha_i K(E_i, W_j) + b$$

where $m$ is embedding dimension, $n$ is the length of testing data, $E_i$ and $W_j$ row vectors of matrix $X$.

6. Using the training data as the input of LS-SVM, according step 2 figure out the $b$ and $\alpha_i$, finally get the prediction equation $y(x) = \sum_{i=1}^{n} \alpha_i k(x, x_i) + b$.
7. Getting the scope of $\sigma^2$ and $C$ by verifying the forecasting data sets, and initializing $\sigma^2$ and $C$.
8. With $f(x) = \frac{1}{R(\sigma^2, C)}$, work out the fitness value.
9. Correcting $\sigma^2$ and $C$ by crossover operation and reciprocated operation.
10. Stop the iterating if the stop condition is satisfied, else continue.
11. Using $y(x) = \sum_{i=1}^{n} \alpha_i k(x, x_i) + b$ with the most optimal $\sigma^2$, $C$ and testing data to solve the forecasting data.
12. Effect evaluation

Calculating the Maximum Error (ME) and Mean Absolute Percentage Error (MAPE) as the valuation criterias for forecasting:

$$ME = \max\left\{\left|\left(X_i' - X_i\right)\right|, i = 1, 2, \cdots, l\right\}, \quad MAPE = \frac{1}{l}\sum_{i=1}^{l} \frac{|(X_i' - X_i)|}{X_i}$$

where $X_i'$ is the predictions, $X_i$ is the real volume, $l$ is the length of testing step.

### 12.3.4 Conclusion

Vast influencing factors are involved in the short-time passenger flow forecasting, while the method provided in my paper is driven with data completely, there's no doubt that when the station makes decision some additional conditions should be taken into account. For instance, in a sunny day, the motor transportation is prosperous, as a result short-haul traffic by train will decrease, and the pressure of railway transportation will be eased, otherwise on the contrary.

## 12.4 Outcomes and Value

Rail transit is a huge, complex system which has the feature of irreversibility, once the line was built on change was available.

Short-term forecasting can be used to provide travel information for passengers, and train operation diagram's automatic generation, which may maximize the benefit, and with the result of short-term forecasting train operation can be dynamically adjusted, which will be efficient and energy saving. And this method can be applied more widely in the operation of urban rail transit, because the start and arrive time of urban rail transit is not settled, there is no necessary to book the ticket. While in the national railway, each time the changing of operation diagram will take dramatically financial and human cost, so if the software can be applied into the practice, it can bring up the enormous economical benefits and social benefits.

Above all, China's high-speed railway and the construction of the rail transit development of China will occupy an important position in the future, this part will occupy a great share of the market, and the government will donate huge financial support into the research, as a result academic research in the area of rail transit will also have a great prospect.

**Acknowledgment** We wish to acknowledge the support of Beijing Laboratory for Mass Transit, the National High-Technology Research and Development Program ("863"Program) of China No. 2012AA112801, The Project of the Fundamental Research Funds for Central Universities No. 2011JB2004.

## References

1. Feng Q, Xiyu L, Yinghong M (2009) Prediction of railway passenger traffic volume based on neural tree model. In: Second international conference on intelligent computation technology and automation. ICICTA'09
2. Xiaogang C (2009) Railway passenger volume forecasting based on support vector machine and genetic algorithm. In: International conference on future computer and communication. FCC'09, p 244
3. Qing C, Cuihong L, Wei G (2009) Railway passenger volume forecast based on IPSO-BP neural network. In: International conference on information technology and computer science. ITCS 2009
4. Qing C, Wei G, Cuihong L (2009) Railway passenger volume forecast by GA-SA-BP neural network. In: International workshop on intelligent systems and applications. ISA 2009
5. Dougherty MS, Cobbett MR (1997) Short-term inter-urban traffic fore-casts using neural networks. In: Int J Forecast 13(1):21–31
6. Ledoux C (1997) An urban traffic flow model integrating neural networks. Transp Res Part C–Emerg Technol 5(5):287–300
7. Shmueli D (1998) Applications of neural networks in transportation planning. Prog Plann 50 (3):141–204
8. Geng Y (2008) Based on support vector machine research and application of least squares power system short-term load forecasting[D]. Shandong University, Shandong

9. Yang YQ (2006) Research on least squares support vector machine power system short-term load forecasting[D]. Sichuan University, Sichuan
10. Yao ZS (2007) Short time prediction theory and method of real-time data traffic on the road network[D]. Beijing Jiaotong University, Beijing
11. Dia H (2001) An object-oriented neural network approach to short-term traffic forecasting. Eur J Oper Res 131(2):253–261
12. Xia Q (2011) Research on fluctuation of holiday railway passenger flow and application on forecast of passenger flow[D]. Beijing Jiaotong University, Beijing

# Chapter 13
# Research on Transfer Design of the Subway Station Buildings

**Yuxiang Dong, Xin Wang, and Yuanyuan Ding**

**Abstract** The distribution of transfer station affects the operation of the rail transit network, thus it affects the whole city traffic. Through analysis and research on multiple international metropolitan underground rail transit transfer station cases, this article starts from the interchange station layout form, the transfer mode and transfer efficiency. It discusses the method of urban underground rail transit transfer station design and finds that the layout form of interchange station decided to the passenger transfer way, for different forms of transfer way lead to different transfer time and the transfer way influent the transfer efficiency.

**Keywords** Subway station buildings • Transfer mode • Transfer flow • Transfer efficiency

## 13.1 Construction and Development of Urban Underground Rail Transit Transfer Station

With the development of city and the increasingly perfect of the urban rail transit, the subway operation mileage, the subway line and the number of subway station's increasing, the number of transfer station is also increasing. In 2004, the number of transfer station in London is 65. It rises to 35 % and the number of the transfer station sum up to 100 seats in 2012. So far, the largest number of transfer station city is New York, the number of the transfer station sum up to 111 seats in 2012.

Y. Dong (✉)
School of Architecture and Art Design, Beijing Jiaotong University, No.3, Shangyuancun, Haidian District, 100044 Beijing, China
e-mail: yxdong@bjtu.edu.cn

X. Wang • Y. Ding
School of Architecture and Design, Beijing Jiaotong University, Beijing, China
e-mail: 11121941@bjtu.edu.cn; 10122047@bjtu.edu.cn

H. Xia and Y. Zhang (eds.), *The 2nd International Symposium on Rail Transit Comprehensive Development (ISRTCD) Proceedings*,
DOI 10.1007/978-3-642-37589-7_13, © Springer-Verlag Berlin Heidelberg 2014

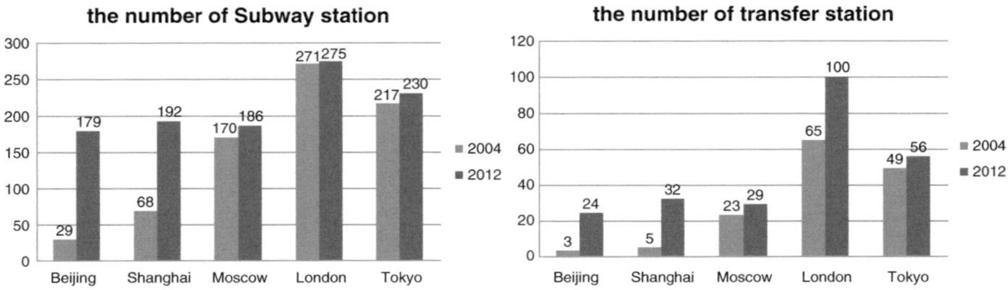


**Chart 13.1** Related statistics on part of the urban subway construction all over the world (*compared with 2004 and 2012*)

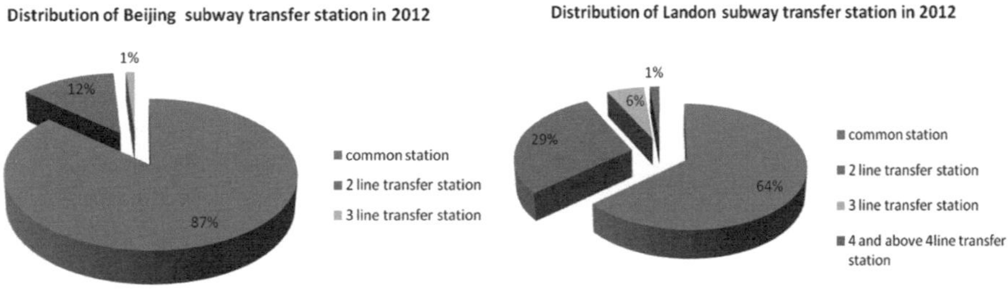


**Chart 13.2** Related statistics on transfer station in Beijing and Landon (*compare with 2004 and 2012*)

Chinese subway construction is in the development stage, so the speed of transfer station increasingly is particularly significant. By the year 2004, number of transfer station from Beijing and Shanghai is insufficient ten seats. In 2012, the number of transfer station in Beijing rises to more than 20, and that in Shanghai rises to more than 30. Transfer stations are greatly increasing in recently, the number of transfer station from Beijing and Shanghai increased seven times and 5.4 times. By the year 2015, Beijing subway line will reach 19; wire network transfer station will reach 74 seats, more than that in 2012 years and increase about 2.1 times in 3 years, and increase 35 times than that in 2004. With the increase of urban subway lines, the subway by two line transfers, gradually developed into three wire transfers, even four wire transfers, some cities even have five line transfers, transfer station number increased sharply. With the number of transfer station rising, subway interchange stations play an important role in the subway network (Charts 13.1 and 13.2).

## 13.2 The Distribution of Underground Rail Transit Transfer Station Based on the Overall Urban Planning

With the aggregately construction of the subway network, the line between network nodes – transfer station – continues to increase, and transfer station in the rail transit network is playing more and more important role. Interchange station acts on improving the efficiency of urban residents traveling [1].

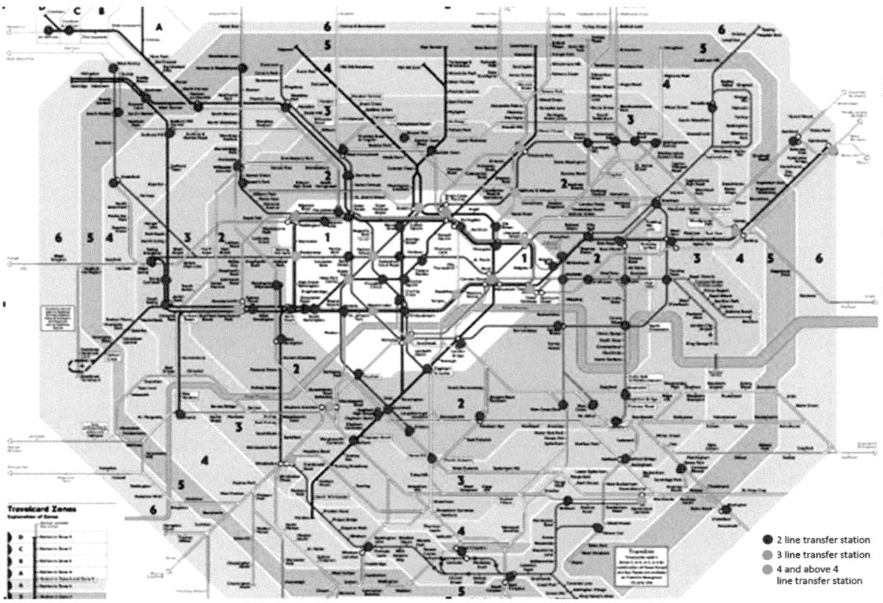


**Fig. 13.1** Distribution of transfer station in London

The distribution of transfer station bases on the planning of urban design, and transfer stations form to node which is intent distributing in the city's important area and the core zone such as the urban traffic, commerce, culture, information, entertainment center and so on in urban underground traffic network. Underground rail transit construction abroad started early, and the distribution of transfer station closely based on the urban planning [2]. Subway in London form into loop radial and 12 subway lines across the city's six areas. Transfer stations were constructed at the intersection and contacts the rail transit network and covers the whole city (Fig. 13.1).

The construction of Underground rail transfer station in China bases the urban planning. Take Beijing as an example, Beijing subway carried out "a ring and second lines" construction planning in the early time. The transfer stations were distributed in DongCheng district and XiCheng district of the city center, and were set in line interchange, connecting with the urban areas. In the twenty-first century, Beijing subway is developing rapidly. At this time, the construction of transfer station in Beijing is closely attached to the "underground space development and planning of center area in Beijing (2004–2020)". The subway line will reach "three-ring, four cross, five longitudinal and seven radiation" totally 19 lines, including 73 seats transfer station by 2015.

Transfer can realize the transfer, contacting by surface cover replace the point line. Reasonable transfer station will promote the urban traffic system operating efficiently.

## 13.3 Layout Form of Urban Underground Rail Transfer Station Decided to Transfer Mode

The layout and graphic design of the underground space of transfer station is very important to rail transit function in the network, at the same time, different layout decides to transfer distance and transfer time even influences transfer efficiency.

### *13.3.1 Layout of the Underground Rail Transit Line Influence the Layout form of Transfer Station*

Layout of urban underground rail transfer station should consider the rail transit network planning, the construction sequence and mixed form, transfer station transfer passenger flow and the traffic, business development and the location of the project summary and so on various factors. Among them, the layout of the underground rail transit line plays an important role to transfer station layout form. The layout of underground rail transit line should combine with city function, using of city resources reasonability, coordinating of urban space layout, making rail transit and city combine as an organic whole, and then to get reasonable transfer station layout form.

### *13.3.2 The Transfer Station Layout Form*

For single transfer station, the mode number of transfer station relates with the number of subway line. the layout form of two line transfer station profile to "+" type, "T" type, "L" type and "=" type in accordance with the relationship with platforms (Fig. 13.2). For three lines transfer station, there are a variety of other

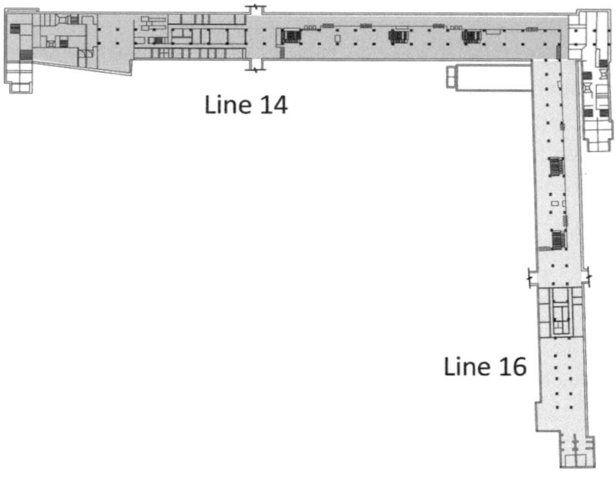


**Fig. 13.2** SanLuJu transfer station scheme

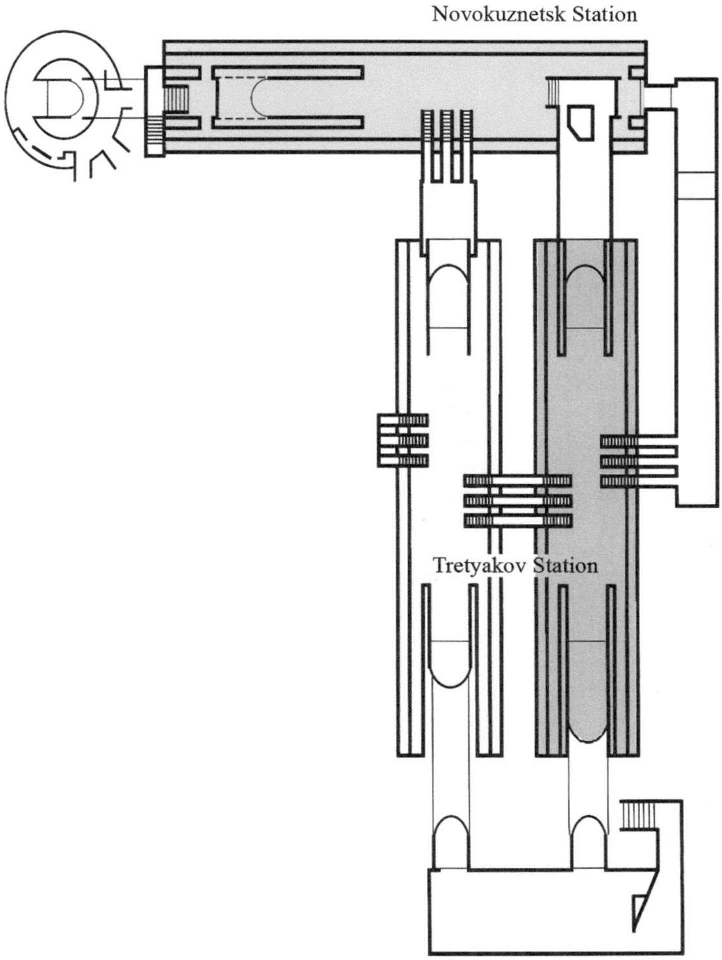


**Fig. 13.3** YaKeFu transfer station scheme

forms basing the two lines, such as "△" type, "*" type, "F" type (Fig. 13.3), "H" type (Fig. 13.4), and four line transfer station "□" type (Fig. 13.5) and so on.

### *13.3.3 Influence of Transfer Station Layout Form to Transfer Modes*

The transfer station layout form decided to transfer mode. Different transfer mode is suitable for different transfer station layout.

Fig. 13.4 SongJiaZhuang transfer station scheme

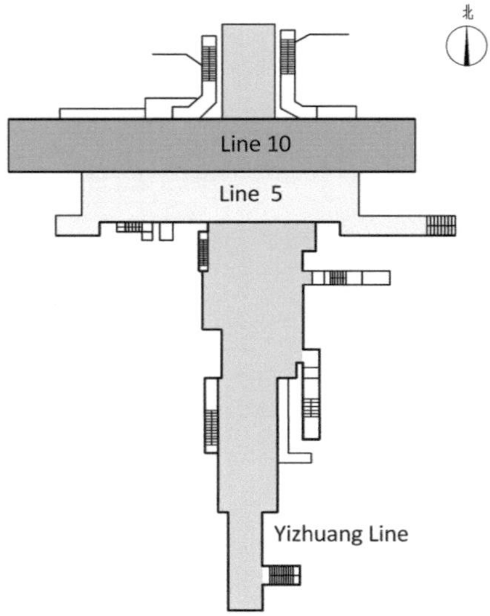


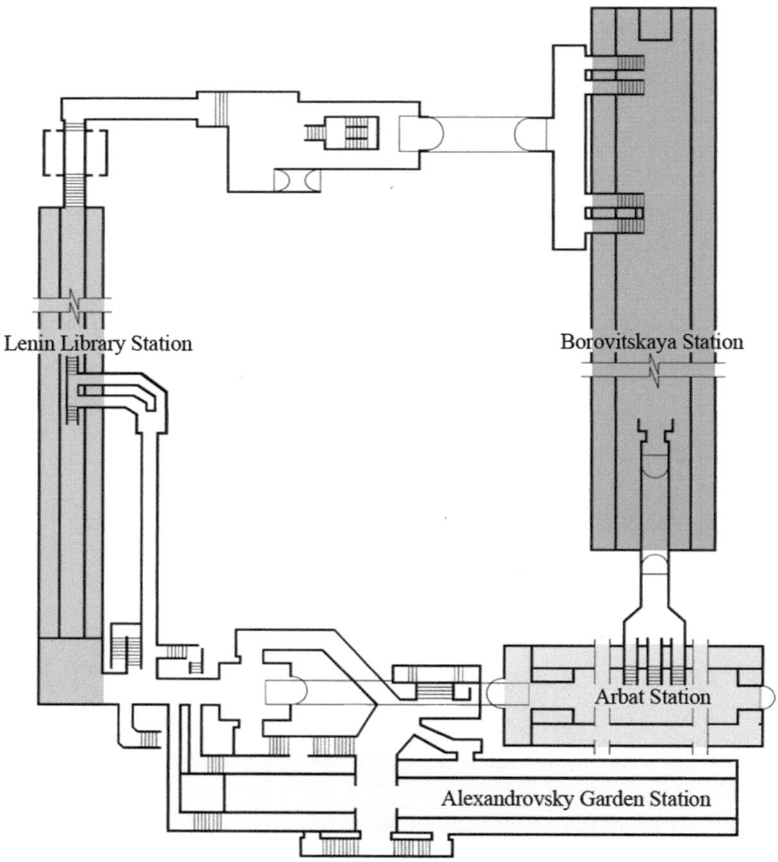


Fig. 13.5 Moscow Alexander Garden transfer station scheme

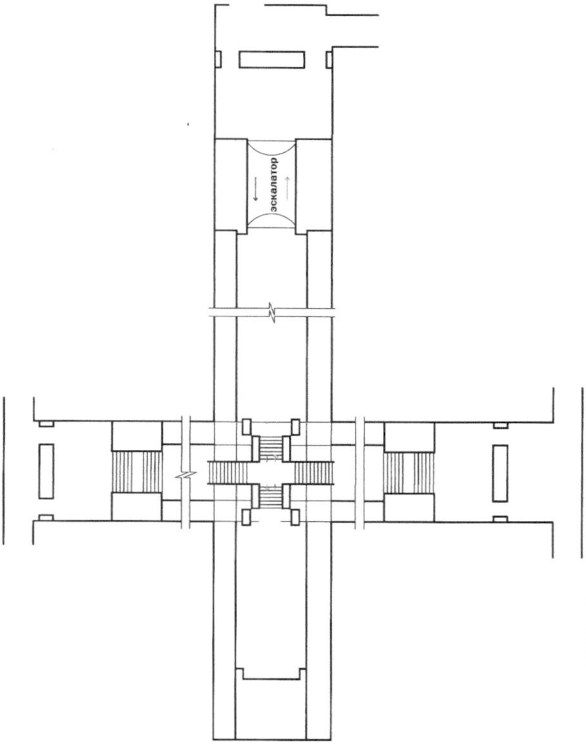


**Fig. 13.6** Central hall transfer layout plane

Transfer mode can be divided into different kinds of forms corroding different angles. On the basis of transfer position, it can be divided into the central hall transfer (Fig. 13.6), the hall end transfer (Fig. 13.7), middle and end combination transfer (Fig. 13.8) and According to the transfer form, it can be divided into same hall transfer (Fig. 13.9), bridge transfer, porch transfer and combination with middle bridge and porch transfer. Bridge transfer means transfer through the bridge crossing the orbit, which is a kind of common transfer forms that widely used in Moscow subway stations. Its transfer distance is shorter, causing less transfer time. Porch transfer are used for vertical transfer, which using the end of the channel mutual to transfer in two line hall, and the transfer distance is a bit long, casing transfer time less. Combination with middle bridge and porch transfer is a more complicated transfer form, which generally applied in three lines and four line mutual transfers.

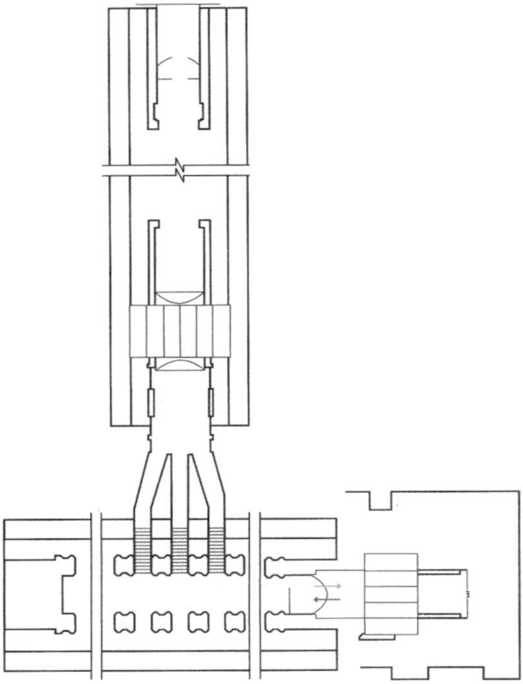

**Fig. 13.7** Hall end transfer plane layout

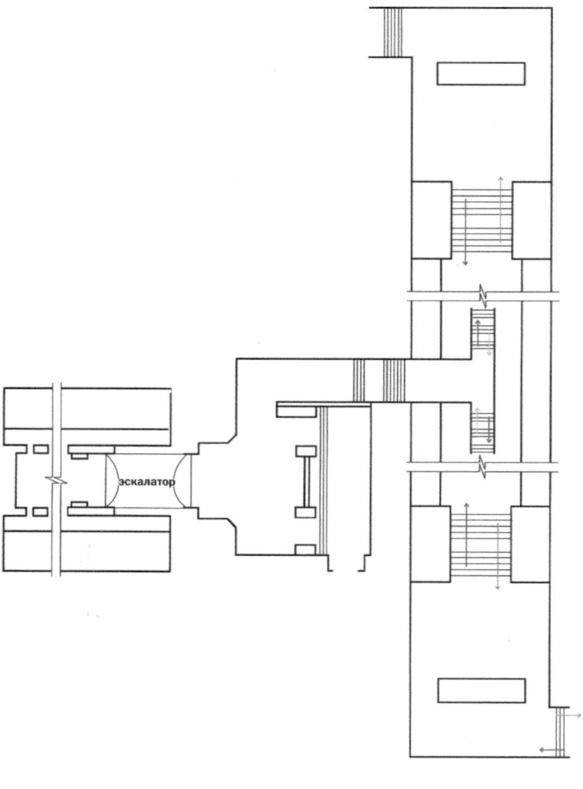


**Fig. 13.8** Middle and end combination transfer plane layout

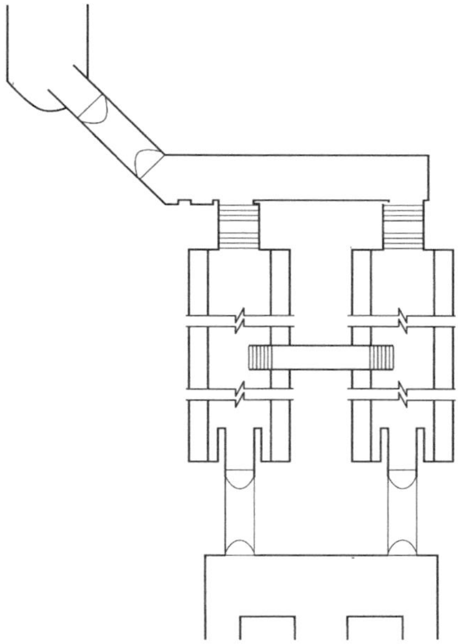

**Fig. 13.9** Combining with hall change layout

## 13.4 Influence Factors of Transfer Efficiency

In the railway traffic transfer station design, improve the transfer efficiency is one of the main points, while there are many factors affects the efficiency of transfer, such as the transfer station space layout form; Change the channel, the staircase and ramp forms, scale and direction; Transfer guide logo design and layout; The design of the transfer line; Transfer space design, etc [3].

### *13.4.1 Transfer Station Space Layout Form*

Traffic hub of the space layout form decided to interchange way, decided to change to the mouth position. Change to the mouth in the central hall general central or ends. Measure transfer station layout form, interchange way to transfer the stand or fall of position and mouth standard is transfer time, layout form, interchange way and the transfer of the mouth is designed to position to minimum distance, change to the shortest time. Common transfer station layout form, interchange way and the relationship between the transfer time see Table 13.1.

Through the Table 13.1 we can see, different layout forms combined with different directions and the ways have important influence on the length of transfer time [3]. In general, the two wire transfer, as a result of "=" type "+" type can form part of the passenger flow, so under the situation of the same transfer from equilibrium time traffic, its transfer efficiency in order to "=" type, "+" type, "L"

**Table 13.1** The layout form of metro interchange station, the interchange way and interchange time

| Line | Layout form | Name | Transfer mode | Transfer line | Transfer time | Evaluation |
|---|---|---|---|---|---|---|
| Two line transfer | "=" | Kitay-gorod in Moscow | Transfer with hall | Line 6 – line7 | 15 s | Shortest |
| | | | Central bridge | Line 6 – line7 | 40 s | Short |
| | | | Porch | Line 6 – line7 | 2 min 40 s | Shorter |
| | "+" | Lama Temple in Beijing | Fringe porch | Line 2 – line 5 | 40 s | Short |
| | "T" | Kuznetsky Most in Moscow | Fringe porch | Line 7 – line 1 | 3 min | Shorter |
| | "L" | Oktyabrskaya in Moscow | Fringe porch | Line 6 – line 5 | 2 min | Shorter |
| Three line transfer | "F" | Kiyevskaya in Moscow | Fringe porch | Line 5 – line 4 | East 2 min 30 s | Shorter |
| | | | Fringe porch | Line 5 – line 4 | West 5 min 20 s | Longest |
| | "F" | Kiyevskaya in Moscow | Fringe porch | Line 3 – line 5 | East 1 min 40 s | Short |
| | | | Fringe porch | Line 3 – line 5 | Bridge 2 min 30 s | Shorter |
| | "F" | Kiyevskaya in Moscow | Middle stair | Line 4 – line 3 | East 4 min 40 s | Longer |
| | | | Middle stair | Line 4 – line 3 | West 5 min | Long |
| | "H" | Song JiaZhuang in Beijing | Fringe porch | Line 5 – Line YiZhuang | 1.5 min | Short |
| | "H" | Song JiaZhuang in Beijing | Fringe porch | line 10 – line YiZhuang | 1.5 min | Short |
| | "H" | Song JiaZhuang in Beijing | Central bridge | Line 5 – line 10 | 0.5 min | Shortest |
| Four line transfer | "□" | Alexandrovsky Sad in Moscow | Central bridge | Line 3 – line 9 | 2 min 10 s | Shorter |
| | | | Fringe porch | Line 3 – line 4 | 3 min 10 s | Shorter |

type and "T" type, in the three lines above transfer, while, the passenger transfer agreement, the transfer efficiency of the shortest transfer time is the highest.

### 13.4.2 Transfer Mode and Transfer Time

Transfer station transfer modes with the platform mainly include transfer, node transfer, station hall transfer, channel transfer and mixed transfer and several ways. With the platform is to transfer the passengers off directly after the other side to walk to the platform, waiting for the train. Node, the station hall and channel transfer respectively is refers to the transfer passengers get off use floor escalator (nodes), transfer hall, transfer channel to another platform waiting. And hybrid transfer is a fusion of several transfer way in front of the comprehensive change.

In a variety of transfer mode in the comparison, with the platform transfer can greatly simplify the transfer line, is the ideal transfer way, but to wire network planning, circuit design, train operation organization the demand is higher, and the cost is expensive. Node, station hall, channel transfer is the present domestic rail transit commonly used transfer organization way, convenience from high to low arrangement with the platform is: One-platform-interchange > node transfer > channel transfer > station hall transfer.

### 13.4.3 The Influence of the Transfer Time

In general, the interchange station layout form, transfer modes, transfer line between various factors are closely related. The transfer between items reflects on the influence of time. And the length of the transfer time has a direct link with the transfer efficiency. Under construction in Beijing BaiShiQiao transfer station scheme for patients (Figs. 13.10, 13.11, and 13.12) [4], the designer is based on different BaiShiQiao transfer station layout form, transfer mode, transfer streamline design, this paper combined with the city's comprehensive factors to consider, to form the perfect design.

Of course, the interchange station layout forms need to be affected by many factors in practice, including travel passengers characteristics, transfer facilities, transfer station operation situation, the transfer station space environment quality, etc. But the transfer station itself, transfer station space layout, transfer mode, transfer time, its layout form of passenger transfer efficiency of the influence is no doubt. In these main factors determine the premise, combined with other factors can be determined from comparison of various transfer station integrated transfer evaluation.

Fig. 13.10 BaiShiQiao transfer station "+" type scheme

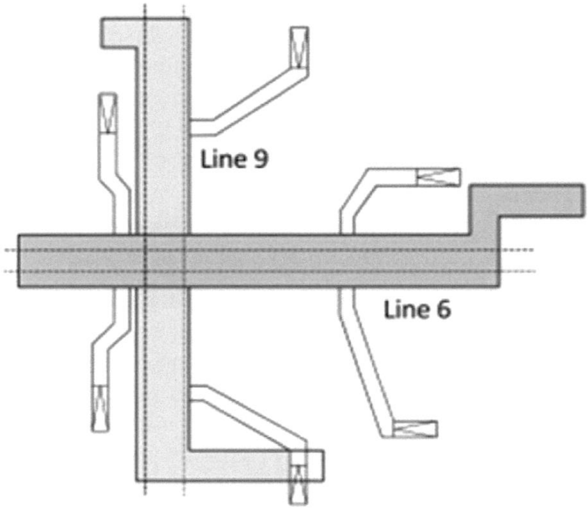


Fig. 13.11 BaiShiQiao transfer station "T" type scheme

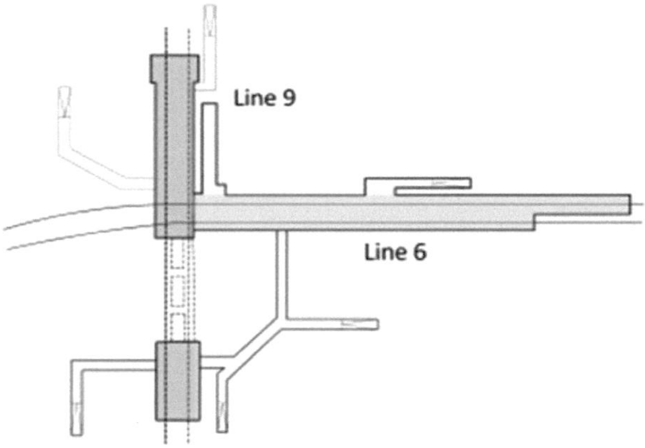


Fig. 13.12 BaiShiQiao transfer station "L" type scheme

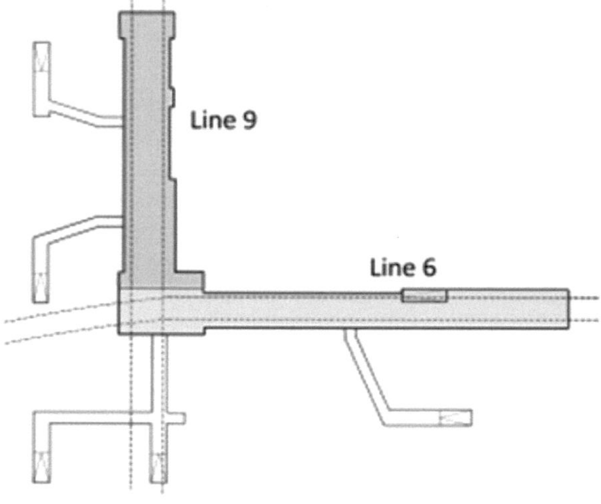

## 13.5 Conclusion

In the metro transfer station construction practice, we can see the metro transfer station quick and efficient transfer design mainly embodied in the following: (1) the interchange station location and layout closely combined with city planning. (2) the way to optimize the design, make full use of the same hall, the same transfer and other high efficient transfer form, the transfer distance is short, transfer time less optimization scheme; (3) transfer station transfer streamline design is reasonable and change to the flow of a single, clear and do not cross.

There are various layout forms and transfer modes in rail transport hub station . But no matter what methods should be to meet the demand for passenger travel, it must ensure there is enough transfer efficiency and convenience. Paying attention the transfer efficiency of passenger flow in rail transport hub is one of the important directions in urban rail transit construction transfer station design and planning.

## References

1. Wang Bo, Li Xiaoxia An Shuanzhuang (2010) Analysis on transfer passenger flow characteristics and station design of urban rail transit [J]. Urban Rapid Rail Transit 02:55–57
2. Dong Yuxiang (2012) Research on Russia Metro Transfer Station underground space design [J]. Huazhong Archit 01:6–10
3. Bao Ning (2009) The study of the urban subway transfer station design [D]. Beijing Jiaotong University, pp 16–23
4. Kang Yuehua (2009) Study on transfer schemes on Baishiqiao Station of Beijing Metro [J]. Railw Stand Design 10:38–40

# Chapter 14
# Research on Traffic Assignment Problem in a Multimodal Hyper-Graph

**Lixiang Zheng, Bingfeng Si, and Ying Wang**

**Abstract** Based the theory of hyper-graph, a multimodal hyper-graph is given in this work. Taking into account many factors such as route travel time, transfer times and comfort, route generalized cost function is established. On this basis, the meaning of effective route is redefined, and combined with the search algorithm in graph theory, an improved Dial algorithm is proposed. In this work, the improved Dial algorithm is used to solve the problem of urban multimodal traffic assignment. At the end of the work, an example is given to verify the feasibility and effectiveness of the algorithm. The results show that the improved Dial algorithm is suitable for solving the urban multimodal traffic assignment problem, and avoided the abnormal results that may appeared when use the original Dial algorithm to solve the problem.

**Keywords** Hyper-graph • Generalized cost function • Dial algorithm • Traffic assignment

## 14.1 Introduction

In recent years, more and more domestic and foreign scholars study how to modeling the urban multimodal travel networks, and establish some route search methods. Lozano and Storchi [1] used hyper-graph modeling the urban multimodal travel network, and propose an algorithm to solve the shortest viable hyper-path problem in multimodal networks based on the pareto optimal theory. Bielli and

L. Zheng (✉) • B. Si
School of Traffic and Transportation, Beijing Jiaotong University, 100044 Beijing, China
e-mail: mofeixiangge@163.com

Y. Wang
School of Economics and Management, Beijing Jiaotong University, 100044 Beijing, China

H. Xia and Y. Zhang (eds.), *The 2nd International Symposium on Rail Transit Comprehensive Development (ISRTCD) Proceedings*,
DOI 10.1007/978-3-642-37589-7_14, 

Boulmakoul [2] established an object-oriented multimodal travel network, in which nodes and arcs are all regard as objects. Fernandez [3] studied the dual-modal travel network and given an algorithm to solve the shortest path problem in dual-modal travel network. But all these studies are limited to modeling the urban multimodal travel networks, and do nothing with the urban multimodal traffic assignment.

According to the behavioral characteristics of travelers and features of multimodal travel networks, Hu [4] studied the route choice behaviors of travelers under the multimodal travel information service conditions. According to the traveler's choices of the path and transfer site, Huang and Li [5] studied the mixed equilibrium model and solution algorithm in transportation networks with combined modes. Si [6–8] studied the defects that may occurred when use Dial algorithm to solve the traffic assignment problem, and proposed an improved Dial algorithm. The improved Dial algorithm avoids the shortcomings of the original algorithm, and improves the efficiency of the algorithm.

The article considering the above two aspects that have studied, propose a model to solve the traffic assignment problem in urban multimodal traffic networks. In the second part of the work, the author establishes an urban multimodal traffic network based on hyper-graph. Route generalized cost function is been studied in the third part of the work. And the meaning of effective route is redefined in the fourth part, based on the new defining an improved Dial algorithm is proposed. A simple example of the algorithm application is given in the fifth part and the conclusions are shown in the sixth part of the work.

## 14.2 Urban Multimodal Traffic Networks

Urban multimodal traffic network is a complex composite network. Single-modal traffic network can be described as $G = (V,E)$, where $V$ is the set of nodes and $E$ is the set of arcs. Multimodal traffic network is composed by many single-modal networks and can be described as $G = (V,E,M)$, where $M$ is the set of modes. Let $M = \{M_1,M_2, \ldots,M_k, \ldots\}$, then different modes $M_k$ have different network structure $G_k = (V_k,E_k,M_k)$. Modal transfer arcs are added to connect different modes networks, and the transfer mode is represented by $M'$.

Through an example below shows how to construct an urban multimodal traffic network. A multimodal network is shown in Fig. 14.1. Assume that the multimodal network has one OD pairs, 9 stations, 17 arcs and has 6 traffic modes: bus($M_1$), taxi ($M_2$), metro ($M_3$), car ($M_4$), walking mode ($M_5$), modal transfer ($M'$). From the Fig. 14.1 we can see that the multimodal network is composed by five single-modal networks, and different single-modal networks are combined with mode transfer arcs. Station 1, 2, 3, 4 are bus stations, station 5, 6, 7, 8 are metro stations and station 9 is taxi station.

In a multimodal traffic network, travelers' travel choices mainly include two aspects: one is the travel mode choices, and the other is the route choices. Usually, there have many factors that influence travelers' choice to travel modes and travel

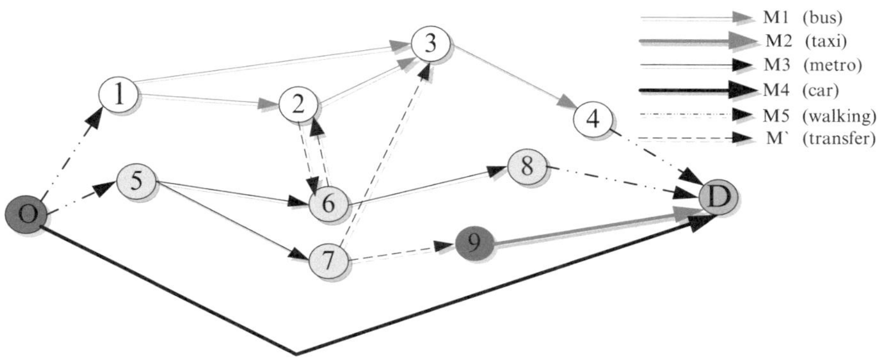


**Fig. 14.1** Urban multimodal traffic network based on hyper-graph

routes, such as personal attributes, travel time, transfer times, comfort and security. This article focuses on considering travel time, transfer time and comfort three factors' influence on travelers' travel choices.

## 14.3 Route Generalized Cost Function

This paper assumes that all travelers are as a single type of travelers that all travelers have the same perceived costs to transfer times and comfort. Route's generalized cost includes four parts: boarding time, travel time, transfer time and leaving time. Then analysis these four parts of cost respectively.

### *14.3.1 Boarding Time*

Boarding time is mainly generated from travelers' walking time and waiting time. Walking time refers the time that traveler walk from start point to the first station. Waiting time refers the time that traveler wait for the mode arriving.

$$T^b = T^{b0} + T^{b1}. \tag{14.1}$$

Where, $T^b$ is boarding time, $T^{b0}$ is walking time, and $T^{b1}$ is waiting time.

### 14.3.2 Travel Time

Travel time refers the time that travelers spend on the traveling. In the case of considering the crowded impact, traveler's perceived travel time is increased with the increase in actual travel time.

$$T^t = \alpha \cdot \widetilde{f}(c) \cdot T^{t0}. \tag{14.2}$$

Where, $T^t$ and $T^{t0}$ are traveler's perceived travel time and actual travel time respectively; $c$ is traffic flow and $\widetilde{f}(c)$ is an increasing function that is relevant to c; $\alpha$ is an undetermined parameter.

### 14.3.3 Transfer Time

Transfer time includes transfer walking time、transfer waiting time and transfer penalty time. Transfer penalty time is caused by the inconvenience of transfer, and increased with the increase of transfer times.

Transfer walking time refers to the time that travelers spend on transfer from one mode to another mode or transfer from one line to another line. And transfer waiting time refers to the time that traveler waiting for the arrival of mode or line. Assumed that the transfer times is N in one route, then the transfer time can be shown as:

$$T^c = \sum_{i=1}^{N} \left(T_i^{c0} + T_i^{c1}\right) + T^{cp}. \tag{14.3}$$

$$T^{cp} = \pi \cdot N^{\beta} \tag{14.4}$$

Where, $T^c$ is transfer time; $T_i^{c0}$, $T_i^{c1}$ and $T^{cp}$ are transfer waiting time, transfer walking time and transfer penalty time respectively. $\pi$ and $\beta$ are undetermined parameters.

### 14.3.4 Leaving Time

The time that travelers spend on leaving travel modes and walking to the destination is the leaving time. And be represented by $T^l$.

To sum up, route generalized cost function can be represented by:

$$T_{rs} = T^b + T^t + T^c + T^l \tag{14.5}$$

## 14.4 Traffic Assignment Model of Multimodal Traffic Networks

This paper adopts Dial algorithm to solve the traffic assignment problem in multimodal traffic network. Consider travelers' choices of travel modes and the influence of transfer on travelers' travel behaviors, the article redefine the meaning of effective route.

The new effective route required to meet the following three points: (1) the route does not include loops; (2) the travel cost of the route $T_{rs} \in [0, (1 + h)T_{rs}^{\min}]$; (3) transfer times must less than K, K is the upper limit of transfer times. $T_{rs}^{\min}$ is the minimum generalized travel cost between origin r and destination s, h is non-negative parameter.

Based on the route generalized cost function and the new definition of effective route, the paper proposed an improved Dial algorithm to solve the traffic assignment problem. The procedure has the following steps:

**Step 1:** Search the effective routes between OD pairs and the effective arcs. Use traversal algorithm searching all of the effective routes. If the arc $(i, j)$ is belong to one of the effective routes, then it is an effective arc, and set $k_{ij} = 1$, otherwise, $k_{ij} = -1$.

**Step 2:** Initialize.

1. $T_{ri}^{\min}$: The minimum generalized cost of the routes between origin r and node i.
2. $T_{is}^{\min}$: The minimum generalized cost of the routes between node i and destination s.
3. $U_i$: Set of nodes, includes all upstream nodes of the node i.
4. $D_i$: Set of nodes, includes all downstream nodes of the node i.
5. Calculate likelihood value $L(i, j)$ of every arc $(i, j)$:

$$L(i,j) = \begin{cases} \exp\left\{\theta\left(T_{rj}^{\min} - T_{ri}^{\min} - t_{ij}\right)/T_{rs}^{\min}\right\} & k_{ij} = 1 \\ 0 & k_{ij} = -1 \end{cases} \tag{14.6}$$

**Step 3:** Calculate the weight of arcs. Sort the nodes by ascending order of $T_{ri}^{\min}$, then calculate the weight of arc $(i, j)$ in turn, where $j \in D_i$.

$$W(i,j) = \begin{cases} L(i,j) & i = r \\ L(i,j) \cdot \sum\limits_{m \in U_i} W(m,i) & i \neq r \end{cases} \tag{14.7}$$

Where, $W(i, j)$ is the weight of arc $(i, j)$.

**Step 4:** Calculate the flow of arcs. Sort the nodes by ascending order of $T_{js}^{\min}$, then calculate the flow of arc $(i, j)$ in turn, where $i \in U_j$.

$$v(i,j) = \begin{cases} q_{rs} \cdot \dfrac{W(i,j)}{\sum\limits_{m \in U_j} W(m,j)} & j = s \\ \sum\limits_{m \in D_j} v(j,m) \cdot \dfrac{W(i,j)}{\sum\limits_{m \in U_j} W(m,j)} & j \neq s \end{cases} \tag{14.8}$$

Where, $v(i, j)$ is the flow of arc $(i, j)$.

## 14.5 Example

Based on the multimodal traffic network has been proposed in the second part of the work, calculate and verify the traffic assignment model. For ease of calculation, the generalized cost of every arc is given in the multimodal network. And in the network, only origin、destination and transfer nodes are shown. The network is shown Fig. 14.2 below.

Assign the traffic flow according to the original Dial algorithm, let $\theta = 20$,then we can get four effective routes between OD pairs, and the traffic assignment of every effective route shown in Table 14.1.

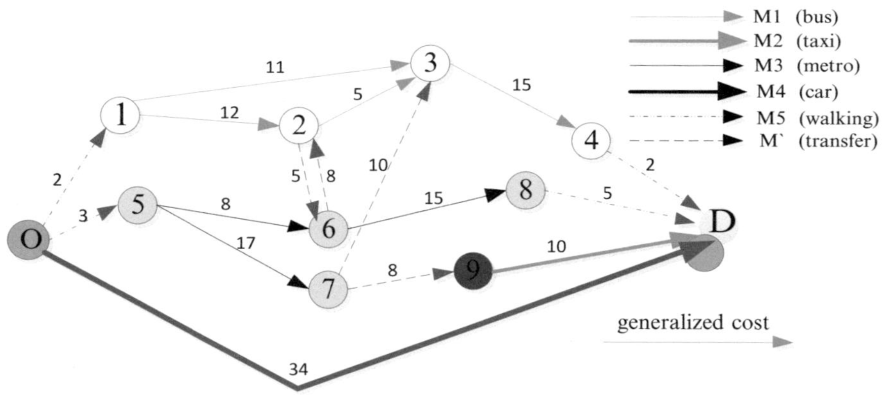


**Fig. 14.2** Urban multimodal traffic network with route generalized cost

**Table 14.1** Result of traffic assignment based original Dial algorithm

| Number | Route | Generalized cost | Transfer times | Traffic ratio |
|---|---|---|---|---|
| 1 | $O \to 1 \to 3 \to 4 \to D$ | 30 | 1 | 0.62983 |
| 2 | $O \to 5 \to 6 \to 8 \to D$ | 31 | 1 | 0.32336 |
| 3 | $O \to 5 \to 7 \to 9 \to D$ | 38 | 1 | 0.00304 |
| 4 | $O \to D$ | 34 | 0 | 0.04376 |

**Table 14.2** Result of traffic assignment based improved Dial algorithm

| Number | Route | Generalized cost | Transfer times | Traffic ratio |
|---|---|---|---|---|
| 1 | $O \to 1 \to 3 \to 4 \to D$ | 30 | 1 | 0.624525 |
| 2 | $O \to 1 \to 2 \to 3 \to 4 \to D$ | 36 | 2 | 0.011438 |
| 3 | $O \to 5 \to 6 \to 8 \to D$ | 31 | 1 | 0.320642 |
| 4 | $O \to D$ | 34 | 0 | 0.043394 |

According to the original definition of the effective route, route O→1→2→3→4-D is not an effective route, so the traffic on the route is 0. But the generalized cost of the route is 36, lower than the generalized cost of the route 3. Obviously, it is not a reasonable assignment result.

Then, based on the new definition of the effective route, use Dial algorithm to assign the traffic in the multimodal network. Let the upper limit of the transfer times is K = 2, and h = 0.25, $\theta = 20$, then the results of effective routes and traffic assignment are shown in Table 14.2.

From Table 14.2 we can find that, in the case K = 2, h = 0.25, route O→1→2→3→4-D replace the route O→5→7→9→D as an effective route. It is obvious that the result is more reasonable. From the result can also find that the improved Dial algorithm take into account the influence of the transfer times. The more transfer times in a route, the less traveler will travel on it. It can be reflect from two aspects, one is that with the increase of transfer times, the generalized cost of the route also increase. The other is when the transfer times of a route is high than K, it would not be an effective route, i.e. traffic flow assigned to the route is 0.

## 14.6 Conclusion

Establish multimodal traffic network based on hyper-graph theory, can let different modes be studied under one network environment. In order to simulate the effects of internal transfer in one travel mode and external transfer between different travel modes on the travelers' travel choice behaviors, this paper establish a route generalized cost function considering three factors which are route travel time, transfer times and comfort level. To solve the problem that may appear when use the original Dial algorithm to assign traffic flow in a network, the paper redefined the effective route and propose an improved Dial algorithm. At the end of the paper, the result of the example shows that the model and the algorithm are correct and effective.

**Acknowledgment** This work is supported by the National Natural Science Foundation of China (71071016).

## References

1. Angélica Lozano, Giovanni Storchi (2002) Shortest viable hyperpath in multimodal networks. Transp Res Part B 36:853–874. Elsevier Science
2. Maurizio Bielli, Azedine Boulmakoul, Hicham Mouncif et al (2006) Object modeling and path computation for multimodal travel systems. Eur J Oper Res 175:1705–1730. Elsevier Science
3. Fernandez E, Cea J, Florian M, Cabrera R et al (1994) Network equilibrium models with combined modes. Transp Sci 28:182–193. Operations Research Society of America
4. Hu Hua, Teng Jing, Gao Yufeng (2009) Research on travel mode choice behavior under integrated multi-modal transit information service. China J Highway Transp 22:87–92. Chang'an University Magazines Company
5. Huang Haijun, Li Zhichun (2006) Mixed equilibrium model and solution algorithm in transportation networks with combined mode. J Syst Sci Math Sci 26:352–361. The Chinese Academy of Science Institute of Mathematics and System Science
6. Si Bingfeng, Yang Xiaobao, Gao Liang (2010) Urban multimodal traffic assignment model based on travel demand. China J Highway Transp 23:85–91. Chang'an University Magazines Company
7. Si Bingfeng, Gao Ziyou (1999) The multimodal equilibrium assignment model for urban mixed traffic network. J Highway Transp Res Dev 16:44–48. The Traffic Department of Transportation Highway Academy of Sciences
8. Si Bingfeng, Zhang Haozhi, Gao Ziyou (2009) Improved Dial's algorithm for logit-based stochastic traffic network assignment problem. China J Highway Transp 22:78–83. Chang'an University Magazines Company

# Chapter 15
# Research on the Legibility of Urban Rail Transit Square

**Yunan Zhang, Ye Zhou, and Xiao Liu**

**Abstract** Based on improving the city's legibility, the paper puts forward the URT (urban rail transit) square environment design on the requirements of the location, direction, culture and time, Combined with Beijing's instance, it explore how to handle the legibility of the plaza through spatial layout, design details of buildings and material colors.

**Keywords** URT • Square • Legibility • Interface

**Forward.** If the URT (the urban rail transit) is a sign of a contemporary symbol of urban civilization and urban developed public transportation, the rail transit square is an important conversion and exchange place for the urban system and huge URT system underground. Through the design of their environment, not only the evacuation capacity of the station can be highlighted and strengthened and to provide protection for the large crowded dispersal, but also enhance the overall sense of place and legible of the urban spaces, so that people can clearly caught the urban space and strengthen the impression of the place during the hurried commuter process.

## 15.1 The Requirements of Legibility

With the accelerated speed of urban renewal and the complicate of urban ground and the underground transport system, it is very difficult to build a clear impression of the city. This resulted in the lack of recognition on the urban spaces and a lot of

Y. Zhang (✉) • X. Liu
School of Architecture and Design, Beijing Jiaotong University, 100044 Beijing, China
e-mail: ynzhang1@bjtu.edu.cn

Y. Zhou
School of Economics and Management, Beijing Jiaotong University, BJTU, Beijing, China
e-mail: ydzhou@bjtu.edu.cn

H. Xia and Y. Zhang (eds.), *The 2nd International Symposium on Rail Transit Comprehensive Development (ISRTCD) Proceedings*,
DOI 10.1007/978-3-642-37589-7_15, © Springer-Verlag Berlin Heidelberg 2014

inconvenience in the city living. It seems that the legibility itself, including the dialogue through space and time annotation between the creator and the viewer. And during the urban space design process, it's important to make sure the observer (commuting public) can clearly and easily interpret the following basic information environment of urban space through environmental design.

### *15.1.1 The Legibility of the Area and Location*

The legibility of the area and location mainly identify the spatial situation of the observer in the city. Because of standardized design, people in the complex subway system are very easy to loss of location-aware capacity during the travel. Compare with the streets of the ground, underground system lack of the outdoor asymmetry in the larger scene markers, such as the sun, moon and stars and landmarks landscape tips. Especially people may often be a momentary negligence and lost the resolution of location during the tired subway commuter. Sequencely, the similarity of all the city and street and architectural image is also easy to make it difficult to identify the location.

### *15.1.2 The Legibility of the Direction*

In addition to the need to understand their own situation, People also need to point to more identifying to indicate the target and determine the direction of travel. Generally, these meaningful marks include not only texts and signals, but also more easily described urban spatial information's, for example an special imply of square, street, buildings, trees, flower beds, billboards and even a particular wall which can be easily observed by people on the path to the destination. People need such marks as a guide to obtain more accurate spatial information.

### *15.1.3 The Legibility of Culture*

In addition to the identification of the spaces, the city itself should also actively create a positive dialogue of spirit between space and observers. Designers need to consider more about the unique characteristics of local culture. This cultural characteristic need not only show the coordination of environmental and geospatial which formed due to characteristics of place, but should form the consensus on the usage of spaces and value orientation between the creators and the observers.

### 15.1.4 Legibility of Time

If culture is given to the pursuit of the context inheritance and value stability, the characteristics of the society and the changing times is need to imply by the modern technology development and cultural trends change. Especially in modern time design should advocate the characteristics of the times, marked the ability that advancing with the times which shown in cultural and scientific and technological progressing of the conversion and absorption. It provides opportunities for people to interpret the history of the city through the stone and other building materials epic.

## 15.2 The Legible Requirements of the URT Square

### 15.2.1 The Characteristic of the Location's Legibility – Grading Symbiosis of Networks, Lines, Nodes

That "URT station can be used as an important urban space node" [1] is highlighted in the book The Image of the City which was written by the American urban researcher Kevin Lynch. By above on legible issues and combination of the characteristics of URT, of the unique characters of rail transit system and other urban space can be distinguished easily. Location identification has the characteristics as follows:

In the first place, the URT system, as the main network in the city's public transportation, should show unified features in its station square, and the square also distinguished with other spaces such as in the motor vehicle road, the pedestrian street and the ordinary urban residential premises. Therefore, this should distinguish between a rail transit system differs from other urban spaces.

In the second place, the transfer of people is realized by route, and series spaces surrounded rail transit corridor lead lines and organized by route as a whole system. In spite of the overall network characteristics, transit line is an important factor to strengthen the spatial correlation which makes the spaces closely linked. Therefore, the characteristics of rail transit lines in urban design should be highlighted.

In the third place, the city space and urban functions around URT stations should be reconstructed around the square in front of the station as an important place to local situation. In a way, the square in front of the station to become a center of district, it should be reflected in the environmental design.

The identification exist in the whole network system, corridor and nodes, this makes rail transit stations not only has a clear distinction between the other city system, but also combine the line's features with the area's characters, formed the combination of "the same character a line" and "one scene of each station" [2].

### 15.2.2 The Legibility of Direction – Accurate Mark of the Structures and Spaces

Urban space requires a clearer identification in addition to rail transit system general location, especially for people to determine the specific urban locations away from the underground urban space to the surrounding areas above ground. Usually the urban spaces on the ground are divided into four relatively independent quadrants by the crossing roads, each corresponding to an entrance plaza. Most of the main roads intersection is impact by the rapid traffic stream, the contact between those squares is difficult. Thereby a clearer identification is required to enhance the sense of direction of the city and to reduce the mis-judgment of direction and more accurate situation.

### 15.2.3 The Legibility of Culture – Regional Differences of the Urban Spaces

Urban rail traffic through not only the new district of the city, but also many old town areas. Geographical identification and lots combination of the culture is important for people to get the environmental features at the conversion during the import and export process. After all, the whole impression of people on the first and the last environmental experience to the environment cannot be ignored.

Therefore the square and the entrance of the URT should adapt to the surrounding environment, Different sites of the rail transit station should have the corresponding style and character with the area. While rail transit square design with rich cultural characteristics, the passengers are able to clear understanding of the area and the site's direction. It also reflects the characteristics of the site, and enhances people's memory about local space, helps people integrate lots scattered impression.

### 15.2.4 The Legibility of Times – The Modern Transit Technology and the Brand of the Times

Despite the history of the rail transportation is no more than 200 years compare with the long history of urban construction, especially in China is about half the century. However though the 100 years of construction time, technology of construction and operation of the rail transit has been continuously developed and improved, there is rarely any kinds of construction of the city can be seen in the growth process of building construction system. Therefore the mark of the times is shown in the process during construction is able to reflect in the long period of urban history, it can also emerged in the verified characters in different line of the URT. And also in each different node of a line, although position in space is different, relatively speaking, it is also reflects the whole level of science and technology, and cultural characteristics of the construction period. The design should based on the line, and explore the features of the transport facilities and social or public space which given by the times.

## 15.3 The Specific Styles to Strengthen the Legibility of URT Station Squares

### *15.3.1 Spatial Design*

#### 15.3.1.1 Processing of Spatial Form

It should first be showed in landscape spatial forms processing in strengthening the legibility of the Station squares. Station square spatial form should not only be beneficial to the evacuation but also reflects a system of urban public character and by the enclosed space. At the same time, it needs to be a system of geographical landmark landscape through enclosure walls, flower beds and some architectural pieces of the square. In addition, the design with some bicycle parking space, cooling tower, or wind pavilion surrounded, we can cover some places which are easy to be mixed and disorderly by enclosure and shelter.

For example, in the square designs of Ping'anli and Beihai North stations of the 6th line, which has been recently completed, squares of station was enclosed by the interfaces formed by grey walls, only form an opening direction to the main street, Ping'an street, some low green fence and steps in the opening part both play a certain role of direction guiding and define open interface with its boundary elements. This spatial form echo urban public spatial legibility. In the design of Garden Expo Square landscape of 14th line, the sculpture was put on the oval area in the north of the square, contrast with the surrounding floor coverings and green landscape. Sculpture is made by bending colored steel pipe, which is designed in accordance with the scale of the situation into the landscape of the shape and the image of the track. Garden Expo Station Square design ingenuity in the conception, reflected the traditional Chinese garden landscape with modern sculpture techniques and a strong sense of the times. Installations and illuminational lights play an important role in guiding the crowd and dividing the space.

**Table 15.1** D/Hand the viewer's visual relationship [4]

| The ratio | Vertical viewing angle | Observed effects | Spatial feelings |
|---|---|---|---|
| D/H = 1 | 45° | Observed detail | Close to the sense of local and oppressive |
| D/H = 2 | 27° | Observe the main body | Closed space without pressure |
| D/H = 3 | 18° | Observed overall (comfortable angle) | Weakening of spatial relationships, absolutely no pressure |
| D/H = 4 | 14° | Observation outline | Not a general spatial interaction |
| D/H = 5 | 11° | Observe the relationship between squares and environment | Not a general spatial interaction |

#### 15.3.1.2 Design of Spatial Scale

In addition to fit the functional requirements of the traffic organization, the square and its scale of the enclosure interface should maintain a certain degree of coordination relations to form a suitable degree of enclosure. In many urban space research of the past, the degree of spatial enclosed is often take placed of the ratio of D and H [3], it means the ratio of Plaza distance (D) and the height (H) of around objects. Most of pedestrians walk through the square, so it can be drawn from the past urban research, as shown in the table as followed. After all, the suitable ratio of D and H should between 2 and 3. In this way, the degree of the square enclosed is available and easy to form a positive feeling of space in the city (Table 15.1).

Dongsi and Beihai North station squares consider the relationship between the walls and the square, they are more opening than ordinary Beijing courtyard and more closed than public plaza. The D/H of these plazas are about 3. As Garden Expo Square, it is more learn from the city's public spatial ratio and scale, more open.

### *15.3.2 Details and Design of Structure*

#### 15.3.2.1 Differences of Detailed Features

During the design courses, URT squares on the same line always have similar style. Squares should changes in unified and look for combination of "the same character a line" and "one scene of each station". It embodies a local regional and cultural legibility. The master color of URT squares should be similar to the nearby landmarks' color. Contrasting color can be put to highlight the sense of the times in certain degree. These colors form a special texture of material. The colors and textures can make a strong affection, and the texture of color formation is a manifestation of the local regional and cultural identity.

**Fig. 15.1** Ping'anli Station square design

For example, when Square interface is established by brick wall, the texture of the land and combinations pattern can be used as a more delicate handling to make some differences under the whole style. In addition, if the change of paving mechanism, the form of the flower beds, landscape plant species selection, structures skit, stone of a rock garden or image can be combined appropriately, which can play a "unity of opposites" [5] role. It should be noted that, subway square is crowded in many cases, so the arrangement of these marks should be placed in a reasonable spatial position. Preferably, some mark should be appropriately higher than the eye level of average person, so that it can be easily observed and understood. In some cases, the altitude difference between the square base surface and the road base surface can make people more clearly awareness of the significance of the station square as a place space.

#### 15.3.2.2 Combined with Other Structures

Rail transit system not only has environmental landscape structures but also many of the structures with the characteristics of its industry requirements, such as subway logo, entrance signs, bulletin boards, prohibit marking, warning signals, road boot logo and advertisement windows. The design of these furniture is usually not belong to the scope of rail construction company, it belongs to the rail operating companies in accordance with its operational requirements (of course, this way deserve to debate and explore). As we can draw above, the location of these furniture in environmental needs to be integration design. These integration efforts need to be further strengthened to reduce the contradictions between the various systems, and improve its overall spatial image and enhance the quality of environmental design (Figs. 15.1 and 15.2).

**Fig. 15.2** The Garden Expo Station Square design

## 15.4 Conclusion

Environmental design requires to fit with both material and spiritual needs. The legibility play a very important role in the urban construction, it can quickly and accurately guide the people informed of its location and direction in the complex network of rail transit. Legibility can also improve the overall identity sense of people to city, and provide more comfortable and interesting life to people in the city.

Rail transit plaza, as a portal of a place, is the place for more and more long-haul commuter people to form the first and last impression, also in the future will become important locations for more people to meet with friends, wait for their loved ones, which can form more people's memory about the city and living events. For designers, the environmental design should be proceeded from big direction, put the impression of the geographical location, direction, culture and age into themselves. The rail transit square needs to start from the pavement and materials, space and scale, conception and practices to strength its legibility as an important part of the rail transit station and urban squares. Only by this way, people can realize the requirements of the different levels of the city and dreams in materialized spaces.

## References

1. Image of the City (USA) (2001) Kevin Lynch forward (trans: Fang Yiping). Huaxia Publishing House
2. Yunan Zhang, Minghua Mu (2011) Small volume, big building. Huazhong Architecture Publishing House

3. Town design (UK) (1983) Gilbert, China Building Industry Press
4. Dehua Li (2011) Urban planning 3rd. China Architecture & Building Press
5. The principles of construction of formal beauty (U.S.) (1982) The Torbert, Hamlin (t. hamlin) forward (trans: Zou Nong), China Building Industry Press Publication

# Chapter 16
# Comparison & Analysis Among the Development Modes of Ground and Underground Urban Rail Rolling Stock Depots

**Yingdong Hu, Xinran Zhang, Yuxiang Dong, and Honghong Zhang**

**Abstract** Combined with examples of the urban rail rolling stock depots at home and aboard, this paper has made macro & micro comparisons on the development modes of underground and ground depots under TOD, and explored domestic development status, significance and issues.

**Keywords** Urban rail rolling stock depot • Underground • Ground • Development modes

## 16.1 Development Modes

With the acceleration of urbanization and resource constraints of urban land, depot as an important part of subway system, the contradictions as low intensity land use and large area are rising, and ways have been adopted to face the challenges. One is to construct economical depot, reduce operational costs through multiple lines shared one depot base; the other is through joint development of depot bases, to make up the funding gap by fully excavating resource potential of space [1]. With practices

Y. Hu (✉)
Associate Professor of Beijing JiaoTong University, Beijing, China
e-mail: ydhu@bjtu.edu.cn

X. Zhang
Beijing Institute of Architectural Design, No.62 Nanlishi Road, Xicheng District, 100045 Beijing, China

Y. Dong
School of Architecture and Art Design, Beijing Jiaotong University, No.3, Shangyuancun, Haidian District, 100044 Beijing, China
e-mail: yxdong@bjtu.edu.cn

H. Zhang
Professor of Beijing JiaoTong University, Beijing, China
e-mail: hhzhang@bjtu.edu.cn

H. Xia and Y. Zhang (eds.), *The 2nd International Symposium on Rail Transit Comprehensive Development (ISRTCD) Proceedings*,
DOI 10.1007/978-3-642-37589-7_16, © Springer-Verlag Berlin Heidelberg 2014

**Table 16.1** Several models of depot development

| | Classification | Cases | Schematic diagram |
|---|---|---|---|
| Ground development mode | Carpet mode | Sihui Depot, Beijing, MTR Tsuen Wan Depot, Hongkong, Lohas Park in Tseung Kwan O Depot, Hongkong, Guogongzhuang Depot in Beijing, parking development, Jiading, Shanghai | 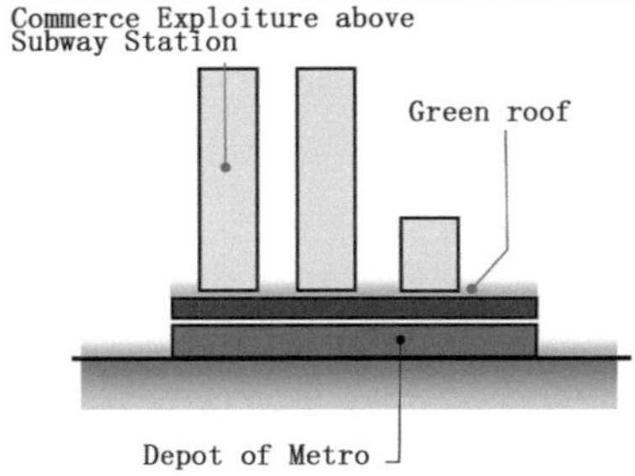 |
| | Elevated mode | Nanjing University City Depot | 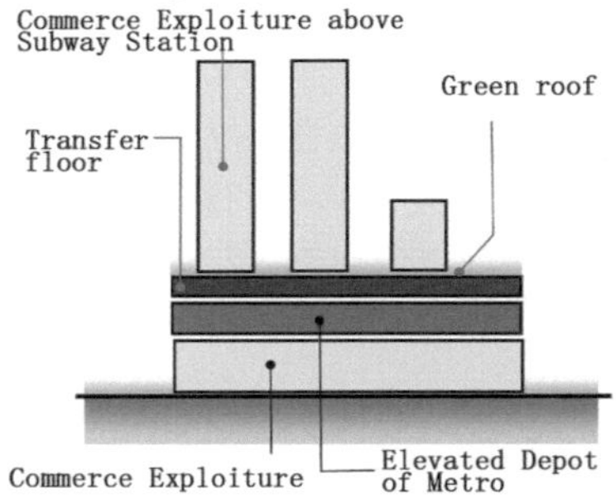 |
| Underground development mode | Underground covering soil mode | Tseung Kwan O Depot in Hongkong, Jinquan Depot in Singapore, Light Chiu Depot in Japan, Misaki Depot in Japan, Underground Depot in White town, British | 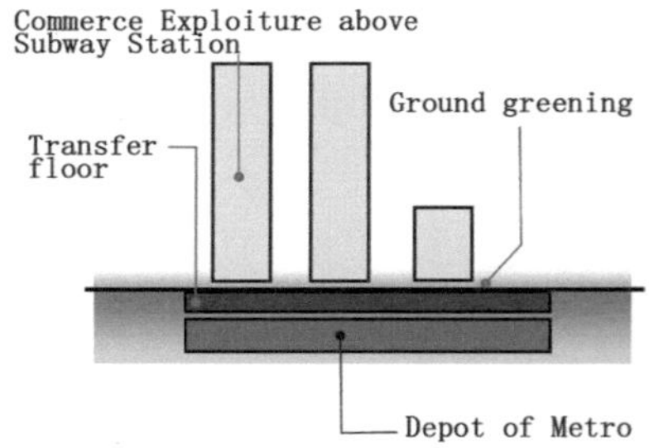 |

accumulated at home and abroad, it has been paid more and more attention by governments, subway departments and investor, and the transit-oriented land development mode TOD (Transit-Oriented Development) has also been widely accepted, which shows good prospects.[1] Therefore, it is of great significance and urgency to discover the construction and development modes based on our national condition.

Based on the mutual spatial relations, it can be defined as three types: carpet, elevated and underground,[2] and the above first two can also be classified as ground development mode (Table 16.1).

[1] Domestic Depot covers about 15~30 ha, much larger than the subway site, with the land appreciation produced by the accelerated urbanization, the potential of property development value should not be underestimated

[2] Zhongling Xiao Exploration on property development models of Metro Depot and integrated base [J]. Urban Rapid Rail Transit, 2010, 23(6):48–53.

**Fig. 16.1** The platform and superstructure development of Sihui Depot (*L*) (Source: Self-shot); West Depot in Shekou, Shenzhen in the construction, Superstructure development has not started yet. (*R*) (Source: Self-shot)

Carpet mode is to arrange depot horizontally in ground elevation, accordingly to form an overall platform as construction land for the property development, such as Sihui Depot in Beijing (built in 1999) (Fig. 16.1), Lohas Park in Tseung Kwan O Depot in Hong Kong (in 2008).

The elevated mode is a form of leaving the lower space to layout property development when depot are arranged about 10 m above the ground. Commercials are arranged in the lower space, and the upper can be also planned for vertical development, such as Nanjing University City Depot (built in 2010).

The underground covering mode appeared later, which is to arrange deport facilities in the ground, and property development on the ground. For example Jinquan Depot in Singapore (built in 2009).

## 16.2 Comparisons of Depot Development Models Between Ground and Underground

### *16.2.1 Policy and Decision-Making Levels*

Presently, the corresponding regulations and policies of land, development and design specifications have not kept pace. And the construction and development of depot often introduces corresponding guidance documents or implementation measures regarding a proposal.[3] Besides, due to policy gaps of ownership of underground, underground development faces more regulations missing than the ground. The mode of two-level transfer bidding makes the ownership of land use undetermined, which limited on the unified planning and integrated comprehensive

[3] "Hypsometric rights, Respectively for the land", first implemented in the development of Bay Depot in Shenzhen (in 2008), will be carried out to subway operation and MTR Station through specified policy, and issues of land ownership of the Depot comprehensive development will break up from the practical level.

**Table 16.2** Several models of Depot development

| Name of depot | Development model | Total land area ($hm^2$) | Total building area (million $m^2$) | Property development area (million $m^2$) | of total floor area ratio (%) |
|---|---|---|---|---|---|
| Sihui Depot in Beijing [2] | Carpet mode | 34.04 | 103.4 | 60 | 58.0 |
| Parking development in Jiading, Shanghai | Carpet mode | 41.5 | 136.0 | 122.3 | 90.0 |
| Willow Depot in Beijing | Underground covering soil | 17.4 | 6.8 | 0 | 0.0 |
| Coking Factory Depot in Beijing | Underground covering soil | 31.7 | 65.6 | 48.62 | 74.1 |
| Nanjing University City parking development | Elevated mode | 13.84 | 33.2 | 27.8 | 83.7 |

Part of the data sources: Beijing Urban Construction Institute

development. During the carpet and underground mode, it should be avoided if developers pursue their own profits at the expense of structural rationality, which affect operational efficiency of depot.

### 16.2.2 Economic Level

It can be seen from Table 16.2, due to differences location, land size, period and social factors, different modes have diversified in the proportion, and it's difficult to directly draw the conclusions from development ratio. But what can be determined is that later cases of same mode tend to higher proportion, such as Jiading parking is up to 90.0 %, about 32 points higher than Sihui Depot.

In order to have a comprehensive consideration on economic benefits, It should calculate the factors including hydrogeological conditions, construction costs and so on. For example, the integrated engineering costs are larger in underground space or elevated set will add (Table 16.3).

As shown in the table, in the three existing programs, the investment of ground depot program is least (8.5 million), and the developed area is also least; the investment of underground is the largest (19.76 million), and the area is also the largest (65.6 million $m^2$); traditional ground mode is balanced. To draw scientific economic evaluation, it remains to make a comprehensive evaluation of the additional value and investments.

### 16.2.3 Planning Level

The planning of rail traffic including joint development is an important part of urban spatial planning.

**Table 16.3** The comparison of schemes of development models of Coking Plant Depot in Beijing

| Development model programs | Development area (million $m^2$) | Development occupies (million $m^2$) | Depot occupies (million $m^2$) | Depot construction area (million $m^2$) | Roads and squares | Urban Landscape | Investment |
|---|---|---|---|---|---|---|---|
| Ground Depot | 40.9 | 12.9 | 18.8 | 11.6 | 1,440 | 29,360 | 850 million |
| The ground large cap development | 48.5 | 9.3 | 22.4 | 11.6 | 1,580 | 26,560 | 1,030 million |
| Underground Depot development | 65.6 | 14.72 | 16.98 | 12.4 | 29,092 | 65,950 | 1,976 million |

Data sources: Preliminary design and related outcomes of "North coke plant Depot of Subway Line No. 7 in Beijing and the surrounding development", Beijing Urban Construction Institute

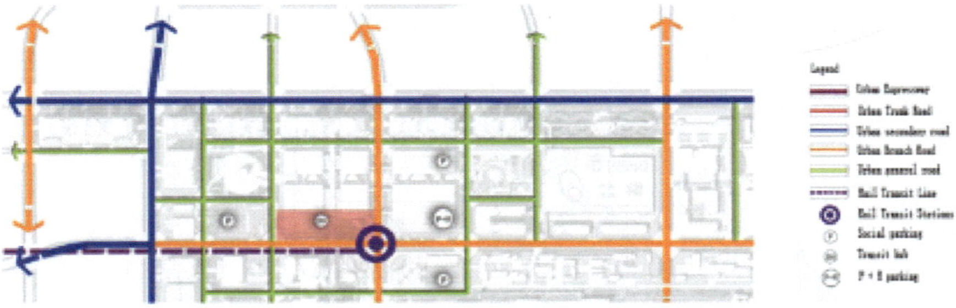

**Fig. 16.2** Coking Plant underground depot in Beijing has not produced barrier to urban road network (Source: Beijing Urban Construction Institute)

#### 16.2.3.1 City Location

Depot bases are usually at large-scale and produce noise and vibration pollutions, so they are usually set in the urban edge in the early. With urban sprawl, land prices soaring and the economic development value also be reflected. In this case depot development requires higher utilization of land and commercial value-added, more inclined to use underground mode.

#### 16.2.3.2 Traffic Impact

Due to the large-scale (up to 1 km), the traditional ground depot has blocked links of municipal roads and destroyed the proper neighborhood scale of the city. Elevated mode and underground mode have overcome the above shortcomings and made a replacement for urban space on the ground. Coking plant depot adopts underground mode, which can form city blocks of about 150–300 m (Fig. 16.2).

### *16.2.4 The Level of Urban Design*

#### 16.2.4.1 Hierarchical Use of Space

The development of underground depot is more reasonable in vertical functional partition, conducive to link the development and utilization of part of the underground space with rail transit station. Underground space development, guided by the rail transportation company, is clearer on nature, ownership, development interests and other aspects; but with the development extending upward.

#### 16.2.4.2 Traffic Organization of People and Vehicles in the Area

Development functions of elevated mode are located on the ground floor, pedestrian and dealers having the most direct convergence with city, but the height of the lower space is restricted. The main problem facing carpet mode is the transportation alternatives up and down as well as fire evacuation are larger issues. For example, in Sihui Depot, daily travel are affected, which restricts quality. The underground model can make full use of underground space. The urban space of the ground is returned to the public, thus makes traffic organization of people and vehicles in the area closer to urban general area.

#### 16.2.4.3 Urban Spatial Form

Mode has a closer relationship with the surrounding environment and mutual influence. To have better integration, the best condition is to take full advantage of some ups and downs of the terrain, which makes depot complex coordination and symbiosis with surrounding and continues urban spatial form and landscape. Comprehensive property[4] of Tseung Kwan O Depot in Hong Kong is a model.

### *16.2.5 The Technical Level*

#### 16.2.5.1 Structural Transformation

Traditional carpet mode development, structural transformation is located above ground elevation, seismic requirement is high, and structural transformation is difficult (Fig. 16.3); while underground depot, due to the conversion part (weak layer) moved under the ground elevation, the overall stability of underground structures is far better than the terrestrial transition, correspondingly the engineering costs decreased. The research of large platform structure, seismic and down is still in the initial stage, there is a need to identify reasonable and optimized structure types through research on the structure of selection, calculation and analysis, tests and so on to provide a theoretical basis and guidance.

[4] The Superstructure property development of Tseung Kwan O Depot is a rise residential area "Lohas Park", and the most prominent feature is the separation of people and vehicles. The Depot facilities accounted for only one-third of the land have the organic connection with peripheral land in the horizontal and vertical directions through Footbridge, wind and rain corridor and green platform, and the Depot is covered below in property development. Property platform combined with the adjacent mountainous terrain makes Depot overall better convergence with the surrounding environment.

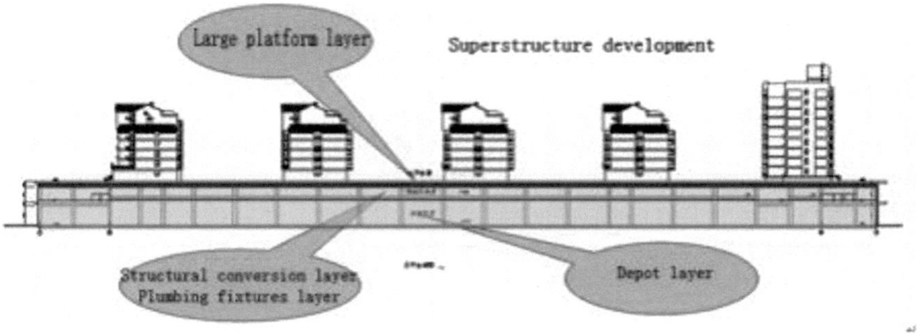


**Fig. 16.3** Conversion layer of the cross-sectional structure and the development of the upper part of Sihui Depot in Beijing (Source: finished according to the information)

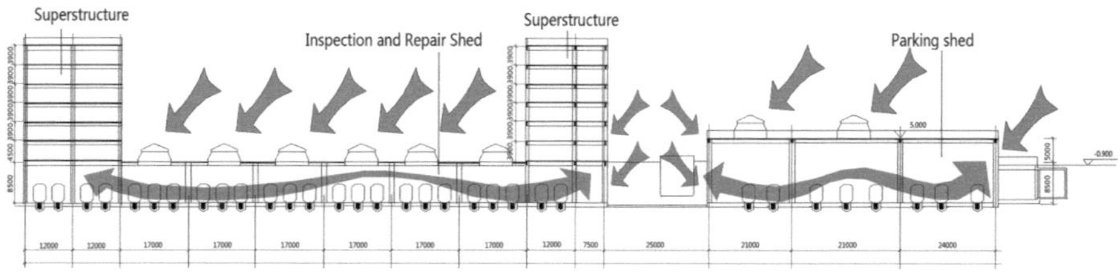


**Fig. 16.4** The diagram of top and side lighting ventilation of underground depot of Beijing Coking Plant (Source: Beijing Urban Construction Institute)

#### 16.2.5.2 Lighting and Ventilation

Due to hot weather, southern cities often use carpet development mode and have an open design of part of the depot houses located in the first floor, making use of natural ventilation to reduce the section consumption. Because of the poor conditions of natural lighting and ventilation, on the development mode of underground depot, the ventilation and lighting well is necessary (Fig. 16.4).

#### 16.2.5.3 Vibration Isolation and Noise Reduction

Operation of transit inevitably brings vibration, noise and dust pollution. Underground depot has advantage in reducing noise (sound wave propagation), but there's still lack of clear evidence on vibration (solid dissemination) to explain the pros and cons of underground and ground. Therefore, it needs further study on what extent underground depot is conducive to reduce noise and isolation.

## 16.3 Conclusions

It can be seen through the above analysis that the depot development modes can not only affect the development content in spatial relationships, but also produce different effects on overall economic efficiency, the type and scale of development, Urban and regional transportation system and responding technological measures. Through more than 30 years of development, the ground mode is gradually matured, reflecting the good adaptability, clear interface development, property development operational and so on; the underground depot shows characters of better convergence with urban spatial form, conducive to coherence and smooth of the regional urban road network, strength of development, the economics of safe structural transformation, reducing noise pollution and so on, which has developed advantages in the regions of high commercial value.

**Acknowledgement** This research is supported by: National Natural Science Foundation of China (51178038); longitudinal research projects funded projects of Beijing Jiaotong University (2011JBM180,2010RC030,2012JBM120); Fundamental Research Funds for the Central Universities (2009JBZ023); Ministry of Education "New Century Excellent Talents" (NCET-10-0217); Longitudinal project (2012-R1-11); Technical consulting projects (A12X00040); This research is also part of cooperation in scientific research with Beijing Urban Construction Institute.

## References

1. Linlin Qian, Haishan Xia (2011) Depot development mode of rail transit in the Sustainable Urban Perspective Comparative Study of transition and reconstruction. In: China Urban Planning Annual Conference Proceedings, Southeast University press, 2011, pp 6216–6225
2. Yongchao Wang, Changjun He, Wei Wang (1999) Bawangfen Depot of Metro Line in Beijing in the construction and superstructure development [C]. In: The tenth symposium of professional committee of the Underground Railroad of Tunnel and Underground Engineering Branch of China Civil Engineering Society, Haitian Press, 1999, pp 660–665

# Chapter 17
# The Discussion of Beijing Rail Transit Development Scale

**Hongwei Gao, Meiqing Zhang, and Weiwei Hao**

**Abstract** Drawing on the basis of the relevant literature, the paper constructs the index system for the analysis of urban rail transit reasonable scale. On the data of 14 typical cities in the world and domestic urban respectively, fitting the rational lower limit and upper limit scale model of Beijing urban rail transit. Based on multivariate analysis and comprehensive model, the conclusion is: contrast the developed cities of the world, according to the actual situation in Beijing, the current scale of Beijing urban rail transit is inadequate, the planned scale for 2015 is more reasonable. After 2015, Beijing's urban rail transit still has some room for development.

**Keywords** Urban development level • Urban rail transit development scale • Multivariate analysis

## 17.1 Introduction

Based the cities building plan about 2015, China's urban rail transit operating mileage will reach 2,500 km that is 4.15 times more than 602 km at present. The total investment is expected to more than 1 trillion yuan. Facing such construction speed, some people believe that the construction of urban rail transit will stimulate domestic demand and promote economic growth, so urban rail transmit need rapid development; others believe that the construction speed is too fast, the huge construction costs and the continued accumulation of operator and maintenance costs will increase the city financial burden. For this opposing view, the industry has started a lively discussion. The core issue focuses on what is the reasonable scale of urban rail transit.

H. Gao (✉) • M. Zhang • W. Hao
School of Economic and Management, Beijing Jiaotong University, 100044 Beijing, China
e-mail: HWGao@bjtu.edu.cn; meiqing_zhang@sina.com; wwhao@bjtu.edu.cn

H. Xia and Y. Zhang (eds.), *The 2nd International Symposium on Rail Transit Comprehensive Development (ISRTCD) Proceedings*,
DOI 10.1007/978-3-642-37589-7_17, © Springer-Verlag Berlin Heidelberg 2014

There are many research perspectives, such as the study of the relationship of a reasonable size, density of road network and the system capabilities [1], demonstrated that the total urban travel, trip distance distribution, as well as travel mode structure, are the important factors which affect the urban rail transit reasonable scale [2], Passenger traffic was a decisive factors to determine the scale of the rail transit [3]. From a purely technical investigate the relationship between the scale of urban rail transit and demand of distribution, the shortest path [4], given suitable scale of city line network based on traffic demand, load intensity of urban rail transit, line density [5], studied relationship between urban GDP and urban rail transit mileage [6]; Determined the network size by the level of socio-economic development of the city, the population and land size [7], considered the urban population, the area of rail transit share rate and other factors to estimate the network size [8].

Viewed those research literatures, the micro-level studies are abundant, the macro level studies are sometimes lack and simplistic. This paper argues that china cities are in a stage of rapid development, the micro indicators can not overall grasp of the scale of the urban rail transit for its volatility character. So it is more reasonable to measure macroeconomic indicators before prediction used microeconomic indicators, so can avoid the dilemma that planned to meet the volume of city traffic demand after the next 20 years will unable to meet the demand of not over 5 years. In short, quantitatively analyses the relationship between Beijing urban rail transit scale and urban development level is a lot more meaningful, especially in the large-scale Beijing rail transit construction period.

The paper constructs the index system for the analysis of urban rail transit reasonable scale. Based on the model, the paper predicts the reasonable development scale of Beijing urban rail transit.

## 17.2 Multivariate Analysis of the Reasonable Scale of Beijing Rail Transit

Rational scale of urban rail transit is affected by many factors, before making a comprehensive analysis of these factors; the first work is to examine these factors individually to measure the reasonable scale of Beijing rail transit. Due to data limitation, the paper carries out intuitionist analysis comparing with the world's representative cities situation.

1. **Total population**. Based on Beijing resident population of 19,612,000, the calculated result is in Table 17.1, the conclusion is: compared with the average scale of the five cities, from demographic factor, the scale of Beijing rail transit should reach 867.4 km.
2. **Urban area size**. Based on Beijing urban area of 1,040 km$^2$, the calculated result is in Table 17.2, the conclusion is: with the average level of developed cities in the world, by the urban area size factor, the scale of Beijing urban rail transit should reach 402.8 km.

**Table 17.1** The development scale of urban rail transit and population size in the world's representative cities

| City | Length of urban rail transit (km) | Urban population (million people) | Length of urban rail transit per 10,000 citizen (km/10,000) | Compared scale of Beijing rail transit (km) |
|---|---|---|---|---|
| London | 410 | 718.8 | 0.57 | 1,118.8 |
| Paris | 306 | 600 | 0.51 | 1,000.2 |
| Moscow | 203 | 867.5 | 0.23 | 458.9 |
| Tokyo | 356 | 835.5 | 0.43 | 835.6 |
| New York | 343 | 728.4 | 0.47 | 923.5 |

**Table 17.2** The development scale of urban rail transit and urban area size in the world's representative cities

| City | Urban rail transit scale (km) | Urban area ($km^2$) | Urban rail transit mileage per square km ($km/km^2$) | The compared scale of Beijing urban rail transit (km) |
|---|---|---|---|---|
| London | 410 | 1,580 | 0.26 | 269.9 |
| Moscow | 203 | 878.7 | 0.23 | 240.3 |
| Tokyo | 356 | 575 | 0.62 | 643.9 |
| New York | 343 | 780 | 0.44 | 457.3 |

**Table 17.3** Urban rail transit passenger volume and scale in the world's representative cities

| City | Passenger traffic by urban rail transit (100 million passengers) | Mileage of urban rail transmit (km) | Per kilometer passenger traffic (100 million passenger/km) | The compared scale of Beijing rail transit (km) |
|---|---|---|---|---|
| New York | 11.3 | 390 | 0.029 | 635.04 |
| London | 8.15 | 390 | 0.021 | 880.49 |
| Tokyo | 27.21 | 230 | 0.118 | 155.53 |
| Paris | 15.3 | 201 | 0.076 | 241.73 |
| Moscow | 32.8 | 275 | 0.119 | 154.27 |

3. **Passenger traffic by urban rail transit**. At the end of 2010, Beijing rail transit length was 336 km, the total amount of passenger traffic by rail transit was 1.84 billion. Analogy of the average value from the five representative cities in the world, calculated according to Beijing's annual passenger traffic of 1.84 billion, the reasonable urban rail transit scale is 402 km. See Table 17.3.
4. **The rate of rail traffic length by per 10,000 citizens**. Beijing's rate is far below the rate of representative cities in the world. The average index of the rate in the representative cities of the world in 2010 is 0.37, it is reasonable for Beijing to construct 750 km rail transit scale, much higher than the actual scale of 336 km at present. See Table 17.4.

**Table 17.4** The rate in the world's representative cities (2007) and Beijing

| City | Urban population (million people) | Network scale (km) | The rate of 10,000 people have rail traffic network length (km/million people) | The category ration of Beijing rail traffic mileage (km) |
|---|---|---|---|---|
| New York | 757.60 | 368.00 | 0.49 | 925.54 |
| Moscow | 900.00 | 282.70 | 0.31 | 615.97 |
| Tokyo | 845.70 | 293.20 | 0.35 | 679.87 |
| Paris | 620.90 | 213.5 | 0.34 | 674.30 |

**Table 17.5** World in some big cities the proportion of public transport structure

| City | Year | The ratio of trolley, bus in total volume of public transport (%) | The ratio of the passengers transported by rail transit in the passengers transported by public vehicles (%) |
|---|---|---|---|
| New York | 1984 | 20.00 | 80.00 |
| London | 1985 | 9.00 | 91.00 |
| Paris | 1982 | 29.00 | 71.00 |
| Moscow | 1985 | 49.00 | 51.00 |
| Tokyo | 1985 | 10.00 | 90.00 |
| Osaka | 1987 | 14.00 | 86.00 |

5. **The ratio of the passengers transported by rail transit in the passengers transported by public vehicles**. According to Table 17.5, the ratio has a big different compared with foreign cities'. The planned rate in 2015 is that Beijing rail transit will bear 50 % of public transportation volume. According to this ratio, the development scale of Beijing rail transit is about 600 km.

In addition to the above indicators, the paper also analyses other indicators, such as the GDP indicator, the disposable income per capita, the urbanization rate, vehicle ownership, as well as the local fiscal revenue. Due to the data of compared cities is difficult to obtain, these indicators are not included in the text.

## 17.3 Integrated Measurement of the Reasonable Scale of Beijing Rail Transit

Examine the individual elements can make us clearly to understand a reasonable scale of the Beijing Rail Transit. Above analysis shows that each reasonable scale corresponding its factor is not the same. What will it be when we consider these factors together? (Table 17.6).

**Table 17.6** Scale of urban rail transit and other data in the world's representative cities

| City | Rail transit length (km) | GDP (100 million dollar) | Urban population (10,000 people) | Urban area (km) |
|---|---|---|---|---|
| Tokyo | 356 | 4,043 | 835.5 | 575 |
| New York | 343 | 4,030 | 728.4 | 780 |
| London | 410 | 3,360.4 | 718.8 | 1,580 |
| Paris | 306 | 1,884.5 | 600 | 1,496 |
| Mexico City | 202 | 1,240 | 920 | 1,526 |
| Berlin | 152 | 1,039.5 | 339 | 833 |
| Seoul | 287 | 1,039 | 963.7 | 605 |
| Singapore | 91 | 959.3 | 264.1 | 98 |
| Moscow | 203 | 522.5 | 867.5 | 878.7 |
| Bussan | 71.6 | 365 | 203 | 181.1 |
| Taipei | 67.2 | 342 | 278.1 | 269 |
| Sao Paulo | 57.6 | 305 | 1,010 | 162.4 |
| Cairo | 43 | 280 | 520 | 214 |
| Hong Kong | 109 | 1,794 | 735.7 | 1,098 |

Data from: The International City Statistical Year Book, published by the World Bank, data of the world's representative cities. Collated data shown in Table 17.6

This paper selects the urban land area, the gross national product and the total urban population indicators as representative indicators and 14 typical cities be chosen as the sample cities to build the following model:

$$L = kS^{a}P^{b}E^{c} \tag{17.1}$$

Where L is a total length of urban rail transit, units km; S is urban area, units square kilometers; E is GDP, units 100 million U.S. dollars; P is total urban population, units million; k, a, b, c are parameters to be estimated. See Table 17.6.

Make a Logarithmic transformation, the results of the model are as follows,

$$\begin{aligned} \ln L &= -1.2797 + 0.3011\ln S + 0.1465\ln P + 0.5121\ln E \\ t &= (-1.4042) \quad (2.6253) \quad (0.9117) \quad (5.0656) \\ R^2 &= 0.9155, \quad \overline{R}^2 = 0.8873, \quad F = 32.4994, \quad DW = 1.2659 \end{aligned} \tag{17.2}$$

The overall linear relationship significantly, higher degree of fitting. The parameters all passed test. Revert equation (17.2) to the exponential form as:

$$L = 0.2781S^{0.3011}P^{0.1465}E^{0.5121}. \tag{17.3}$$

Although the typical cities in the world have developed into a relatively well-developed and mature stage, and have a lot of difference with Beijing which is in the fast development now, but the three selected indicators are the basic indicators and has higher reference to. So this paper takes the model as upper limit model of Beijing reasonable scale.

Domestic model selected is different with above model for there are more indicators can be chosen. Using the data of Beijing and Shanghai Metro (Data source: "Beijing Statistical Yearbook", "Shanghai Statistical Yearbook", "China City Statistical Yearbook", as well as the website of the subway company of Beijing and so on), deal with multicollinearity by the step-by-step regression method, the final selected indicators are the urban area (AREA) and local general budget revenue (BUDGET). The model and calculated results are as follows:

$$\begin{aligned} L &= 11.9153 + 0.0312AREA + 0.0718BUDGET \\ t &= (1.1838) \quad (1.5469) \quad (6.8935) \\ \overline{R}^2 &= 0.8942 \quad F = 123.3620 \quad DW = 1.2719 \end{aligned} \tag{17.4}$$

The model's F-test and t-test are significant. This paper makes the model as the lower limit model of Beijing reasonable scale.

After getting world cities' and domestic models, we use urban area in 2010 in Beijing (1,368 km$^2$), the total urban population (22 million), GDP (13,777 billion yuan) and local general budget fiscal revenue (223 billion yuan), to forecast the 2015 year's data, the result is that Beijing urban construction land area is 1,437 km$^2$, the total urban population is 23.92 million, GDP is 22,189 billion and local general budget revenue is 359.1 billion yuan, finely we input these forecast data to the upper limit model and the lower limit model to predict the reasonable interval scale of Beijing urban rail transit in 2015. In order to avoid large errors in the use of the world's urban model, we use PPP commuted to replace the currency units (assuming conversion coefficient is 0.5), i.e., the Beijing GDP (PPP translated) in 2010 and 2015 values were 6,888.5 and 11,094.7.

After taking the data into the lower limit and upper limit models, we get the results: in 2010 Beijing urban rail transit reasonable size range: 273–699 km, the average is 436 km; in 2015 Beijing urban rail transit in a reasonable scale range: 405–915 km, the average is 610 km.

The above results show that when consider three foundational factors comprehensively, regarding the average of world developed cities as upper limit and average of domestic large cities as lower limit, the scale of Beijing rail transit in 2010 (336 km) entered a reasonable lower limit scale, but away from the average of 436 km. In 2015, Beijing Railway scale should reach more than 610 km, this value is more than 561 km of the Beijing 2015 planning mileage. However, the Beijing 2015 planning mileage is within the range of reasonable scale, just slightly lower than average value.

## 17.4 Conclusion

By multivariate analysis, the results show that, from the population size, the reasonable scale of Beijing' in 2010 is 867 km; from the city area, is 402.8 km; from annual passenger volume, is 402 km, from rate of 10,000 people, is 750 km, from the total public transport from rail transit proportion of total volume, is about 600 km. These contrasting indicators were significantly more than the actual level

336 km in Beijing 2010. This clearly shows that Beijing rail transit scale is not in excessive situation. By comprehensive analysis, the results showed that in 2010, the scale of Beijing rail transit still has a gap of 330 km with the average level of developed cities in the world. Beijing's 2015 planning scale is also not beyond the average scale of the developed cities' in the world.

The conclusions of this paper are: to compare the developed cities of the world, the current rail transit scale of Beijing is insufficient, the planning scale for 2015 is reasonable. After 2015, Beijing rail transit still has some development space.

It needs to emphasize that the total predicted value is based on macro indicators. Predict or judge just considers about the urban economy, the financial level, the population, the urban area and other factors, predictive value is only match these indicators. Therefore, after in the grasp of the total level, we also need to forecast it on micro-level indicators. Given the total size of urban rail transit, how attain optimize spatial layout, how to meet the needs of urban public traffic continuously and the requirements to lead the urban areas with economic expansion, still need us to do a lot of thorough and meticulous research.

**Acknowledgement** The chapter is supported by Beijing philosophy and social science planning major program "the research of sustainable development of urban traffic in Beijing" (11ZDA03) and the program from Beijing Education Commission "the research of the interaction and coordination between urban traffic and urban space structure of research in Beijing" (B12H100040).

## References

1. Wang Zhongqiang, Gao Shilian et al (1999) Urban rail transit system planning. Southwest Jiaotong University, Chengdu, 27–47
2. Quan Yongshen (2000) A Discussion on Some Critical Issues of Metro Network Planning. Vol 2. Engineering Science 75–82
3. OuYang Zhijian, Ma Xiaoyi (2004) The analysis of influencing factors on scale of passenger flow of urban rail transit. Study Urban Rail Trans 7
4. Current JR, Re Velle CS, Cohon JL (1985) The maximum covering/shortest path problem (1985): a multiobjective network design and routing formulation. Eur J Oper Res 21(2):189–199
5. Hu Chaofan et al (2006) The study of the reasonable scale of Beijing rail transit. Urban Transport 10:52–54
6. Liu Yin, Lu Jian (2006) The study of the construction scale of the urban rail transit. Urban Transport 3:16–20
7. Lin Lifan, Zhang Weihua (2008): A Study on the Method of the Network Scale of the Urban Rail Transit. Vol.10. Technology & Economy in Areas of Communications: 110–112
8. Jing Jian et al (2006) The mechanization and model of reasonable scale of urban rail Transit. J China Railw 12:16–20

# Chapter 18
# Development Tendency and Solution Research to China's Large Airport Transportation

**Jinjin Sun and Xianming Zhang**

**Abstract** There is small, scattered, weakness and other issues in China's large-scale airport development process. Refer to this, based on analysis of China's current airport traffic status and functional position, research of the current collection and distribution profile of the international large hub airport's ground transportation, ground traffic organization mode. Learned from its advanced experience, this paper proposes a development model suitable for China's large-scale airport, which is a future direction of China's large-scale airport's development that should be combining railway transportation, ground public transportation, airport transportation and urban outbound traffic together as one comprehensive transport hub.

**Keywords** Large-scale airport • Transportation development • Comprehensive transport hub

Until end of year 2011, the number of domestic large-scale airports whose passenger throughput over 10,000,000/year has reached 27, as the passenger traffic volume continually increased, it stimulates the development demand of airport transportation, but on the other hand, it also exposes some problems which need to be solved as quickly as possible in airport business, like transfer way limited, inconvenient Interchange etc. So based on this situation, this paper conducts some research and analysis on china's airport transportation status and learn from advanced experience of international large-scale airport management mode, dedicates to put forward an transport development model and solution suitable for china's airport situation and hopes to offer an decisional support to china's airport transportation development [1].

---

J. Sun (✉) • X. Zhang
Civil Aviation University of China, Binhai Airport, Tianjin, China
e-mail: vivibp@126.com; zhangxianming1314@126.com

H. Xia and Y. Zhang (eds.), *The 2nd International Symposium on Rail Transit Comprehensive Development (ISRTCD) Proceedings*,
DOI 10.1007/978-3-642-37589-7_18, 

## 18.1 China's Large-Scale Airport Transportation Status Analysis

### *18.1.1 Airport Function Position Analysis*

The function of approaching airport traffic is comprehensive; it always takes many functions at same time like transit, urban citizen's public transportation, and urban external links, therefore it is hard to guarantee the reliability of airport ground transport distribution. And since the rise of construction boom at the airport economic zone around the airport, the non-air transport demands grows, so how to build an functionally-clear and well-facilitated transportation infrastructure within the limited space around airport becomes a big problem for each party in airport construction. there is another main problem that image are more important than function in terminal plan and construction, this leads to air and ground transport unwell-connected, the conflict between passenger flow continually increasing in peak hours and shortage of ground transportation capability shows up gradually [2].

### *18.1.2 Position of Airport in Transportation Network Analysis*

Currently China's airport does not notice its position in generalized mass transportation network, or just simply works as an end link. Normally it only has input or output traffic flow, but no cross traffic flow. Airport's ground transport system usually only includes parking building, bus stations and station platform of railway as last stop, lines are always single-way, but no intersection. So come to conclusion, most of airports could not position itself correctly in transportation network in current big integrated transportation development time.

## 18.2 International Large-Scale Airport Management Analysis

We can see from decades of Europe & USA airport development course that hub airport will eventually become the hub integrated multiple-transport ways in its located region. As a regional transportation hub, hub airport should consider to build up a multi-dimensional, multi-transport ways of transportation network system, the critical point is building up a correct rail transit to connect airport with city town and how to make rail transit play its role properly.

### 18.2.1 Example Heathrow International Airport Overview

London Heathrow International Airport is located in 25 km far southwest of London, it is top largest airport of Europe, and its passenger throughput was 69,433,565 in year 2011, 3rd worldwide, four Terminals are under using now. There are two rail transit lines connecting with Heathrow Airport, line London Metro Piccadilly and Heathrow Express (Hex). Both of the two lines have two stops in the airport, which can directly go to terminal 1, 2, 3 and terminal 4 [3].

Heathrow Airport is not only the most important aviation hub, but also the biggest long distance bus station in southeast UK area, now 10 % of long distance buses of whole UK go through Heathrow Airport bus station, you can go directly to 1,200 destination cities of whole UK via Heathrow Airport bus station, another 1,200 cities only via 1 stop exchange.

### 18.2.2 Inspiration to China's Large-Scale Airport Construction

Beijing Capital International Airport, Shanghai Pudong International Airport and Guangzhou Baiyun International Airport are all aimed at working as Asia most important airport pool, compared with oversea famous airport, China's airport ground transportation mainly rely on road transportation, rail transit as supplement, and has not established its own traffic hub system. So we can get some hint from the inspiration of international airport as follow:

1. Increase the competition ability of public transport in comprehensive transportation network. As to increase the competition capability of public transportation in comprehensive transport network, what we need is to put the public transportation at the first priority in the airport overall construction planning process.
2. Solve the conflict between system and management level in airport construction. There should establish a management system which takes government as a leading role, airport authority as main body in every airport construction phase.
3. Comprehensive concept implement in transportation network construction [4]. In airport construction plan phase. Airport ground transportation design should be put in city comprehensive transportation network scale, not work as isolated terminal, and it should consider the connection between different traffic ways and its extensibility as a transportation hub.

## 18.3 China's Airport Future Development Tendency Research

### *18.3.1 Airport's Future Development*

The future airport will not only be one link of air transport, but also a connection place of various transport-ways, which is comprehensive transportation hub. The comprehensive transportation hub defines integrating various transportation nodes in comprehensive transportation system. The key point is the interchange between different traffic ways, and using this node to reach multi-mode transportation of various transit ways, to reach optimal effect, break the block among different traffic way, finalized the efficiency of comprehensive transport.

The future airport should be surrounded by traffic ways as much as possible, in order to reach multi-mode transportation, like long distance bus station, Train Station, Highway Station, Metro, road transport and water transport etc. already existed and future traffic ways, we can boldly guess that future airport terminal will be an architecture complex which vertically includes various traffic ways, and executive seamless interchange between each traffic ways, so as to reach out optimal travel way for passenger.

The traffic in hub airport is continuously increasing, and airport scale also increasing, so the future of hub airport should not only be a place for passenger transport and cargo transit, but also a comprehensive city function area, a transportation hub which collects the function of passenger flow, cargo flow, and capital flow together, the future large-scale airport would face a huge challenge, which is transferring its function from single isolated hub airport to multiple-functional and comprehensive transportation hub.

### *18.3.2 Comprehensive Transportation Hub*

The future airport comprehensive transportation hub would have three main function areas: airport terminal; Railway hub and city traffic center, mainly responsible for travel and interchange of air passengers, train passengers and other ground transport passengers.

Airport Terminal: it serves those who travel via aircraft, so optimizing the passenger service process should be put more effort, like rebuild the workflow of inter-transit, shorter the inter-transit time, more satisfaction from client, thereafter increase the proportion of passengers who make inter-transit in airport.

Railway Hub: using location advantage, build advanced railway transportation network, inter-cities express, train, and city metro together as an integrated railway hub.

City Traffic Center: as to collect and distribute the passenger from airport and railway station, it should plan a city public traffic center between them. And this center should include city metro, taxi, long distance bus station and other various traffic ways together to build a comfort, convenient, modern and large city transportation hub [4].

## 18.4 China's Airport Transportation Development Solution & Suggestion

Through the above analysis, we could see that future of china's airport should be an integrated comprehensive transportation hub using airport as center. From the specific transportation structure side, it should be using railway network as backbones, taking airport transport support, collecting bus stations, taxi station, parking and long distance bus station together or even put them in the same building, to compose an interchange hub with both internal and external functions. The hub can make transfer between express and regular speed traffic way, between external and internal, between walking, parking and vehicle flow, at the same time, it can also contain some shopping malls, service centers or entertainment facilities and other public facilities through extending hub's functional exploitation. This already increase the exploit level of development, so it can not only match people's demand for interchange in peak traffic time, but also give part of passengers a chance to finish daily work or fulfill shopping demands in the middle of interchange. In some degree, it does not only reduce passenger's pure waiting time or pure shopping demand out, but also promotes diversification of public service, therefore it can increase the attractiveness of public transport and passenger traffic volume [5].

## 18.5 Conclusions

There are still some issues in the development process of china's airport construction, like mis-position, disordered traffic connection and so on. Based on the analysis of china's airport transport organization status, and learning advanced experience of function position and traffic connection in comprehensive transportation system from international hub airport, this paper put forward the future development direction of China's airport, which should be a comprehensive transportation hub model centered by airport itself, meanwhile this paper also offers the solution and suggestion on development and construction of transport hub. It can work as brain reference for China's large-scale airport development.

## References

1. Galster G (1996) William Grigsby and the Analysis of Housing Sub-markets and Filtering [J]. Urban Stud 33(10):1797–1805
2. Zhang Guohua (2011) Planning and design technique system of large-scale and comprehensive airport transport hub [J]. City Plann Rev 35(4):61–68
3. Xue Yuan, Zhao Yali et al (2007) Research on the public ground access network of London Heathrow Airport [J] Traffic Transport 2007(12):12–20
4. Huang Zhigang, Rong Chaohe (2007) The development tendency and its reasons of foreign city. Large passenger transportation hub [J]. J Transport Syst Eng Inf Technol 17(12):2–4
5. Lu Ximing (2003) Integrate transport for metropolis [M]. Shang Hai: Shanghai Scientific and Technical Publishers, pp 100–200

# Chapter 19
# "Mix and Match" in Urban Design

**Kehui Shi, Xiaoqian Chen, and Xuesong Hu**

**Abstract** As the buzzword nowadays, "Mix and match"'s power should not be overlooked. It spread to all areas of life, and has its own ways of expression and characteristics. The rises of "mix and match" boom in the field of urban design and in the most directly affect the lives of the city level, so it comes with its own feature.

This article studies mix and match city in terms of why did it arise, operating means, urban design and its feasibility. With the current practice of building unprecedented, the mix and match city concept has penetrated into every aspect of from design to practice. And more participation in the broader sense, designers need to learn more about the research on the city of mix and match's content, which makes urban design innovative, practical and create a pleasant urban space, and make urban design truly implemented to improve the people's living environment.

**Keywords** Mix and match • Urban design • Methodology

Whether a fashion people or not, you can feel mix and match's life has become the most stylish trend of today's living. It not only be a design method, but also integrated into everyone's lifestyle and work habits, as a collective, conscious or unconscious phenomenon.

This mix and match trend spontaneously rises in the field of urban design.

K. Shi (✉)
School of Architecture and Design, Beijing Jiaotong University, Beijing, China
e-mail: khshi@bjtu.edu.cn

X. Chen
Zongxu Planning & Architecture Design Co. Ltd, Beijing, China

X. Hu
School of Architecture, Beijing University of Civil Engineering and Architecture, Beijing, China

H. Xia and Y. Zhang (eds.), *The 2nd International Symposium on Rail Transit Comprehensive Development (ISRTCD) Proceedings*,
DOI 10.1007/978-3-642-37589-7_19, 

## 19.1 The Definition and Outline of "Mix and Match"

The basic meaning of "mix and match" is combining the different styles, different types, and different worth into a completely personal style, according to personal tastes. This trend spread to many other areas of music, video, design later, becoming the mainstream gradually from the 1990s. For influencing many areas, attracting the audience widely, the term "mix and match" has been more profound meaning and numerous manifestations.

The informatization makes the materials increase exponentially. Service-based society makes people free to be maximized. Within the range of conditions, they can choose different types of information, services, products, etc. Greatly enriched existing resources and personal preferences different and people hate the old nature decided to mix and match in this way can be obtained more freshness and infinite creative play space, more and more able to reflect it in the context of contemporary society adaptability.

The acceleration of the development of the times make the mix of elements innovative ways has been generally accepted. Rising extensively in various fields can show that mix and match is not limited to the disciplines of the different subjects as a way of thinking operating and working methods. The Mix and Match has wide range of applications and enriched expression in the field of urban design gradually.

The postmodern trend is also the design style of collage in 60–90' to some extent. Mix and match of contemporary architecture is not surface design techniques, more in design early research, and cooperation of other areas 'teams and a variety of professional. Standing on a different angle to do one thing, broking the universal thinking routines generally engaged in the design, which can easily lead to results completely different from the traditional and promote the development of the entire industry to a more open, pluralistic, richer connotation and denotation greatly expanded.

## 19.2 Feasibility of Mix and Match in Urban Design

Firstly, the computer provides a virtual perspective of the city, playing the huge scale of things between in the applause, increasing the overall effect of the control and design flexibility. Its interoperability and easy to grasp take other material collage feasibility objectively.

Secondly, original design field continues to expand and compatible, multiple and complex and overlap makes starting from a different professional perspective on urban design, complex body, mergers and Urban Design presents a wide range of operating.

Landscape designers, architects, urban designers operate simultaneously, pulsing with a variety of other professionals. Personal or individual professional's role in the entire process is nothing more than a functional organ of an organism, and they are complementary, comprehensive, mutation, selective mixed finally forming an effective organic whole.

So, a variety of different stimulation multiple solutions for planning is inevitable after a design topic brainstorming.

Thirdly, as for a complex system such as city, there is no single quantifiable standard; the evaluation system is also different because of the multiple points of view. So it's often not simply to define whether an urban design is appropriate or inappropriate, good or bad. The results screening out from different evaluation system constitute the complexity of the city. And because of the professional continuously strengthen, research depth constantly having new progress and time continued, the dimension of design concept is more and more widely, and the evaluation system is also more and more complex, which makes design no pure concept of said, but comes from the mixing and change of one or several segments in different original subjects.

## 19.3 "Mix and Match" in Urban Design

### *19.3.1 The Multifarious Mixed of Form and Prototype*

Spatial form, for example, the emergence of mixed settlements and a large number of large-scale commercial complex, get rid of the concept of partition in the simple sense of modernism. That makes different functions of the space mixed in one building, with no clear boundaries and split in more cases, so that the mixing showing a freer and more complex trend. On one hand, it is a way to solve a variety of social issues through the city building, on the other hand is a response to the complicated of urban itself. Such comprehensive building is not abruptly making rules based on chaos, but to solve the problems exist in the macro chaotic with the chaos of the organic system, to produce more positive results.

In the design, Koolhaas wants to show the convenience of communication and exchange of the inside of the building, forming a three-dimensional ring, to strengthen contact. "Most of the skyscrapers are a waste of space, vacated part in the middle of the CCTV building (Fig. 19.1) like a capsule, making the ground vitality and bringing life to the building." Its inter part has a very complicated system, nothing less than a medium city.

Also, the building configuration in city blocks is always rectangular matrix arrangement, while the mobility curve layout and free-form shape appear, breaking the graphics and layout. The change of plane figures can just be seen as graphical changes, but the graphic itself does not make sense. Comparing with putting the graphic into the three-dimensional space, they produce a new impetus and impact, providing more possibilities.

**Fig. 19.1** CCTV building

**Fig. 19.2** SOHO city

Zaha Hadid designed the "SOHO city" (Fig. 19.2) in the planning project of Beijing Logistics Port. Zaha said she designed a streamlined city, which seems like waves, from low to high, and forming a climax in the middle.

There are three parts in the building. The circle part is a courtyard, has a long tail is the High-rise building, the bend part form the floor, the following green part is that the park. The green part is like branches as she described. Something going down the oblique, like touching something, actually give the building an open visual effect, bringing in the front landscape, the more we going down the lower we are. Below the tower there is a tail, like a flower, actually is shops or public area.

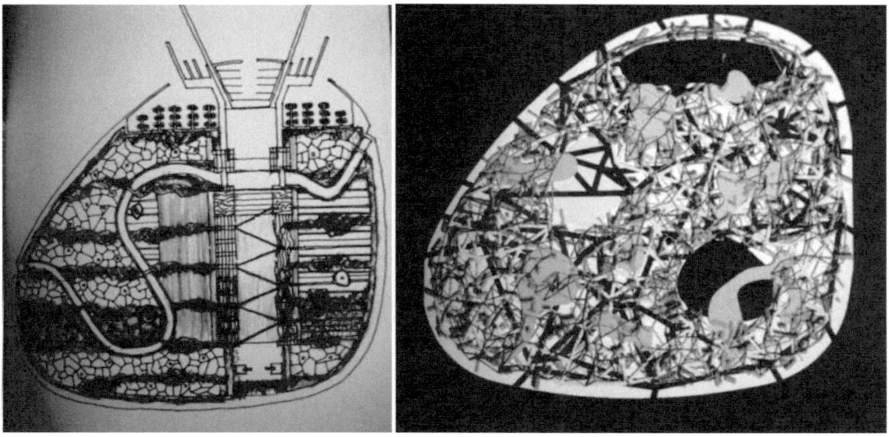

**Fig. 19.3** Simulation experiment of blocks generated

The programs amazing the world, is not only like fish or plant growth form when looking when looking down, but also force on the top of wave when looking far from in the city, then modeling is slow stretches, then going to the corner, each different direction has a different state.

Therefore, the mixing of the various forms has its appeared inevitable and the effect of solving the problem, apart from that the complication and mixing is just budding, a variety of methods and a variety of elements of the hybrid also have the very big development space.

#### 19.3.1.1 The Realization of the Theoretical Model

In the angle of area Urban Design is a macro large-scale, you can have an infinite number of different level segments and spaces; however, it is also subject being affected most by various intricate factors, so there are also plenty of regularity.

Many architects have unique deep understanding of the city, they have made a lot of meaningful urban design theory model. Many of these theoretical models are to stay in the level of city research, not having the feasibility of coming out, but the way of thinking and the research of urban problems provide us with a template and framework, we can extract useful part, combining with the reality, to form an urban mix and match the picture. So the reference and continue of the existing form and theoretical prototype is particularly important in the new urban design. Retention and extension of the existing urban fabric,, or the unlimited replication of one urban form are all based on this.

Such as the Zhuhai Haishi planning (Fig. 19.3) made by Arata Isozaki, which is an experiment of drawing the unknown city blueprint with different thinking. Arata Isozaki said in the Design Description of Haishi Planning: in the high-speed

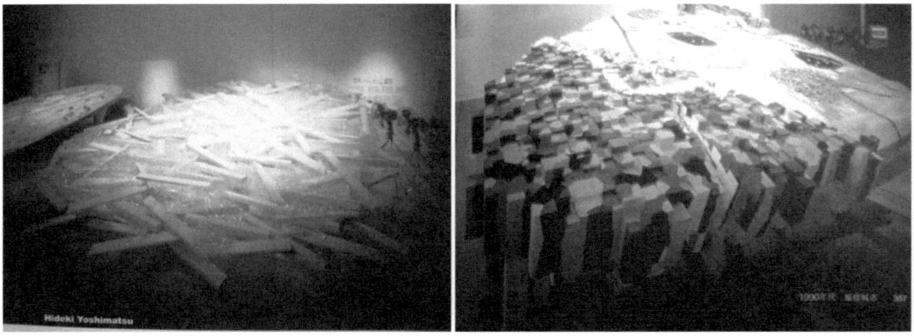


**Fig. 19.4** Open systems models

operation of the electronic world, in fact, there is no concept of distance, which is just a topology world forcing on the relationship between things. On the other hand, urban planning is controlled by Euclidean geometry. However, the real city is high-speed melt from the classical space into topology space. The Haishi plan is showing the generation process of this topological space.

"Haishi" proposed to be built on an artificial island in the South Bay of the Hengqin Island. The program began in 1993, commissioned by the Zhuhai municipal government. As for Arata Isozaki, the "Haishi" has two meanings,, one of which is literally, the city on the sea, another layer of the meaning is a mirage, a world completely cut off with current political, social recognized systems.

Facing such a new beginning, Arata Isozaki used Fengshui theory, in order to promote the circulation of the "air flow", simulating "whirlpool", such a growing and changeable the shape, building up the basic framework of the plan. Arata Isozaki, however, does not indulge in this fixed urban form, also opposed that the urban design mode finished by a few architects or planners, because he believed that "the city will not have the final form, it is a process". Thus, in 1997, on the "Haishi - and a utopian" exhibition, Arata Isozaki try to invite a number of others to participate in this future planning of the city.

The exhibition founded on the basic framework of Arata Isozaki, matching based on Rome recovery plan, and then let the architects, as the "signature person", to exchange the building and let the 12 architects, as the "visitors" to participate in the previous results during the 12-week exhibition time. During the exhibition call from all walks of life though the Internet to participate in this planning. The results of this simulation fell into chaos. "Haishi" (1997) Eventually completed though the Internet, is very similar to the form of "the ruins of the city" that Arata Isozaki put forward in 1960s. As the city is not a subjective decision by a body, it has become a very complex system, showing an intricate structure.

These models, similar as Haishi, themselves do not have a fixed interpretation and program, but just floating utopia. They are open systems, with varying degrees of adaptability to other elements in the city, there is abundant room for interpretation and develop, provided a large number of elements that can be mixed and matched, waiting for the more valuable comprehensive (Fig. 19.4).

Typical design is familiar, accepted and flexible applied by people, which makes the overall control ability of the outcome strengthened, and can manage a variety of complex design method and integrated application.

Aimed at different urban problems, there are a lot of dedicated research group, provides a variety of solutions. In reality there is a wide range of issues that are intertwined, so using the existing research to make the appropriate adjustments and changes.

Apart from a certain type solution, the mix and match with other subjects forming a new cross-cutting area, making the conventional method a new look, a new direction.

#### 19.3.1.2 Era Sensation of City

Urban development is a long-term process which needs to continuously adjust and compromise, current designs also need to do lots of adjustments as time goes by, as well as the combination and co-ordination between city's old town and the new zone which are to fix heterogeneous things to the complicated system. Not only form the aspect of viewing the whole region of a city, but also from the aspect of a single region, buildings and the surrounding environmental factors also present the idea of collage and the effects.

In 1975, Rowe and Kotter have used the word "collage" in fields of urban and architectural design in the book <collage city>. They think that "the discontinuous structures, various, up-and-down passion and a series of travel memories" gather together to present the so called "collage". In fact, historic cities are always displayed as a wonderful "collage" of vestiges of every period of history.

### *19.3.2 Urban Design Increasingly Show Its Sociality*

A city is a complex thing that can't be decided by a group of people. Its development has a lot of uncertain factors and roles. Designers are very limited in this process.

Urban construction can't always be carried out in a more reasonable way from a professional view; the reason is that every leader in power holds the absolute right which becomes more powerful with other various forces participating in. Finally, the result is a palette with mixed colors rather than a simple proposal.

Koolhaas said two years ago that "Building has become an abstract metaphorical symbol, the significance of practice becomes more and more small, and the metaphor meaning becomes more and more important." Now this viewpoint has gradually become a reality. The CCTV new building is such a media through which it presents a gesture of openness, tolerance, and even the worldwide vision beyond its adventurous and individual new figure. In a way, the brand new CCTV building is not only a representation of the image of CCTV's culture, but a new card of Beijing in this era of media.

### *19.3.3 Multidisciplinary Participate in Urban Studies*

Urban design has escaped from the original design methods that start from the traffic organization and the city's architectural form, and then it has produced a plural analytical method that intersect with large-scale discipline such as human geography, anthropology and so on, also from areas that seems to have nothing related like philosophy. This method of analysis determines the connotation, extension and direction of design by itself, and its impacts on design have even beyond itself.

It is decided by the social division and the increasing contact of various fields that the existence and development of various forms mix.

The in-depth development of every industry make people to choose profession and in-depth study in the early without building the foundation of whole knowledge frame, so it is very easy to fall into a vicious circle that work behind closed doors in a field. However, a lot of innovations appear in Interdisciplinary. We hope that all kinds of fields can cooperate with each other, after all, one person's learning capacity and energy are limited and it is not easy that he have mature skill in just one field. Not only is teamwork important, it's also a significant method that can quickly get experiences from various fields.

For example, the wind tunnel trial that is used to analyze wind environment has already exceeded the limitations of climate research, and many results of the research have been applied in urban design, even applied in architectural practice, landscape design and so on. Multidisciplinary sciences impact on every aspect of urban design.

It is very common that a team have different members from different professional backgrounds, for instance, in a project, there are architects, civil planners, landscape planners, estate agents also play a catalytic role, and even artists are involved in. After break through the barriers of professional field, all kinds of thought collide together in the same place to produce various interesting results and more forward-looking ideas, which have positive effect no matter for the construction of future or for the development of discipline.

It is not enough to keep urban-mix going only have different field staff collaborate. The way of working, that is what makes all aspects of intelligence involved in the entire process, it is the key in design or even directly determines the effect that can be achieved in the end.

For example, the research of architectural practice has become an indispensable tool that has been used to interpret the Capital programme by contemporary architect, thereby guiding and inspiring people to read their works. In virtue of it has an amazing position and proportion in MVRDV's work, the research is inevitable topic even become the most attractive part when we talk about their works. It is hard to imagine that the research has played such a significant role in their work. The research operation transform the date and fact to the prospect of city and

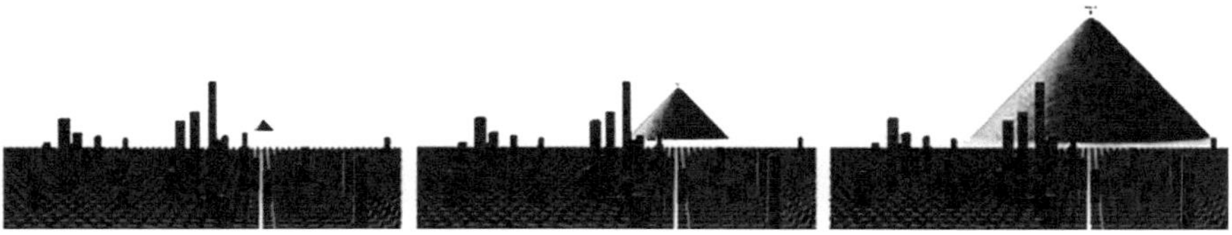

**Fig. 19.5** MVRDV's research works

building, which show a new method of building research and the role change of architect in society, then provide new possibilities for architectural design.

MVRDV has incredible efficient and practical ideas. In order to force they work accurate, rather than out of intuition, they through the practical analysis to help understand the rule and logic in the complicated system. On one hand, in the process of transformation between research and practice, MVRDV's research process is the foundation of production practice, which combines research and practice more tightly. Thus, making the research part as the starting point of design concept generally holds a larger proportion than the practice part which is just construction after research. On the other hand, MVRDV put architecture research in a more grand scope, which have strengthen the combination of architecture and humanity research such as social, economic, ecological and so on.

It makes the research becomes one of the indispensable part. However, it is not changeless that the research account for the proportion and the relationship between it and the practice. Sometimes, they work start from the analysis, and then go up to the level of the building via their research. We can clearly see the same strain between Farmax and WoZoCo's, 3DCity and Pinault, Functionmixer and Silodam. Sometimes, MVRDV products programme one by one via a series of abstract and conceptual proposal, but seldom stay in how to build. In 2002, the housing scheme that has been built in Ypenburg is only a geometrical practice and successful community planning, a set of abstract graphics, a highly synthetic Consolidation. Sometime, their research seems an extreme idea, completely without construction. In the Metacity/Datatown published in 1997 and exhibitions, MVRDV analyzed and compared a series of maps and big cities around the world, imagining the future cities via accurate calculation. The exaggerated sized city scene presented by giant tower and garbage mountain are like new work created by surrealism artists. MVRDV's research works as important part in the early practice, as well as narrative construction readings that been separate tasted (Fig. 19.5).

## 19.4 Summary

In the large-scale domestic construction, the mode of thinking and design method of Mix and Match City have shown their characteristic. Along with the further development of the industry, this method not only have been gradually accepted by the designer but also will lead the entire design develop to the direction that is broad, diverse, complex and integrated. This kind of benign design pattern is really needed and will development as well. What designers have to do is to make their own thinking more open.

**Acknowledgement** This research is supported by: longitudinal research projects funded projects of Beijing JiaoTong University 2011JBM180.

# Chapter 20
# Optimization Model of Land Use Surrounding Rail Transit Stations Based on TOD

**Yan-ping Liang, Shu-juan Cui, and Xue Hou**

**Abstract** This paper formulates a multi-objective linear optimization model of land use surrounding rail transit stations based on TOD theory. With FAR value of different land use types as decision variables, this model is to maximize rail transit ridership, maximize service facility level and maximize total added economic values from land. And Xizhimen rail transit station is selected as a study case. The data used in the case come from investigation on-the-spot, images of Google Earth. At last the model is applied to optimize land use of Xizhimen rail transit station surroundings, and some suggestions for future development are given in this area.

**Keywords** Traffic planning • TOD • Rail transit stations • Land use • Optimization model

## 20.1 Introduction

High intensity exploitation in rail transit station surroundings based on TOD (Transit-Oriented Development) theory will bring benefits of economy and passenger flow. Rail transit station attracts large number of passengers along the rail transit line because of its convenience, and thus it activates the economic elements to gather together. Furthermore, more service facilities attract more passengers. But over-developed intensity would certainly lead to traffic jams, environmental deterioration, and lower life quality and so on. Ridership, level of service facilities and economic benefit should all be taken into account and be balanced in the system,

Y. Liang (✉) • S. Cui
MOE Key Laboratory for Urban Transportation Complex Systems Theory and Technology, Beijing Jiaotong University, 100044 Beijing, China
e-mail: ypliang001@163.com

X. Hou
Qingdao Metro, 266000 Shandong, China

H. Xia and Y. Zhang (eds.), *The 2nd International Symposium on Rail Transit Comprehensive Development (ISRTCD) Proceedings*,
DOI 10.1007/978-3-642-37589-7_20, 

when planning to design and construct a rail transit station. In order to guide metropolis to form a balanced developing mode effectively, it is necessary to study how to keep land use surrounding rail transit stations in a beneficial cycle.

Different studies have been done in the field of land use around rail transit stations. Ming Ruili [1] investigated land use data including seven different-type rail transit stations from four cities of Tokyo, Seoul, Hong Kong and Singapore, and analyzed the characteristics of land use, developing intensity and land layouts in different distance circles around rail transit station. Yang Liya [2] proposed two chain combination prediction models of Gray system cloud-Markov chain and BP neural network-Markov, which had been used to predict spatial distribution patterns of land use and prices of real estate along rail transit line.

Aiming at looking for a suitable developing intensity of land use around rail transit stations, Yao Wenqi [3] gave a propositional FAR (Floor Area Rate) value of Shenzhen when the city was first planned, according to the FAR value of Tokyo and Hong Kong. Li Linglan et al. [4] proposed a proper FAR value in TOD district of Suzhou by referring to the development intensity of land use surrounding rail transit stations in Hong Kong, Shenzhen and Shanghai. The FAR value of comprehensive development land is 4, and 2 is suitable for residential area.

Some researchers studied the issue of land planning and optimization. Mo Yikui et al. [5] studied the issue from three aspects of centralized development, balancing use and environmental improvement; J.J. Lin et al. [6] discussed it from another three aspects of efficiency, environment and justness in land development intensity. But both of the studies do not consider the aspect of economic benefit. Although Lu Huapu et al. [7] took the economic benefit into account, the upper and lower limits of FAR were set neither clearly nor reasonably.

Different from the studies above, this paper proposes an improved optimization model of land use based on TOD theory, and then applies the model to optimize the land use of Xizhimen rail transit station of Beijing.

## 20.2 Establishment of Optimization Model

### *20.2.1 Objective Function*

There are three functions in this model:

① The first objective function is to maximize the passenger flow of urban rail transit station. Improving utilization rate of urban rail transit station can enhance traffic efficiency of rail transit line and utility of land resources. Passenger flow of urban rail transit station can be calculated by its share rate and traffic generation rate of unit building area of different land use, which is expressed with formula (20.1).

$$\max Z_1 = \sum_i \left(k_i^r T_i^r L_i^r\right) + \sum_j \left(k_j^b T_j^b L_j^b\right) \tag{20.1}$$

Where,

$X_i^r$—FAR of residential building;
$X_j^b$—FAR of commercial and office building;
$k_i^r$—the share rate of rail transit in residential area (%);
$k_j^b$—the share rate of rail transit in commercial and office area (%);
$T_i^r$—the generation rate/attraction rate of traffic in residential area (person times/100 m$^2$);
$T_j^b$—the generation rate/attraction rate of traffic in commercial and office area (person times/100 m$^2$);
$L_i^r$—area of residential building (100 m$^2$);
$L_j^b$—area of commercial and office building (100 m$^2$);
$r$—residential building;
$b$—commercial and office building;
$i$—the type of residential building ($i = 1, 2, 3$);
$j$—the type of commercial and office building ($j = 1, 2, 3$).

② The second objective function is to maximize service facilities level. Good service could improve life quality, and the level would be improved only when the proportion of land used for public service facilities is increased. This model only considers total gross of buildings in each type of land use, and land spatial layout is not included. So the level can be expressed by the area ratio of service facilities land with the total area of residential, commercial and office buildings. The larger the value is, the higher service facilities level in a rail transit station will be.

$$\min Z_2 = \frac{L_D}{\sum_i X_i^r L_i^r + \sum_j X_j^b L_j^b} \tag{20.2}$$

Where,

$L_D$—area of service facilities land (100 m$^2$);
Subscript $D$—the type of service facilities land;

Other parameters have the same meanings with formula (20.1).

③ The third objective function is to maximize the total added economic value of land. P is a percentage of average added economic value in certain type land use.

$$\min Z_3 = \sum_i \left(P_i^r X_i^r L_i^r\right) + \sum_j \left(P_j^b X_j^b L_j^b\right) \tag{20.3}$$

Where,

$P_i^r$—percentage of the added value of unit residential building area (%);
$P_j^b$—percentage of the added value of unit commercial and office building area (%);

Other parameters have the same meanings with formula (20.1).

### 20.2.2 Constraint Conditions

There are three types of constraint conditions in this model:

① The first are upper and lower limits of FAR. This paper aims to study how to improve development intensity of land use around rail transit stations based on TOD theory, thus FAR value should not be lower than a certain limit, and it also should not overflow an upper limit to avoid over-developed [2]. Referring to the results of reference [3], and at same time taking the restrict regulations of FAR in China into account, this paper chooses the upper limit value of current rule as the lower limit value, and its 1.4 times is used as upper limit value.

$$f_i^r \le X_i^r \le 1.4 f_i^r, \quad \forall i \tag{20.4}$$

$$f_j^b \le X_j^b \le 1.4 f_j^b, \quad \forall j \tag{20.5}$$

Where,

$f_i^r$—upper limit value for FAR of residential building in current rules;
$f_j^b$—upper limit value for FAR of commercial and office buildings in current rules.

② The second type of constraint condition is quantitative relationships among different type of land use.

$$\sum_j X_j^b L_j^b \ge \alpha \sum_i X_i^r L_i^r \tag{20.6}$$

Formula (20.6) means that area of commercial and office buildings must meet the demand of residents around a station under certain demand rate.

$$\sum_i X_i^r L_i^r \ge \beta \left( \sum_i X_i^r L_i^r + \sum_j X_j^b L_j^b \right) \tag{20.7}$$

$$\sum_j X_j^b L_j^b \ge \gamma \left( \sum_i X_i^r L_i^r + \sum_j X_j^b L_j^b \right) \tag{20.8}$$

Both of the above formulas tell that the areas of residential building, commercial and office buildings must satisfy the development demand of station surroundings respectively. For example, when a station focuses on residential function, it will make $\beta$ lager and $\gamma$ smaller; when it's a commercial station, it should let $\beta$ smaller and $\gamma$ larger.

③ The third type of constraint condition is the area in the station surroundings must satisfy the demand of residents in that district.

$$\sum_i X_i^r L_i^r \leq \theta L_D \tag{20.9}$$

In the Formula (20.9), α, β, γ, θ represent different quantitative relationships among different types of land, and they can be adjusted when necessary.

### 20.2.3 Establishment of Model

Turning the second nonlinear objective function into a linear one:

$$\min Z_2' = \frac{\sum_i X_i^r L_i^r + \sum_j X_j^r L_j^r}{L_D} \tag{20.10}$$

So, the optimization model proposed in this chapter is a multi-objective linear optimization model:

$$max\ Z_1; min\ Z_2'; max\ Z_3$$
$$\text{s.t.}(3-4)-(3-9);\ \ X_i^r, X_j^b \geq 0, \forall i, j;\ \ \alpha, \beta, \gamma, \theta > 0$$

## 20.3 Case Study: Xizhimen Rail Transit Station in Beijing

### 20.3.1 Introduction of Xizhimen Rail Transit Station

Xizhimen rail transit station locates in the northwest corner of Xizhimen Bridge in Beijing. It is a comprehensive passenger transport hub. There are three rail transit lines transferring here, while combining the National railway (Beijing North Railway Station), road bus, taxi and other modes of transportation into together.

Total land use planning in Xizhimen is 5.99 ha, with a total construction area of 264,000 $m^2$, commercial and office construction area of 182,000 $m^2$, and traffic building area of 67,000 $m^2$.

### 20.3.2 Data Collection of Different Land Use

With the supporting of images in Google earth, the authors investigate on-the-spot actual data of land use in Xizhimen rail transit station district, and then draw out

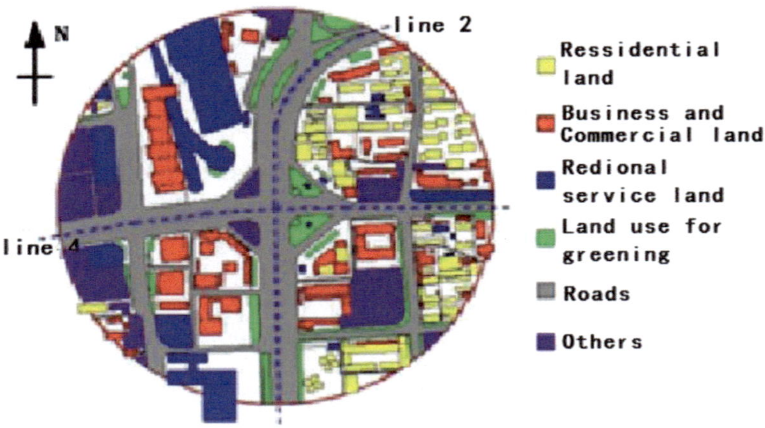


**Fig. 20.1** Various land types in Xizhimen station

**Table 20.1** Total area of various land types (100 m$^2$)

| Type | Area |
|---|---|
| The first class residential land use R1 | 0 |
| The second class residential landuse R2 | 1,081.14 |
| The third class residential land use R3 | 180.29 |
| Land for commercial facilities B1 | 733.24 |
| Land for commercial facilities B2 | 512.02 |
| Office land use A1 | 606.20 |
| Public service land D | 4,096.76 |

**Table 20.2** The known parameters in the model

$k_1^r = k_2^r = k_3^r = 12.56\%;\ k_1^b = k_2^b = k_3^b = 12.56\%$

$T_1^r = T_2^r = T_3^r = 12.4;\ T_1^b = T_2^b = T_3^b = 12.4$

$L_1^r = 0;\ L_2^r = 1081.14;\ L_3^r = 180.29$

$L_1^b = 733.24;\ L_2^b = 512.02;\ L_3^b = 606.20;\ L_D = 4096.76$

$P_1^r = 0;\ P_2^r = 20.3\%;\ P_3^r = 15.4\%;\ P_1^b = P_2^b = P_3^b = 23.01\%$

$f_2^r = f_3^r = 2.8;\ f_1^b = f_2^b = f_3^b = 4.5$

$\alpha = 0.34;\ \beta = 0.30;\ \gamma = 0.50;\ \theta = 25$

Fig. 20.1 with AutoCAD. The total areas of each type of land use are listed in Table 20.1.

Other parameters data are shown in Table 20.2 above.

### 20.3.3 Values of Current FAR

The FAR value of certain type of land use is equal to the ratio between its total area of buildings and its total land area (Table 20.3).

**Table 20.3** Values of current FAR in different land use

| Type | Current FAR |
|---|---|
| The second class residential land use R2 | 3.1 |
| The third class residential land use R3 | 0.75 |
| Land for commercial facilities B1 | 2.0 |
| Land for commercial facilities B2 | 5.1 |
| Administrative land use A1 | 3.6 |

**Table 20.4** Contrast of FAR before and after optimization

| Types of land use | Current FAR | Optimized FAR |
|---|---|---|
| The second class residential land use R2 | 3.1 | 3.92 |
| The third class residential land use R3 | 0.75 | 3.92 |
| Land for commercial facilities B1 | 2.0 | 6.25 |
| Land for commercial facilities B2 | 5.1 | 6.13 |
| Administrative land use A1 | 3.6 | 6.30 |

### *20.3.4 Optimization*

Applying optimization toolbox of MATLAB software for calculation of the model with linear weighted method, a set of non-dominated results is gotten:

$$X_2^r = 3.92; \quad X_2^r = 3.92; \quad X_1^b = 6.25; \quad X_2^b = 6.13; \quad X_3^b = 6.30$$

Comparing the FAR values after optimizing with the current values (Table 20.4); it shows that there is a large space to improve in the study area.

① The current FAR of second class residential land use (R2) and commercial facilities land use (B2) are 3.1 and 5.1. Both of them can be improved and the final data are 3.92 and 6.13.

② The current FAR of third class residential land use (R3) is very low and it is 0.75. According to the rule of <The 2011–2015 Beijing State-owned Construction Land Supply Plan>, the FAR value of residential land should not be lower than 1, especially in district around rail transit stations. The third class residential land should be reconstructed into second class and its value should be lifted to 3.92.

③ The current FAR values of land for commercial facilities (B1) and administrative land use (A1) are also lower. Commercial land is wasted much more seriously there, and many of them are small commercials. The FAR value of these two types can be increased to 6.25 and 6.30 separately.

## 20.4 Conclusions

There are three differences from the existing studies: ① The model in this paper takes economic benefit of land development around rail transit station into account, which is expressed by formula (20.3). And at the same time the upper and lower limits of FAR are given definitely. ② The data used in case study come from investigation on-the-spot, and not from urban planning which used to conflict with reality. ③ The land use types around Xizhimen rail transit station are plotted out referring to the new standard <Code for classification of urban land use and planning standards of development land> (GB50137-2011) which was promulgated in 2012 [8].

It is beneficial to implement TOD strategy in urban planning, especially in the rail transit station districts, and it will amend and perfect the urban developing mode to keep urban in sustainable development.

**Acknowledgment** This paper is supported by National Natural Science Foundation of China (51008016) and the Fundamental Research Funds for the Central Universities (2012JBM068).

## References

1. Ruili M (2009) Research on method of combining rail transit and land-use along. Tongji University, Shanghai
2. Liya Y (2007) Theory and methodology of relationships between urban transportation and land use. Beijing Jiaotong University, Beijing
3. Wenqi Y (2005) Research on strategy of combining the land use planning and rail transit planning in Shenzhen. South China University of Technology, Guangzhou
4. Linglan L, Guohua Z, Yunli C (2007) Approaches to land-use planning and adjustment adjacent to rail transit stations: a case study in Suzhou, Jiangsu Province. Urban Transp China 5:30–36
5. Yikui M, Jun D, Jingyuan W (2009) A land use model for urban rail station area planning based on TOD strategy. J Civil Arch Environ Eng 31:116–120
6. Lin JJ, Gau CC (2006) A TOD planning model to review the regulation of allowable development densities around rail transit stations. Land Use Policy 23:353–360
7. Huapu L, Jing Z (2008) Study on planning methodology of TOD for Chinese cities. Highw Eng 33:64–68
8. The People's Republic of China Ministry of housing and Urban–rural Development (2011) GB50137-2011City land classification and land use planning and construction standards. Beijing: China Building Industry Press, pp 6–9

# Part II
# Experience and Practice

# Chapter 21
# Building Complex Integrated with City Rail Transport: Practices from Newly Built Building Complexes in Japan

**Xinran Zhang and Yingdong Hu**

**Abstract** Japan is short of land resource, thus the city construction is extremely integrated. Building complex in Japan is highly developed, usually a marriage of building and urban rail transport, to realize a point to point junction of traffic & property. Based on the writer's research on the newly accomplished Shibuya Station Super high-rise Building Complex (2012), Tokyo Midtown (2007) and Hakata Station transportation Building Complex (2011), the report tries to explore the development and design idea of the building complex in Japan.

**Keywords** Urban rail transportation • Building complex • Japan

## 21.1 Mixed-Use of the Functions

### *21.1.1 Trend of the Comprehensive Functions*

City building complex tends to satisfy the modern citizens' diverse needs by one-stop solution, supported by the integration, completion and mutual amelioration of the functions, It's an incentive effect.[1] Integration of the functions not only satisfies the diverse needs, but also lifts the usage efficiency of transport infrastructure and balance the operation risk. What are more important, integrated functions

---

[1] Dongqin Han, Jinglong Feng. City building integrated design. Southeast University Press. 1999.

X. Zhang
Beijing Institute of Architectural Design, No. 62 Nanlishi Road, Xicheng District, 100045 Beijing, China

Y. Hu (✉)
Associate Processor of Beijing JiaoTong University, Beijing, China
e-mail: ydhu@bjtu.edu.cn

H. Xia and Y. Zhang (eds.), *The 2nd International Symposium on Rail Transit Comprehensive Development (ISRTCD) Proceedings*,
DOI 10.1007/978-3-642-37589-7_21, 

**Table 21.1** Comparison of the economical & technical indexes and functions

| | Shibuya Station HIKARIE super high complex (Fig. 21.1) | Roppongi Hills | Midtown, Tokyo | Hakata station transport complex | Kokura station building complex |
|---|---|---|---|---|---|
| Project location | Shibuya, Tokyo | Roppongi, Tokyo | Roppongi, Tokyo | Fukuoka, Hakata | Kitakyushu Kokura |
| Site area (ha) | 0.964 | 11.5 | 10.2 | 2.2 | 1.5 |
| Construction area (10,000 m$^2$) | 14.4 | 75.91 | 56.9 | 20 | 7.6 |
| Floor-area ratio | 14.9 | 6.6 | 5.58 | 9.1 | 5.1 |
| Building height (m) | 182.5 | 270 | 248 | 60 | 56 |
| Position to the railway infrastructure | Close to | Underground channel | Underground channel | Station integrated | Station integrated |
| Retail | ● | ● | ● | ● | ● |
| Office | ● | ● | ● | ● | ○ |
| Hotel | ● | ● | ● | ○ | ● |
| Entertainment | ● | ● | ● | ○ | ● |
| Culture & art | ● | ● | ● | ○ | ○ |
| Public infrastructure | ● | ● | ● | ● | ● |
| Leisure (green, public space) | ● | ● | ● | ○ | ○ |
| Residence | ○ | ● | ● | ○ | ○ |
| Parking | ● | ● | ● | ● | ● |

Note: "●" for with the function, "○" for without the function

could help reduce the traffic pressure as a tide. The integrated functions include: transport, retail, entertainment, office, hotel, residence, culture, infrastructure, parking [1] (Table 21.1).

Based on the above table, the development modes are of distinct features and of independent system. Combination of the functions is based on the regional infrastructure characters. Small & medium city building complex usually don't include residences, yet include business, hotel and infrastructure services.

### *21.1.2 Public Green, Culture and Public Infrastructure Under High-Density Development*

Modern citizens expect to be closer to the nature. Take the case as Tokyo midtown, which integrates functions as business, office, & hotel, yet also keeps a green of nearly 20,000 m$^2$. So as the case of Roppongi Hills Maori courtyard.

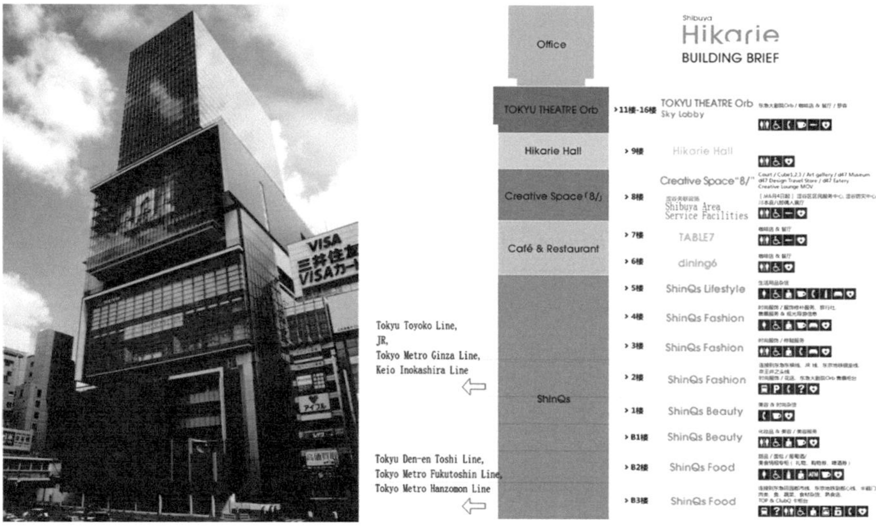


**Fig. 21.1** (**L**) Shibuya Station HIKARIE super high complex (Source: self-shot). (**R**) Floor function profile of HIKARIE (Source: collation according to the official guide)

City complex is usually the regional landmark and culture & info center. Building complex developers prefer to introduce the art infrastructure to lift the quality of property. Such as the Museum "21-21design sight" designed by Tadao Ando at Tokyo midtown. Tokyu Theatre Orb of 2000 seats locates at floor 11–16, skyscraper complex HIKARI. And infrastructures as HIKARIE Hall, Disaster Prevention Center, and exhibition hall are set at floor 9 & 8 respectively.

### *21.1.3 The Match of Function Positioning & Regional Characters*

Besides advantages as transport & multi-function integration, city complex should also be precisely positioned to avoid competition of similar nature. This requires city complex to integrate regional industries and core competence. Shibuya region, Tokyo is of many creative & promotional culture enterprises. It is of high potential of being a center of popular life style and info transit. HIKARIE complex not only integrates a massive culture exhibition infrastructure, but also a high-level college of creation talents. The efforts demonstrate its position & willingness as the cultural core at Hikarie [1].

## 21.2 Lift the City Image

City complex could help build city image, establish space guide and reinforce city distinction.

### *21.2.1 Renovate Regional Traffic Infrastructure & Re-exploration of the City*

Symbol building demonstrates its positive function by the city renovation and transport infrastructure modernization driven by the massive building complex construction. As the case before, HIKARIE as a preface would push the modernization of station and the development of the bi-tower building (Fig. 21.2).

### *21.2.2 City Landmark*

Landmark in a city usually demonstrates its distinctive height, outline and feature. The distinction could also be embodied by its size, shape, material & color, etc. Skyscrapers are guidelines of the city space and the symbols of city culture, spirit and economic power. Midtown Tower, by its distinctive building group, is over passing Roppongi Hills Mori Building to be a new regional landmark.

In small & medium cities, building complex embodies its city life acknowledgement by its distinctive inner building design. The Tokyo building complex introduces huge step-shape space into its building, which reach a perfect marriage of the building function & city space.

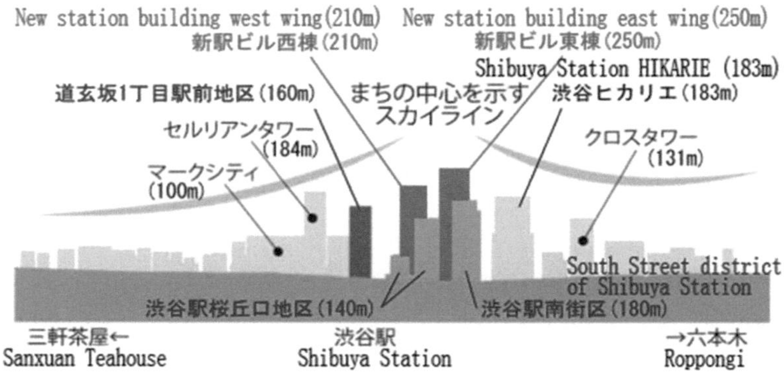


**Fig. 21.2** Project "Dual-Towers Plan" – New Shibuya Station building (Source: collation according to "Infrastructure policy in the central district of Shibuya Station" (October 24, 2010))

## 21.3 Contribution to the Walking Overpass System

Accessibility of the walking overpass is the key success standard of a high-efficiency transport and the development of building complex.[2] The walking system is part & extension of the street.[3] Street system & walking system should be connected from different dimensions, and the connections should be continuous, accessible, cycling, and of good environment. In one words, the new walking system should modernize the original walking system, and help transform the walkers to consumers.

### *21.3.1 Relations with Railway*

The connections between building complex & railway could be defined as three modes: the inclusive mode, the paste building mode and the channel connected mode.

The integration mode refers to the building complex with one or several integrated traffic infrastructures; the advantage is the convenient traffic. The sub-systems of the building complex could distribute & attract the population to lessen the traffic pressure on the main route, which could reduce the peak traffic to"cut peak fill valley effect".[4] Take the case of Hakata Station, passengers could take the central transit passage of JR Station to enter directly Hankyu Department Store.

The attachment mode refers to the horizontal design of the building complex & traffic infrastructure. Which are connected directly with several horizontal interfaces. As demonstrated by the case of the building complex HIKARIE & Metro Fukutoshin Line Shibuya Station.

The channel-connected mode refers to the ground, underground & sky connections between the complex & the transport infrastructure. Since the distribution capacity is always associated to factors as width, time & distance, certain distance to the traffic would maintain clear and stable passenger traffic. Tokyo midtown & Roppongi Hills are in less than 300 m distance to Roppongi Station.

---

[2] TOD theory encourages travelers to take public transportation hubs, pedestrian traffic as the main mode of transport within the region.

[3] "Team10" (1959) considered street and traffic system environment as "generator", which should be restored to the streets as the medium of urban living tissue.

[4] Guo Hong. ɣbn 2.9 to build a new gate to Beijing-Focus on Xizhiwen Hub Project. China investment, version 11, 2002, P73.

### 21.3.2 The Passing Mode Creates Connected Public City Space

To the building complex, the comprehensive economic benefits come from the speed and quality of the passenger flow. So the effective measure to explore the passenger flow is to create the traffic & stay space for individuals of different aims & natures, thus to ameliorate the opening and accessibility of neighboring spaces. Notably, the reason why an integration of complex & transport infrastructure could be realized is because of the unified ownership & operation by the same railway enterprise. As the HIKARIE complex & neighboring Tokyu Toyoko Line belong to the Tokyu Corporation.

### 21.3.3 Multi-layer Walking Network

When activities are too many to develop on two-dimension space, the three-dimension walking system becomes a necessity. A right metaphor is that the stretching branches growing from the transport hub as a "tree", bringing nutrients to the end. A suitable walking system should be convenient, safe, comfortable, continuous, from the moderately attractive, diversion, and others.

Ground pedestrian system is convenient, clear, open and with good psychological feeling, yet the disadvantage is that fusion dangers, and is vulnerable to the weather, motors, tail gas. At some traditional sites of high density development area, the ground walking system has become fragmented.

To ease the short of land in the central area, urban space is extending to the overhead and underground space. Accompanied with the transfer of some transport lines and features (such as public transport lines, stopping the garage) to the underground space, the ground was used to provide an open green space and natural landscape in order to improve the environment on the ground.

Topography, climate, geology, development size, engineering difficulty and habits are all factors impacting the walking modes. In the cold region, the forms as overhead or underground walking systems are always chosen. Such as Canada Montreal (winter up 4–5 months); Japan can accept and use large underground space, yet the Chinese (including people from Hong Kong, Macao and, and Singapore region) are reluctant to accept underground passages and space.

## 21.4 Development of High Strength and Dimensional

Accompanied with the emergence of three-dimensional city and the concept "vertical city", the urban complex tends to grow higher.

### 21.4.1 Public Space for High Floor Area Ratio

Location affects the strength of development. Tiers 1 hub area located in the city center is often of high floor area ratio (such as Shinjuku and Shibuya areas of Tokyo & the central at Hong Kong are at floor area ratio of 10.0–15.0,), whereas the traffic complex in a small city, town or outskirts of big cities may be less than 5 (such as Kokura station complex at floor area ratio of 5.1).

To weaken the negative impressions brought by the over-strength development on the city image, urban complex also adopts positive measures in response. For example, in response to the intensive passenger flow from Shibuya Station (about 50,000 people a day), Shibuya HIKARIE super high complex highlights the degree of openness and the three-dimensional organization of the traffic flow. Since this project contributes through a variety of ways to the city,[5] Shibuya HIKARIE super high complex sacrifices the first floor for public space for higher floor area ratio (14.9[6]).

Small land plots make it difficult to retain more public space. Open platforms on the roof and platform spaces of different heights would be better choices. Its open landscape without being blocked is an incomparable advantages and can also attract the upward visit of the passenger flows. With the case Roppongi hill, to create the themes "vertical Garden City", architects unite the plazas of different heights, connected tiered streets & green to form a "three-dimensional around forest" (Fig. 21.3).

### 21.4.2 Dimensional Development

Due to the high land value, complexes near the railway stations always take vertically stacked and mixed-and-penetrated approaches. The vertically stacked approach is suitable for the ultra-high strength development of properties in city center or above MTR stations. The advantage is land saving, and can form a regional landmark, yet the shortcoming is the heavy burden of large vertical transport, and the complexity to organize flows of multi-functions. Mixed and stacked approach is a common form shifting from two-dimensional to three-dimensional structures. As illustrated by the concept "semi-network structure"

[5] The regional statutory floor-area ratio is of 9.0, but through the application of "Urban Regeneration Law concerning Special Measures", Shibuya Station HIKARIE (recognized urban regeneration project) is reduced restrictions.

[6] According to the provisions of "the Basic Law of the Japanese construction" on "integrated design license": "Project which has a land area of more than a certain size, traffic, security, fire, health, without obstacles, have more than a certain percentage of open space in the base of the urban compensated volume rate of environmental improvement of buildings, floor area ratio and height can be added to the number, to ensure that the proportion of urban public space as determined by the base."

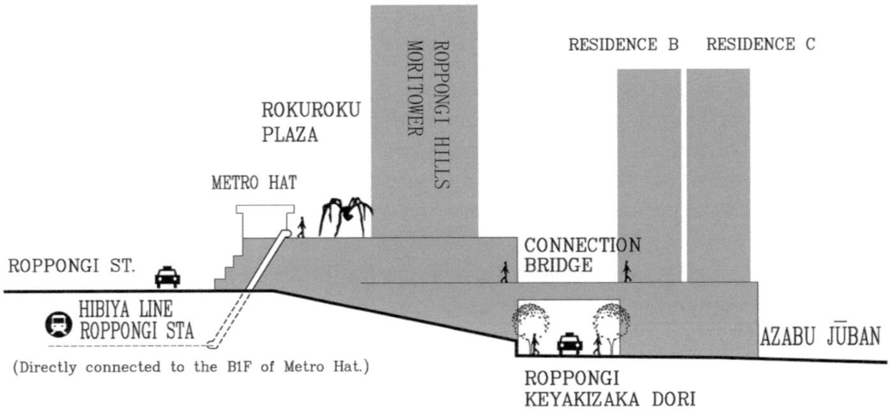


**Fig. 21.3** Stereo greening platform and different modes of transportation in Roppongi Hills (Source: collation according to the official guide)

proposed by the United States architectural theorist Christov Alexandra in his book "A City is not A Tree 1965": The functions should not be simply overlaid or paralleled, mixed & interlaced space would be more dynamic and stimulating.

## 21.5 Concluding Remarks

By the case of Japan, we could seize the trends of building complex closely attached to railway traffic, demonstrates its features of mixed functions, city image, regional walking network, development strength & modes, etc, and expect the research would contribute to China's fast-developing rail transport development and projects.

**Acknowledgment** This research is supported by:National Natural Science Foundation of China (51178038); longitudinal research projects funded projects of Beijing Jiaotong University (2011JBM180,2010RC030,2012JBM120); Fundamental Research Funds for the Central Universities (2009JBZ023); Ministry of Education "New Century Excellent Talents" (NCET-10-0217); Longitudinal project (2012-R1-11); Technical consulting projects (A12X00040).

## References & Websites

1. Yingdong Hu, Xinran Zhang (2009) Research on the commercial design of the comprehensive city hub complex—case Hongqiao Comprehensive Hub, Shanghai. Arch J 56(4):78–83
2. Xiwei Yang (2009) Leading hub–official launch of the large-scale redevelopment project around the Shibuya Station. Mod Arch 26(5):76–77

3. Infrastructure policy in the central district of Shibuya Station (2010, October 24) Shibuya government network: http://www.city.shibuya.tokyo.jp/kurashi/
4. Shibuya Hikarie: http://www.hikarie.jp
5. Tokyu Corporation: http://www.tokyu.co.jp
6. Tokyo National Road Firm: http://www.ktr.mlit.go.jp
7. Japanese Economic News: http://www.nikkei.com/
8. Nikken Sekkei: http://www.nikken.co.jp

# Chapter 22
# The Strategy of the Coordinated Development Between Rail Transport and the Layout of Urban Spatial in Beijing

**Meiqing Zhang, Yaodong Zhou, and Weiwei Hao**

**Abstract** Traffic congestion has become a main factor that restricts the sustainable development of Beijing. Traffic congestion is the phenomenon of urban traffic problem but the fundamental reason is that urban transport and the layout of urban spatial in Beijing are uncoordinated. Especially, the guided and supportive effect of rail transportation on the layout of urban spatial cannot be fully played out. In this paper, an empirical analysis has been done on the inter-relationship of Beijing rail transport and the layout of urban spatial by using the methods of DEA and Elastic Analysis, in order to reveal the incongruity between the two from the perspective of Spatial Aggregation and Function Agglomeration. Then an in-depth analysis has been done on the reasons of incongruity between the two factors and some suggestions have been offered to promote the coordination between railway and the layout of urban spatial on a strategic level.

**Keywords** Rail transport • The layout of urban spatial • Coordinated development • Traffic congestion

## 22.1 Introduction

With the rapid development of urbanization, traffic congestion has become a constraint factor for sustainable development of Beijing Urban Area. The government also issued a series of measures for easing traffic congestion, but the result is

M. Zhang (✉) • W. Hao
School of Economic and Management, Beijing Jiaotong University, 100044 Beijing, China
e-mail: meiqing_zhang@sina.com; wwhao@bjtu.edu.cn

Y. Zhou
School of Economics and Management, Beijing Jiaotong University, BJTU, Beijing, China
e-mail: ydzhou@bjtu.edu.cn

H. Xia and Y. Zhang (eds.), *The 2nd International Symposium on Rail Transit Comprehensive Development (ISRTCD) Proceedings*,
DOI 10.1007/978-3-642-37589-7_22, © Springer-Verlag Berlin Heidelberg 2014

far from the intention, even result in the opposite direction. Traffic jams, however, is only the phenomenon of Beijing urban transport problems, the fundamental reason is that Beijing urban transport and the layout of urban spatial are uncoordinated. Especially, the guided and supportive effect of rail transportation on the layout of urban spatial cannot be fully developed. To this end, how to promote the construction of rail transport and the coordinated development of the layout of urban spatial is an urgent problem to solve.

In recent years, domestic and foreign scholars did great efforts to study the relationship between rail transportation and the layout of urban spatial, such as the evacuation of the population, the urban environment and land use [1]. Another literature which uses input–output mode land gravity model [2] to analysis the role of rail transportation in the urban industrial economy; Liu Yin and Lu Jian [3] point out that the economic level as well as the population density is the most important factor; Guo Xiu-cheng and Lv Shen [4] put forward the mode of big city's layout of rapid rail transport network. However, the vast majority of these literatures view the problem from the point of planning and qualitative to explain the role of rail transportation in the evolution of the spatial structure, and there is still a lack of specific factors and index system methods and also the spatial econometric methods hasn't been used to study the impact between rail transport and the changes of urban spatial, which shows the significance of our thesis.

Therefore, this article bases on the reality of China urban rail transport construction, elaborates the mechanism between the development of urban rail transport and layout of urban spatial by using DEA and Elastic Analysis method, reveals the incompatibility between the Beijing Railway Construction and layout of urban spatial, and proposes some suggestions from a strategic perspective to promote the coordinated development of the Beijing Railway Construction and layout of urban spatial.

## 22.2 The Analysis Methods and Selection of Indicators for Rail Traffic's Impact on the Layout of Urban Spatial

As the city's largest infrastructure projects, urban rail transport has four kinds of influence on urban development and the layout of urban spatial: the transport functional effects, environmental functional effects, economic function effect and social function effect. Regarding the layout of urban spatial as a system, rail traffic is actually a part of the system. Thus, the impact of rail transport on the layout of urban spatial can be viewed as changes in endogenous, mainly involving two methods: elasticity approach and the evaluation method. But this research eventually takes DEA and Elastic analysis method because of the availability of data.

The index system of the layout of urban spatial is divided into spatial indicators and function indicators. Spatial index system aims to analyze the influence between traffic flow which caused by the rail transport site and the degree of site formation gathering. We take the district rail crossings mileage and the number of site to express the importance of the district rail transportation in the entire rail transport system; we use the degree of population agglomeration to show the regional spatial difference. Functional index system is mainly to explore the social and economic impact of rail transport on urban spatial, which includes 18 evaluation indicator: the timeliness benefits of travel, the comfort benefits of travel, travel Security, induced passenger effective and so on.

## 22.3 The Empirical Analysis of the Rail Transport Impact on Beijing Urban Spatial Layout

The empirical analysis of the rail transport impact on Beijing urban spatial layout includes two aspects: the spatial concentration analysis and functional agglomeration evaluation.

### *22.3.1 The Spatial Concentration Analysis*

By examining the overall flow of rail traffic, the regional distribution and site distribution of rail crossing miles in Beijing from 2005 to 2009 and using of elastic analysis methods, we analysis the changes of rail transportation for the degree of concentration of population. We get the following conclusions:

#### 22.3.1.1 Changes in Rail Transport Mileage Doesn't Matter Much to the Overall Distribution of Urban Spatial Agglomeration

In indicator, the Beijing railway prevailing mileage has been greatly improved since 2006, but the relative distribution of population space did not change too much from the last 5 years (Table 22.1). From the Table 22.1, all the indicators were no more than 1, indicating that changes in rail transport mileage has less impact on the population density.

#### 22.3.1.2 Spatial Distribution of Rail Transport Districts Is Unbalanced

From the development process of rail transport in Beijing, it was clear that its construction mainly distributed in urban function expansion areas with a mileage of 168.8 km, accounting for 58 % of the total mileage; especially in Chaoyang and Haidian, where appears a network spatial structure of functional expansion area as the main gathering point. This structure does not match with the existing urban spatial structure.

**Table 22.1** Comparison of rail transport mileage distribution and population density from 2006 to 2009

| | Rail transport mileage | | Population density | | |
|---|---|---|---|---|---|
| | 2006 | 2009 | 2006 | 2009 | Elasticity |
| Urban function core area | 32.1 | 54.3 | 22,308 | 22,849 | 0.0467 |
| Function expansion area | 61.2 | 132.6 | 6,063 | 6,810 | 0.1575 |
| Urban development area | 17.2 | 27.2 | 675 | 781 | 0.0365 |
| Ecological conservation zone | – | – | 202 | 210 | 0 |

Data from: Beijing Statistical Yearbook

**Table 22.2** Principal component score and comprehensive evaluation index of external effects of Beijing's rail transport

| Year | F1 | F2 | F3 | F |
|---|---|---|---|---|
| 2003 | −1.34045396 | −0.88842964 | 0.75570236 | −1.148511624 |
| 2004 | −0.52667642 | 0.43332521 | −1.6020074 | −0.448241492 |
| 2005 | −0.0386538 | 0.93161025 | 0.39241304 | 0.131029198 |
| 2006 | 0.71923463 | 0.77328295 | 0.77456857 | 0.730552366 |
| 2007 | 1.18655084 | −1.24979024 | −0.32067154 | 0.735172654 |

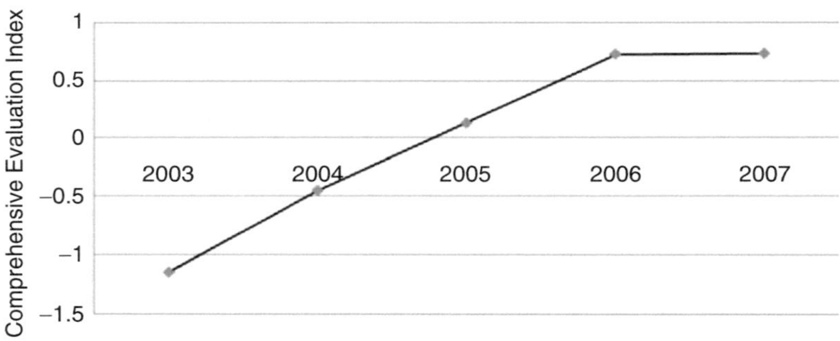


**Fig. 22.1** Comprehensive evaluation index of railway functional spillovers in Beijing

## 22.3.2 *Functional Agglomeration Evaluation*

### 22.3.2.1 Model and Results

According to the Beijing Statistical Yearbook, by analyzing 18 indicators and processing those data from 2003 to 2007, we analyze the results output by DEA as well as the overall evaluation. The final results of principal component analysis operated by SPSS are shown in Table 22.2.

We can draw the curve of Beijing railway functional effects from 2003 to 2007 as shown in Fig. 22.1.

#### 22.3.2.2 Result Analysis

As is shown above, changes in the external effects of rail traffic is a curve with a trend of upward-sloping forward, from which we can sum up the characteristics of Beijing railway functional clustering:

**Firstly, Overall Trend is Up and Functional Spillover Effects Continued.** The composite index of urban rail transport functional spillovers for the period from 2003 to 2007 in Beijing increased from $-1.149$ to 0.735, showing a constantly increasing trend, indicating that the urban rail transport had significant effective overflows in meeting residents' travel needs, environmental protection, spawning along economic and coordination of social development, and the spillover effects continued to enlarge.

**Secondly, the Spillover Effect Gradually Slows Down.** The slope in Fig. 22.1 becomes smaller, which means that during the period, external effect of rail traffic gradually slows down, and the speed of effective diffusion and enlargement also slows down, mainly on the added value of land and housing prices. Rail traffic effect on real estate prices has slowed down with the construction of subway lines, and the overall momentum develops steadily.

**Thirdly, the Traffic Functional Effects and Environmental Effects are Significant.** Observing each initial factor scores by using SPSS, we can see that traffic and environment effect are significant in all external index of rail transport, such as attracting passengers, saving land and fuel energy consumption contribute the most in adding external affection. The main reason is that these two effects are the most basic and straight ones among all external effects.

**Fourthly, Economic and Social Effects Increased Gently.** Through empirical analysis we can see that economic as well as social effects contribute little to external effect of rail transport, initial factors like land value, property prices, city layout etc. fall behind. The reason is that: firstly, economic and social effects belong to indirect effects, where there is a time lag before working; secondly, the real estate in China is affected by many factors, and the lack of efficient market system makes it hard to reflect rail transport's economic effect in price mechanism; thirdly, methods of evaluation, statistical standards and accuracy will influence index scores.

## 22.4 The Analysis on the Causes of the Lack of Coordination Between Urban Rail Transport and the Layout of Urban Spatial

Through empirical studies of the relationship between Beijing rail transport and urban spatial, rail transport plays spatial agglomeration and functional role in the process of urban development, promoting the transfer of industries, population relocation and suburban development to some extent. But overall, the influence of rail traffic on urban spatial structure and layout optimization is not outstanding. Rail transport planning and construction of the city's spatial structure and layout does not match, which cannot meet the requirements of the rapid development of the city, mainly in the following three aspects:

### 22.4.1 Rail Transport Network Does Not Match the Layout of Urban Spatial, and It Is Unable to Support the Structure of Urban Spatial and Layout

Firstly, rail traffic distribution and population density are incompatible. In recent years, distribution of rail transport construction is mainly in the expanding area of urban functions, but central city rail transport network density is very low, only 0.23 km/km$^2$ (and the city center urban rail network density of the rest area of the world are between 1 and 2 km/km$^2$).Secondly, rail transportation planning and urban spatial form planning are Incompatible. Beijing City Master Plan (2004–2020) proposes to build the city spatial structure as "two axes – two belts – multi center". This urban spatial form does not match the urban transport system.

### 22.4.2 The Rail Transport Network Connection Is Inconvenient, Affecting the Rail Transport's Gathering and Diffusion Functions

Transferring to rail transport network connection is inconvenience by reducing travel timeliness. It shows that transfer between rail transport lines is inconvenient, and the performance of poor convergence between rail transport and other modes of transportation, seriously affect timeliness, accessibility, convenience and comfort of residents travel. It reduces the attractiveness and competitiveness of rail transportation. At the same time, transit transfer hub is lack of effective design, reducing the distribution efficiency.

### 22.4.3 Rail Transport Construction and Land Development Does Not Interact Positively, Affecting the Effectiveness of Integrated Development of Land

Development of rail transportation and utilization of land are out of touch, mainly because the development of rail transportation can't keep up with the traffic demand of rapid urban development and high-intensity land development. Meanwhile, rail transport cannot keep pace with the speed of land development and utilization mainly because of the investment and financing channels. What's more, the station around the rail transport hub does not form agglomeration. The mode of land development surrounding rail transportation hub in Beijing is uncorrelated with all residential real estate development, centralized commercial, or they are mainly concentrated office properties. So the crowd, logistics, traffic flow, the flow of information cannot be agglomerated effectively, which is lack of three-dimensional integrated development planning and mechanisms.

## 22.5 Strategy of Coordinated Development Between Urban Rail Transport Construction and Urban Spatial Planning

Alleviating the traffic jams has been regarded as a main task of Beijing Twelfth Five-Year Traffic Planning. It is believed that the most effective solution of traffic jams is to establish an urban spatial structure which led by the thinking of "public transit first". As the backbone of urban public transit system, the urban rail transport plays a great role in the urban spatial structure. Therefore, the adjustment and optimizing of urban spatial planning with the guidance of urban rail transport construction is the answer to traffic jams in urban area of Beijing.

### *22.5.1 Promoting the Coordination Between Urban Rail Transport Construction and City Development by the Integration Strategy*

The concept of integration strategy should be introduced to comprehensive traffic planning, which means the integration between different means of transportation and also the integration between traffic and city master plan. By this kind of integration, the healthy environment for urban rail transport development can be built and the sustainable development of economic can be assured. In a conclusion, coordination and effectiveness express all.

### *22.5.2 City Form Development by the Guidance of Rail Transport and the Realization of Integration Between Rail Transport and the Layout of Urban Spatial*

We should strengthen the support effect of rail transport and establish a Multi-Center city layout. It is important to strengthen the support effect of rail transport in the east and west development zones to drive the formation of industrial cluster along the lines and to lead the establishment of functional zone. Meanwhile, there should be a calculation of resources carrying capacity in each functional zone and adjustment of central city area land function with the Beijing Territorial Planning. As a result, city functional zones can be decentralized and the single center city mode can be changed to multi-centered and the "separation of workplace and residence" problem can be partly solved.

### 22.5.3 Realizing the Integration Between Different Means of Transportation by the Improvement of Transit Network Planning

There are several points should be paid attention to in the work of planning transit network: The first one is to improve the efficiency of connection in one transportation mode and different lines between urban area and suburbs. The second one is to build the integration connection between different means of transportation by the high efficiency of three-dimensional junction terminals. And some other transfer tools should also be arranged nearby, such as bicycle rental and parking lot for cars, to connect rail transport and other environmentally friendly transportation efficiently.

### 22.5.4 Realizing the Integration Between Rail Transport Construction and Land Utilization and the Interaction of Them

We should learn the Tokyo and Hong Kong experience for reference in the land utilization work with rail transport construction. The key point is to bring enterprises in this process but not let the government rule it all. Meanwhile, the designation of transfer terminal is quite important in this task. The combination mode of transfer terminal and commercial center is highly recommended as it reflects the regional agglomeration effect under the integrated layout.

**Acknowledgment** The paper is supported by the key program from Beijing Planning Office of Philosophy and Social Science "Research on Sustainable Development of Urban Traffic in Beijing" (11ZDA03) and the program from Beijing Municipal Commission of Education "Research on the Interaction and Coordination between Urban Traffic and Urban Space Structure in Beijing" (B12H100040).

## References

1. Qin Ying-bing, Du Wen (2000) The effect of rail transportation on the urban structure. J Southwest Jiaotong Univ 35(3):284–287
2. Guo Peng, Xu Rui-hua (2006) The relationship between urban rail transit and urban development based on gravitation field model 24(1):36–40
3. Liu Yin, Lu Jian (2006) A study on appropriate size of urban rail transit systems. Urban Transp China 4(2):16–20
4. Guo Xiu-cheng, Lv Shen (2001) Spatial contribution of the rapid rail transit network in metropolises. Urban Dev Stud 8(1):58–61

# Chapter 23
# Research on Impact of Urban Rail Transit on House Price Along the Line: A Case Study on No. 1 Line of Tianjin Subway

**Suhong Teng, Jianwei Yan, and Shuang Zhou**

**Abstract** The paper choose the No. 1 line of Tianjin subway as a case to research the impact of urban rail transit on house price along the subway line. Authors collected the house price along the subway line and other data on No. 1 line of Tianjin subway for the few years upon and carried out research, trying to figure out what kind of impact that No. 1 line of Tianjin subway act on the house price along the line by using the methods of qualitative and quantitative analysis.

**Keywords** Rail transit • House price • No. 1 Line of subway • Tianjin

The development pattern of high density in Chinese cities decides that only urban passenger travel mainly relies on public transport, can traffic congestion, inconvenience travel, environment pollution and other contradictions be fundamentally solved and demands of health and orderly low carbon development be met. To this end, The State Council Executive Meeting held on October 2012 established the policy of priority development for public transport and placed public transport in the first position in urban transport development. The fast-paced modern city cannot be separated from convenient transport. Rail transit is incomparable than other transports with its fast, safe, comfortable, environment protection and other features, which plays an very important part in construction of metropolitan public transportation system. At present, about 30 cities in China's big cities have or are constructing rail transports. The rail transport to be completed will become the skeleton of public transportation system and important access corridor.

At present, the specific characteristics of large capacity and high reachability held by rail transit also make it possible to concentrate commerce, entertainment, business and cultural facilities, which promote the development of business and

S. Teng (✉) • J. Yan • S. Zhou
School of Architect, Tianjin University, Weijing Road, Nankai District, Tianjin, China
e-mail: sarah_teng@126.com; yanjw22@126.com; zhshuang66@126.com

H. Xia and Y. Zhang (eds.), *The 2nd International Symposium on Rail Transit Comprehensive Development (ISRTCD) Proceedings*,
DOI 10.1007/978-3-642-37589-7_23, 

high-dense development of real estate along the line, thus generating important and far-reaching impact on development and utilization of land along the line.

It is these external value-added features of rail transit that will help understand the interaction between transport and land utilization at the initial stage of construction of rail transit by researching the impact and rules of rail transit on utilization of land along the line as well as promote the integration development of rail transport and utilization of land along the line and marketization development of rail transport, thus achieving the win-win results of maximum social and economic benefits of rail transit construction.

Leaders in Tianjin City enhanced the development strategy of public transport, proposed to further play the roles of underground rail transit in urban public transport and construct urban public transportation system by regarding large-capacity rail transit as backbone and developed network planning of urban rapid rail transit (see Fig. 23.1).

At present, the subway lines being operated in Tianjin include No. 1, No. 2, No. 3 and No. 3 line, where, No. 1 line was put into operation in June 2006, which is the subway line with the longest operating time in central city of Tianjin. During this period, public service facilities of No. 1 line of Tianjin subway have been further improved and each operating indicator has reached a stable level. This article collected the house price along the road and other data on No. 1 line of Tianjin subway for the few years upon being put into operation and carried out empirical study for impact of No. 1 line of Tianjin subway on house price along the line by using the methods of the qualitative and quantitative analysis.

## 23.1 Relevant Research Review

In recent years, with a wide range of construction and operation of rail transit in big cities throughout the country, domestic scholars have conducted relevant research on urban rail transit on house price.

In qualitative research, He Fang and Wang Xiaoli said in (2004) [1] that rail transit would lift the value of real estate at its expansion area and surrounding real estate of station. However, with the increasing distance from downtown or station, the value-added amplitude of real estate of area adjacent to rail transit would decrease, but the noise and other adverse factors of ground rail transit would generate negative impact on values of adjacent real estate. As the positive impact was always dominant, the overall impact was to add the value of real estate. The higher reachability and the stronger regional advantages of regional rail transit station, the smaller impact on value of real estate would be. Zhang Xiaosong, Hu Zhihui and Zhen Rongzhou mainly analyzed the impact of urban rail transit on land utilization from such three perspectives as utilization intensity, utilization type of urban land and vigor of regions along lines by combining relevant cases both at home and abroad in (2006) [2].

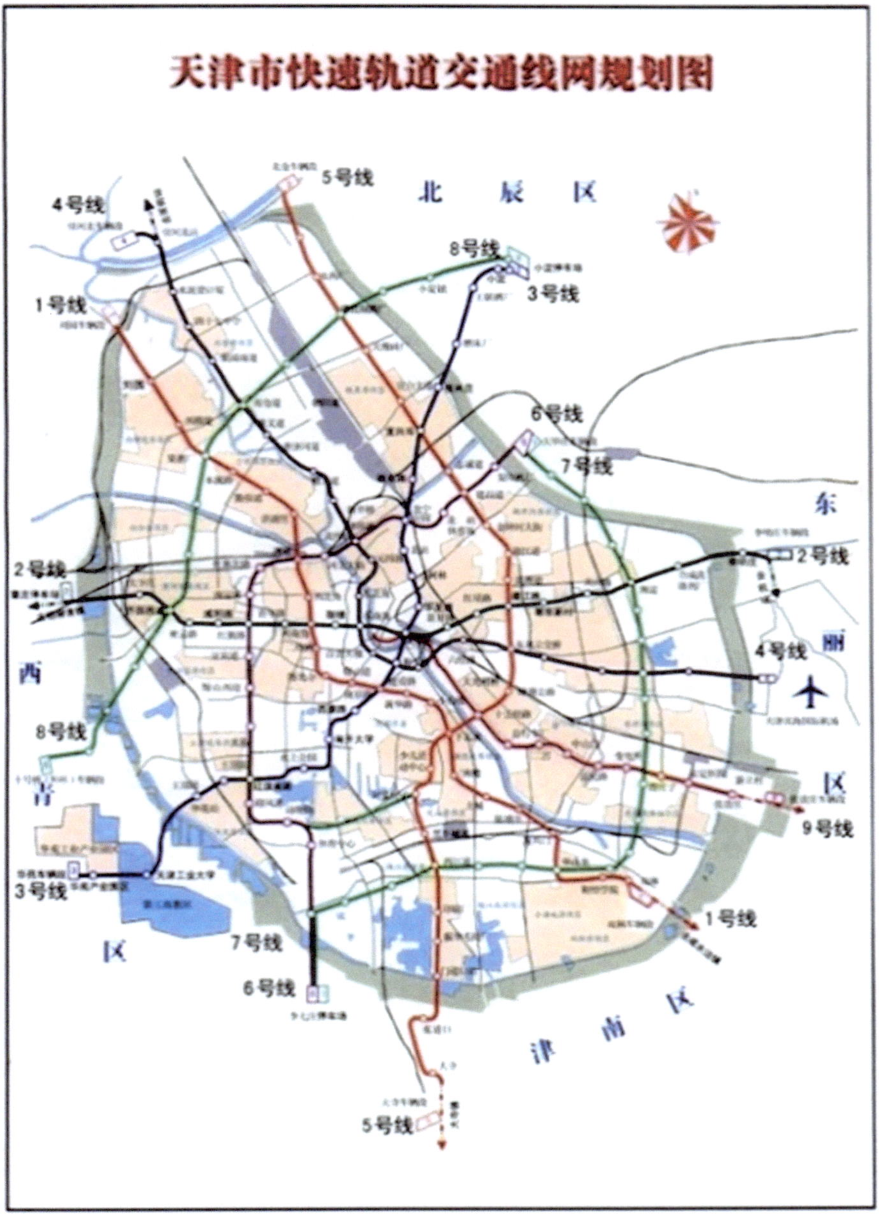


Fig. 23.1 Rail transit planning of Tianjin

In quantitative analysis, Wang Qiong [3] induced the housing price along the subway line and distance between houses and subway station to quadratic equation by regarding Shanghai house price as research object. Finally, the author concluded: In some special areas, the impact of rail transit was very weak due to the impact from regional cultures, region conditions and other factors; but after excluding individual abnormal value, rail transit had significant lifting effect on real estate of adjacent stations, that is to say, the closer of residence to rail station, the higher the house price would be. In addition, with the increasing distance from city center, the impact of rail transit on house price was also increasing.

Wang Xia, Zhu Daolin and Zhang Mingming [4] researched the impact of rail transit on real estate price by taking No. 13 line of Beijing light rail as example. Authors conducted analysis from features of housing price and land price near the light rail stations, non-station regional housing price and land price comparison by selecting the range of 4 km along No. 13 line as research region and came to the conclusion: The impact of light rail station on housing price at city center was smaller; with the increasing distance from city proper, the impact degree and range were increasing. In city center area and peripheral area, the impact on land price always concentrated on the range of within 1 km.

Liu Guiwen and Peng Yan [5] researched the impact of rail transit on house price by taking Chongqing City as an example. Authors measured the amount based on the housing data within 1.4 km from stations, conducted relevance analysis to date according to available degree of rail transit, convenience degree of reaching downtown and substitutability of other transports and by imitating the research results and using multivariate model for analysis and then obtained unigram by excluding independent variable with less relevance; finally, they came to the conclusion: The average house price decreased with the increasing distance from light rail station and there was rebound trend in price when reach to a certain distance. In addition, authors conducted verification through other comparative data and this unigram to analyze the locality and timing of impact of light rail to real estate price.

## 23.2 Research Methods and Objects

This article adopts the methods of combination of quantitative and qualitative analysis, including the following steps: First, determine all factors related to house price and collect relevant data and then obtain variables and their alternation results through alternation and screening to variables; second, conduct multivariate regression analysis to processed data to determine the mathematical relationship between the variables and independent variables and then establish a model impacting on house price to extract and analyze variables related to subway; next, determine its importance through factor analysis and other processes; finally, come to the conclusions [6, 7].

In the researches on the rail transit, domestic and foreign scholars often regard a certain range around urban rail transit stations as main impact area of development benefits based on their experience. Some relevant researches [1] show that European and American scholars generally take 0.5 and 0.8 km around stations and Japanese scholars generally take 2 km. There is a certain range of impact of urban rail transit for subway on the house price along the lines; its price is inversely proportional to the distance from station; the farther the distance between station and city center, the larger direct impact range on the price of adjacent housing will be. Therefore, 57 housing samples within 2 km of radius for stations along No. 1 line of subway are sorted out by collecting the transaction data of Tianjin real estate. There are many factors affecting house price. We believe that average selling price (¥/$m^2$) (Y) of real estate is affected by continuous variables and dummy variables. Continuous variables are the distance between the residential area and the nearest subway station (X1), plot ratio (X2) of residential area, greening rate (X3) of residential area, number of bus lines within 1 km around the residential area (X4) and distance between residential area and municipal commercial service center, etc. Dummy variables are the conditions that whether there are key middle schools within 1 km (X6), ordinary middle schools (X7), parks (X8), hospitals (X9) and whether the residential area is finely decorated (X10), etc. After screening, in this article, we selected variables X1, X2, X3 and X10 for detailed research. As the data of some residential areas in these 57 variables is unavailable, 52 samples are reserved after screening by excluding incomplete information [8].

## 23.3 Data Analysis

### 23.3.1 *Preliminary Qualitative Analysis*

The analysis of results of data statistics shows: The price decreases and then increases with the increasing distance between house and subway station. But the magnitude is not large and the difference is within 1,000 Yuan (See Table 23.1).

In consideration of many factors affecting housing price in downtown, all residential data is divided into two parts according to regionalism, including Nankai District and Heping District, district beyond Nankai District and Heping District. And then, conduct the statistics of change trend of average price with the distance from subway station.[1] Through calculation, the average price of housing in district beyond Nankai District and Heping District increases and then decreases with the increasing distance between housing and the nearest subway station; while the

[1] For the location characteristics of Tianjin City, Nankai District and Heping District through which subway No. 1 line passes belong to city center area with relative dense urban activities and higher population density; while the area beyond Nankai District and Heping District through which subway No. 1 line passes belong to the periphery of center city of Tianjin.

**Table 23.1** Relation between house price and distance from subway station

| Distance from the nearest station | House price (¥/$M^2$) |
|---|---|
| 0–500 M | 12,162 |
| 500–1,000 M | 11,451 |
| 1,000–2,000 M | 11,976 |

**Table 23.2** Relation between statistic house price and distance from subway station according to regionalism

| | House price (RMB/$M^2$) | |
|---|---|---|
| Distance from the nearest station | Nankai and Heping district | Other district |
| 0–500 M | 13,387 | 10,685 |
| 500–1,000 M | 11,666 | 11,042 |
| 1,000–2,000 M | 11,557 | 9,575 |

average price of housing in Nankai District and Heping District continuously decreases with the increasing distance between housing and the nearest subway station. However, the change in house price after the distance over 500 m is not big (Table 23.2).

As mentioned above, there are many factors affecting house price and preliminary analysis does not indicate obvious relationship between house price and distance from subway station. Next, we will further dig the factors affecting housing price through SPSS analysis software.

## 23.3.2 Regression Results and Analysis

### 23.3.2.1 Regression Results

The formula of average house price is obtained as follows:

$$Y = 7663 + 0.387X1 + 702.519X2 + 29.877X3 + 2038.751X10. \tag{23.1}$$

($R^2 = 0.371$, $P = 0.000$, $DF = 51$)

From the output of variance analysis table, the significance level of regression equation is 0.000, indicating outstanding significance of regression equation. The determination coefficient of model is 0.371, which can be explained as the changes of 37.1 %Y (see Table 23.3).

### 23.3.2.2 Result Analysis

As the number of newly built residences along the No. 1 line of Tianjin subway being operated from June 2006 is limited, the sample number is insufficient. Therefore, the determination coefficient of final regression results is only 0.371 and it is impossible to classify in accordance with such condition that whether the

**Table 23.3** Variance analysis results

| Anova[a] | | | | | |
|---|---|---|---|---|---|
| Model | Sum of squares | df | Mean square | F | Sig. |
| 1 Regression | 1.367E8 | 4 | 3.419E7 | 6.904 | .000[b] |
| Residuals | 2.327E8 | 47 | 4951619.671 | | |
| Total | 3.695E8 | 51 | | | |

| Coefficient[c] | | | | | |
|---|---|---|---|---|---|
| | Non-standard coefficient | | Standard coefficient | | |
| Model | B | Standard error | Beta | t | Sig. |
| 1 (Constant) | 7,663.673 | 1983.048 | | 3.865 | .000 |
| Plot ratio | 702.519 | 175.740 | .575 | 3.997 | .000 |
| Greening rate | 29.877 | 44.444 | .107 | .672 | .505 |
| Distance from subway | .387 | .609 | .081 | .636 | .528 |
| Decorated or not | 2,038.751 | 884.265 | .289 | 2.306 | .026 |

[a]Dependent variable: Average house price
[b]Predictive variable: (Constant), Decorated or not, distance from subway station, plot ratio and greening rate
[c]Dependent variable: Average house price

residential area is Nankai or Heping District (center district) and carry out further regression analysis. However, it is possible to make a preliminary judgment to the contribution rate of factors affecting house price and specific impact on house price according to the present regression data.

From the t-test results of coefficient output table, plot ratio (significance level 0.000) has maximum contribution to average price of real estate, followed by the condition that whether the residential area is finely decorated and then by greening ratio (significance level 0.505) and distance between house and subway station (significance level 0.528).

Meanwhile, from Pearson correlation analysis of independent variables and induced variables, the relevance between plot ratio and average price of housing is highest, which is 0.542, followed by the condition that whether the house is finely decorated and then by greening rate and distance from subway station.

If control other variables, carry out partial correlation test to X1 and Y and the obtained results are similar with Pearson correlation results (The correlation degree is only 0.92, $P = 0.000$) (Table 23.4).

According to the linear regression statistics and in combination with Pearson analysis, the dominant factor affecting the house price along subway line is plot ratio, while the impact of distance between housing and subway station on price is relatively low, which even can be deemed as there is no impact of distance between housing and subway station on price.

This result is unexpected, which is contrary to other's conclusions of research on impact of urban subway on residence. The reasons are probably as follows: During research period (2006–2010), No. 1 line is the only rail transit in Tianjin center city, which could not form a network and take the role of backbone of public transport, resulting in big difference on utilization and Passenger flow aggregation between No. 1 line and any single line of Beijing or Shanghai or other cities. This can be

**Table 23.4** Results of Pearson correlation analysis

| | | Average house price | Plot rate | Greening rate | Distance from subway station | Decorated or not |
|---|---|---|---|---|---|---|
| Average house price | Pearson correlation | 1 | .542 | −.309 | −.094 | .343 |
| | Significance (Bilateral) | | .000 | .026 | .508 | .013 |
| | N | 52 | 52 | 52 | 52 | 52 |
| Plot rate | Pearson correlation | .542** | 1 | −.588** | −.300* | .188 |
| | Significance (Bilateral) | .000 | | .000 | .030 | .183 |
| | N | 52 | 52 | 52 | 52 | 52 |
| Greening rate | Pearson correlation | −.309 | −.588** | 1 | .411** | −.383 |
| | Significance (Bilateral) | .026 | .000 | | .002 | .005 |
| | N | 52 | 52 | 52 | 52 | 52 |
| Distance from subway station | Pearson correlation | −.094 | −.300* | .411** | 1 | −.159 |
| | Significance (Bilateral) | .508 | .030 | .002 | | .260 |
| | N | 52 | 52 | 52 | 52 | 52 |
| Decorated or not | Pearson correlation | .343 | .188 | −.383 | −.159 | 1 |
| | Significance (Bilateral) | .013 | .183 | .005 | .260 | |
| | N | 52 | 52 | 52 | 52 | 52 |

**It is significantly relevant above the level of .01 (Bilateral)
*It is significantly relevant above the level of 0.05 (Bilateral)

clearly reflected from the statistics data of passenger flow volume of subway No. 1 line. Therefore, at present, the convenience provided by subway to the residents' travel is limited, and consequently subway is unable to significantly impact the house price.

Another possibility is that the impact of rail transit on real estate price has a strong timeliness. This is to say, so far, No. 1 line has been opened for more than three years, so there is a possibility that the opening time is too long to have a significant impact of subway on housing price.[2]

[2] Worth noting that average price of housing is different from land price. In addition to land price, the determination of price is also related to the residential quality, setting of customer base and other factors, which can be known from Pearson correlation analysis. Meanwhile, the own characteristics of residence sample, such as orientation, number of layers, whether it is forward delivery housing, whether it is rear cubicle and other aspects will impact the determination of housing price. As average price is applied in this research instead of price of single house, these factors are not very important for the research, which cannot be deemed to missing variables.

## 23.4 Conclusion

The preliminary analysis of the data statistics shows that there is a certain relationship between house price and distance between house and subway station, that is, price is decreasing with the increasing distance from subway station. In consideration of impact factors of downtown area to house price, all residential housing data is divided into two categories of urban center and non-center area. From statistics and calculation, the average price of house in non-center area increases and then decreases with the increasing distance between house and the nearest subway station; while, the average house price in center area continuously decreases with the increasing distance between housing and the nearest subway station, but when the distance is over 500 m, the changes in house price in these two areas are no longer significant.

The further regression analysis shows that the operating of No. 1 line of Tianjin subway within the time and section selected in this research has little impact on house price. Because of the limitation in the process of data collection, stage of research and other issues, this research fails to include all variables into regression; it is inevitably to have flaws in the corresponding results. However, it is possible to obtain the conclusion through qualitative judgment of regression results. The results are different from the conclusions from relevant researches in such cities as Shanghai and Shenzhen at the same period. From analysis, the main reasons are as follows: There is only one line of Tianjin subway being put into operation with too low line density to make its roles in urban public transport not comparable to the above cities and lack of enough impact on residents during their residence selecting. Thus, it can be seen that the impact of urban rail transport on the house price along the line require certain preconditions. To play a lot of impacts of rail transit on urban development, it is necessary to establish its main backbone of urban public transport system at first.

With the full opening of No. 2, No. 3 and No. 9 line of subway at the second half of 2012, rail transit in Tianjin City will usher in greater development and rail transit as main corridor of public transport will gradually emerge. High degree of aggregation of passenger will promote the development and utilization of land for rail transit and its surrounding and land along the rail transit. The impact of urban rail transit on house price along the lines will be more significant. These changes will promote us to conduct a further research and explore the interaction impact relationship between them under new trend, thus seizing the changing rules and reasons to better guide actual work.

## References

1. He Fang, Wang Xiaoli (2004) Impact of rail transit on price of real estate along the rail transit [J]. Real Estate Mark (9):13
2. Zhang Xiaosong, Hu Zhihui, Zhen Rongzhou et al (2003) Analysis of impact of urban rail transit on land utilization [J]. Res Urban Rail Transit 26(6):24
3. Wang Qiong (2008) Impact of urban rail transit on price of real estate along the rail transit [J]. Res Urban Rail Transit 2:10

4. Wang Xia, Zhu Daolin, Zhang Mingming (2004) Impact of urban rail transit on price of real estate along the rail transit – the case of No. 13 line in Beijing Light Rail [J]. Urban Issues (6):39
5. Liu Guiwen, Peng Yan (2007) Impact of rail transit on residential real estate price – the case of Chongqing City [J]. Urban Issues (1):65–69
6. Donald WM, Ronald DS (1975) Ridge regression in practice. Am Stat 29(1):3–20
7. Wang Donghui, Guo Zujun (2006) Application of Hedonic Model in analysis of commercial housing price – empirical research of case of Shenzhen [J]. Special Zone Econ 215(12):354–355
8. He Jianhua, Zheng Siji (2004) Can newly built subway enhance house price? – The case of No. 13 line in Beijing Light Rail [J]. Urban Issues (11):36–38

# Chapter 24
# Catalytic Effect of Underground Commercial Space of City and Its Design Method – Take Underground Mall of Hangzhou Wulin Square as an Example

**Zitao Zhang, Haishan Xia, and Jing Sun**

**Abstract** With the constant expansion of the city, the urban rail transit rapid development, but the development of underground space and rail transit construction is not at the same time, this problem seriously restricting the city's comprehensive development process. The paper clues to urban catalyst theory. Taking Hangzhou Wulin Square underground mall project for example, reveals the importance of rail traffic to drive around the development of underground space in the makeup of the district insufficient, and the urgency of underground commercial space of urban development. To explore rail transportation and underground commercial space contact, research planning and design guided by the rail transit urban underground commercial space, and the catalyst effect caused by the underground commercial space.

**Keywords** Rail transportation • Underground commercial space • Catalyst effect • Planning and design

## 24.1 Introduction

Currently, there are some glaring problems of the domestic urban development, the imbalance between urban spatial structure, indiscriminately expanding the scale of construction, clutter image, heavy traffic, and so on. The development of high strength and fixed mode is one of the reasons. Update the current model of development, develop the floor space and increase the utilization of underground

Z. Zhang • H. Xia (✉)
School of Architecture and Design, Beijing Jiaotong University, 100044 Beijing, China
e-mail: 369815332@qq.com; hshxia@bjtu.edu.cn

J. Sun
Beijing Urban Engineering Design & Research Institute CO., LTD, 100037 Beijing, China
e-mail: jingzi588@sina.com

H. Xia and Y. Zhang (eds.), *The 2nd International Symposium on Rail Transit Comprehensive Development (ISRTCD) Proceedings*,
DOI 10.1007/978-3-642-37589-7_24, © Springer-Verlag Berlin Heidelberg 2014

space simultaneously. And there's correlation between the utilization of underground space and rail traffic. The theme of this article is the catalytic effect of the underground commercial space and the design of it, so it will start from the relationship between the urban rail transit and underground commercial space:

## 24.2 The Relationship Between the Urban Rail Transit and Underground Commercial Space

### *24.2.1 The Urban Rail Transit Give Guidance to the Development of the Underground Commercial Space*

Throughout the development of the global rail transportation, most of the major cities in the world have a more mature and complete rail transportation systems, a variety of types, including subway, light rail, suburban rail, streetcar, maglev, and so on. Subway started late in China. The first subway opened in 1969, accordingly, in 1863 in Britain, in 1892 in U.S.A., in 1927 in Japan. But in this short span of a few decades, the length of the subway of Beijing and Shanghai have already over the cities of the developed countries, located in the world's first and third. It shows that the urban rail transport in China is in a period of rapid development.

Annual passenger volume of the metro line per km up to more than 100 million people, the maximum up to 12 million people, the transport capacity is far more than the bus. Urban rail transit can transport large passenger in a short period of time. According to statistics, the metro can deliver 17–20 % of the full-day passengers in 1 h, 31 % in 3 h [1]. The huge flow of people bring huge business opportunities. The status of rail transport in urban underground space decided it will guide the development of the underground commercial space.

### *24.2.2 Underground Commercial Space Become the New City Catalyst*

There is a "Bucket theory" in management, it means the length of the "short board" determine the overall level of development in the whole process. Urban underground space is gradually filled by subway, and the underground business development is relatively slow. It will not be conducive to the comprehensive development of the urban space, and also will affect the continued development of the rail transportation.

With the improvement of quality of life, people increasingly want to shop in a comfortable and convenient environment. A large percentage of the subway passengers is office workers which don't have enough time and energy to go shopping.

**Table 24.1** Expectations of passengers for offering a variety of service projects in the area of metro station

| Project | Bank operation | Utility bills collection | Small supermarket | Public telephone | Fast food | Transport advisory & traveling consultation | Photocopying & fax |
|---|---|---|---|---|---|---|---|
| Number of people | 783 | 558 | 563 | 420 | 293 | 251 | 243 |
| Percentage | 66.3 | 47.3 | 47.6 | 35.6 | 24.0 | 21.3 | 20.1 |

**Table 24.2** The demand of passengers to purchase a variety of commodities in the area of metro station

| Project | Umbrellas & raincoats | Duffel bag & luggage cart | Newspapers & magazines | Gifts & flowers | Food & drinks | Kinds of small commodities |
|---|---|---|---|---|---|---|
| Number of people | 565 | 509 | 493 | 443 | 358 | 361 |
| Percentage | 47.9 | 43.1 | 41.8 | 37.5 | 30.3 | 30.1 |

Note: The above information is summarized in accordance with the demand by decreasing order. In addition, also include clothes dry cleaning, digital audio products, clothing, medicines and other services and commodities demand, and the different nature of the site led to the dissimilarity of the quantity and variety of the demand [2]

So they most hope to take into account their own lives when they take the subway. Tables 24.1 and 24.2 reveals the demand for daily necessities and services during the subway time.

The big commercial demand of people in the area of metro station make underground commercial space become the new city catalyst. Namely, the underground business can bring the development of the whole underground space as catalyst.

## 24.3 Catalyst Theory

### *24.3.1 The Concept of Catalyst*

Urban catalyst is an element shaped by the city. It is formed in an urban environment (the experimental environment), and shape the urban environment conversely with the purpose of updating the urban structure in gradual and continuous way [3]. The most important thing is that the catalyst is not just the end product of a separate, more an element which can inspire and guide subsequent development. Its effects are reflected not only in the economic activation, and in the improvement of urban development, as well as the revival of urban vitality [4]. Underground commercial space of the metro stations, as the new catalyst of urban, repair the disordered state of the development of the city, and become the tipping point of a new development model of the city. The impact of its surrounding environment or things is called "catalyst effect"[5].

### *24.3.2 The Catalytic Effect of Rail Transportation and Underground Commercial Space*

The development of the underground commercial space which is under the guidance of the rail transport is the development model which is an organic integration of functional elements of urban commercial and transport. On the one hand, subway stations gather large crowds of people, which is the source of the customers for commercial space. The special advantage of the traffic, reachability, allows the passengers to complete a variety of trip purpose in the area of metro stations (such as shopping, entertainment, and business). It greatly reduces the city ground traffic flow, and eases regional traffic pressure. On the other hand, the good functioning of the commercial facilities will provide subway passenger traffic, promote the development of the subway conversely, and achieve complementary to the passenger flow between the subway and the underground commercial space. This kind of interactions between urban functions is what we called "catalyst effect". It can effectively improve the operational efficiency of the urban space, promote the high-efficiency function of the city, and make the underground commercial space which guided by the rail transport become the brisk urban active node with local characteristics, and then stimulate urban vitality [6].

## 24.4 The Planning Design of the Underground Commercial Space

There are many aspects of the catalytic effect of the underground commercial space, and we will take Hangzhou Wulin Square underground mall project for example to discuss the planning design of the underground commercial space in the macroscopic and microscopic perspectives.

### *24.4.1 At the Macro Level of Design – The Macro Utility of Catalyst*

#### 24.4.1.1 The Overall Location of the Project

Wulin business district, with Wulin Square at the core, is the core of the traditional business district of Hangzhou, with the intersection of Metro Line 1 and Line 3, arounded by several large stores (such as Hangzhou Department Store, Hangzhou Tower, and Intime Department Store) and several important cultural facilities (such as Zhejiang Exhibition Hall and Hangzhou Grand Theater), have the unparalleled location advantage.

#### 24.4.1.2 Protective Effect of the Catalyst

As an urban catalyst, Wulin Square underground mall play the role of protecting cultural atmosphere, and maintenance of city scenes. It based on integrated into Hangzhou environment, the background cultural, save the original building and the shape of the Square. The contrast of the old and new buildings, a blend of old and new culture, underground mall not only can preserve the cultural, but also can let the people in the square feel the changes of time and space.

#### 24.4.1.3 Promotion Effect of the Catalyst

Wulin Square is the elements of the previous era, has been unable to meet the needs of modern urban development process before the transformation. As an urban catalyst, it can improve the regional business status, activate the vitality. Enhance the brand influence can increase the force of the catalyst, the construction and operation of the underground mall improve the potential of an old brand like Wulin Square.

#### 24.4.1.4 Repair Effect of the Catalyst

The renovation of old facilities performance at two levels while the construction of the underground mall of Wulin Square. The first is the renovation of the reality objects such as building, structure, and materials. The second is the advocation and updating of the culture of the original plots. The parking area, provided by the construction, contribute to the improvement of regional transportation, and also improve and enrich the metro passenger flow line, boost the comprehensive development of rail transportation.

#### 24.4.1.5 Creative Effect of the Catalyst

Wulin Square underground mall, as an urban catalyst, also play the role of creativity and innovation. It introduce a green ecological shopping environment to create a pleasant plaza space, and provide people a stylish, low-carbon, and leisure way to shop [7].

### *24.4.2 At the Micro Level of Design – The Micro Utility of Catalyst*

#### 24.4.2.1 Security

The primary consideration in the design of the environment of underground commercial space is the safety problem. Only to ensure the security, to ensure the

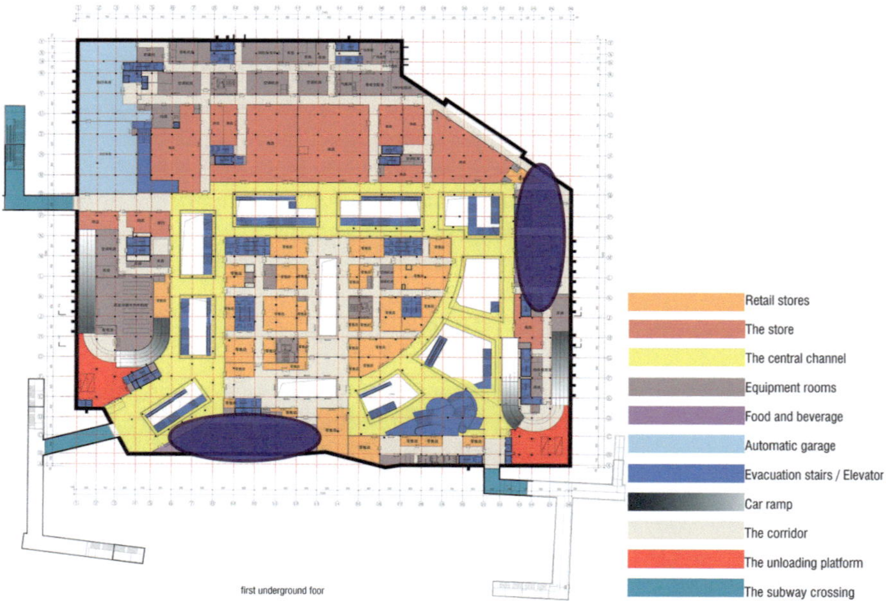


**Fig. 24.1** The kitchen and dining area on the underground floor of Wulin Square underground mall

stability of the catalyst. Due to the closure property, it's very easy for people to lose a sense of direction in the underground space when the fires or other safety incidents happen. The outside light cannot transmit into the underground space. The panic created by the dark environment under the accident will easily lead the stampede. So the security design of the underground commercial space of the metro stations should be taken seriously. The fire-retardant materials were chosen as much as possible in the interior decoration of Wulin Square Underground Mall. And the kitchens which use open flame were arranged at the end of the area to monitor and manage unified. As shown in Fig. 24.1.

#### 24.4.2.2 Comfort

In general, there's no natural lighting in underground space, so it also needs active illuminations even during the day. It makes it difficult to have contact with nature. This project go from underground space to semi-underground space by using these sunken plaza. And it not only can introduce the light, but also the natural landscape. The underground space become an open node which dialogue with the environment. It totally changed the impression of depression. As shown in Fig. 24.2.

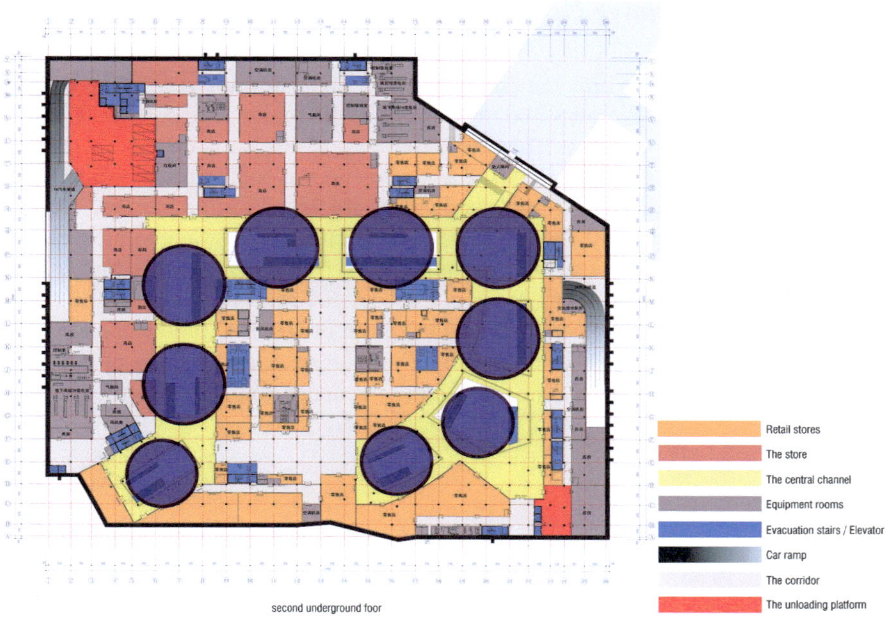


**Fig. 24.2** The sunken plaza on the second floor underground of Wulin Square underground mall

The activities of people are more restrictively, because of the strong closure property and the character of no natural light of underground space. But to use the method of visual signal with lights to guide the flow will solve the problem of the directionless space. The improvement of comfort will improve the shopping desires. This make the catalyst effect of the underground shopping mall get more comprehensive results.

#### 24.4.2.3 Construction

The construction of underground space which with high difficulty, high cost and long duration, is irreversible. So underground space is very difficult to demolish or rebuild like the building on the ground after the completion of it. The utilization of underground commercial space will reduce when the space coincide with the subway. In consideration of sound insulation, the difficulty of the construction will increase when the commercial space parallel to the rail transport. If the depth of Wulin Square Underground Mall is 2 m deeper than the original, the total cost of the parking facilities and the construction will increase by 5,400, and the construction period will be extended for 2 months [8].

### 24.4.3 *Collaborative Development*

The subway station is the important node of the city. The development of the subway area will improve the overall value of the land in the region, and achieve a higher economic and social benefits. Therefore, the corporation should make a subway operators to form a virtuous circle. The emergence of underground commercial facilities in the metro area, to some extent, alleviate the economic pressure brought by the high cost of subway construction. If operating properly, it can create more social and economic benefits, and promote the development of the surrounding business district, the Hong Kong MTR is a successful case. The Hong Kong MTR, the world's most profitable metro, make a net profit of HK $4.0 billion annually [9]. The fare revenue accounted for only a small proportion, most of these profits were from a variety of commercial forms, including the underground commerce.

## 24.5 Conclusion

The jointly of Urban Rail Transit and Underground Commercial Space is not an accidental phenomena. Taking the abundant resources of Urban Rail Transit to exploit Underground Commercial Space, that has a positive effect to stimulate urban vitality, to improve of urban ecological environment, to ease congestion, to ample commercial facilities. As a new city catalyst, Underground Commercial Space is a new tipping point of urban development. We must seize this opportunity, taking use of the good resources of Underground Space, integrating of urban space, improving urban functions, efficient using of land, thus bring people more convenient lives.

To the planning and design of Underground Commercial Space, we recommend to do simulation of the traffic organization, and select the best program by simulation and comparison. Besides, in the aspect of safety, comfort and construction, we must consider the characteristics of underground buildings, and select the program suiting to local conditions. In addition, we should consider the connection between the planning district and the subway station. Combine the planning of upground and underground commercial space by using the vision of forward-looking and developing. In that way, we can creating underground space environment that is safe, comfort and convenient for passengers and consumers.

**Acknowledgment** This research was supported by: National Natural Science Foundation of China (51078022); Doctoral Fund of Ministry of Education of China (20090009120013); Beijing Urban Engineering Design & Research Institute Co., Ltd, Underground Public Space Optimized Simulation Research – Based on Digital Technology (CJ2012-47).

## References


1. Xia Haishan, Li Xiaobei (2012/10) Study on intensive design of urban rail transport hub from the perspective of low-carbon [A]. In: The proceeding of LTLGB2012 [C]
2. Ni Xudong Tao Xiaoma (1999) Commercial value of metro station underground space [J]. Urban Mass Transit 02:12–15
3. Attoe W, Logan D (1988) American urban architecture catalysts in the design of cities [M]
4. Yu Yong, Lu Jiwei (1988/11) Urban catalysts and comprehensive development of subway station [J]. TIME + ARCHITECTURE 04:53–56
5. Jin Guangjun (2006) Catalytic effect of urban planning [J]. Planners 10:22
6. Wang Qianyun (2009) Study of the underground commercial space at subway stations in China [D]. Master degree thesis of Southwest Jiaotong University
7. Zhang Mu (2005) Preliminary discussion on urban catalyst theory [J]. Theory Monthly 03:114–115
8. Jin Bei, Zhu Yanlin (2011) The development and utilization of urban underground commercial space [J]. DaZhongWenYi 6:211–212
9. Chen Jian (2010) Composite design of the subway station underground commercial space [D]. Master degree thesis of Southwest Jiaotong University

# Chapter 25
# Development Strategies on Peripheral Areas of Xuzhou Rail Transit Stations

**Jinghua Li, Bo Yang, and Chunhua Liu**

**Abstract** With the gradual expansion of the city, urban traffic congestion is becoming progressively serious. As an important part of the urban public transport, rail transit plays an important role in ease traffic congestion. Xuzhou, the central city of Huaihai Economic Zone, is facing rapid traffic growth; simultaneously, the city also has the economic conditions to declare the construction of rail transit. In this paper, firstly the necessity of the construction of rail transit in Xuzhou is demonstrated; then the development orientation of the area near the railway station is analyzed, railway stations are classified according to function orientation, transport function and the city zoning it belongs to, finally, the adjustment strategies regarding the land use, development functions and development intensity of the peripheral areas of rail transit stations are proposed.

**Keywords** Xuzhou City • Rail transit stations • Land use development • Strategy

## 25.1 Introduction

Xuzhou is one of the three metropolitan area centre cities in Jiangsu Province, central city in Huaihai Economic Zone, the important transportation hub in the whole country. Along with the increasingly fast urbanization process, the rapid development of city economy, the sharply increased city population; the city scale expands continually, trips of personnel and material exchanges occur frequently, there is a dramatically growth in traffic demand, urban traffic faces with a critical situation.

J. Li
Xuzhou Urban Planning Bureau, Xin'an Road, New City, 221000 XuZhou, China

B. Yang (✉) • C. Liu
Xuzhou Planning and Design Institute, Xin'an Road, New City, 221000 XuZhou, China
e-mail: yang-bo85@163.com

H. Xia and Y. Zhang (eds.), *The 2nd International Symposium on Rail Transit Comprehensive Development (ISRTCD) Proceedings*, DOI 10.1007/978-3-642-37589-7_25, 

Currently, Xuzhou has met the requirements to declare the rail transit construction in indexes such as Local Finance General Budget Revenue, Gross Domestic Product, city population, passenger volume in the planning line. The initial tentative plan is to build three track lines by 2020: line No. 1, No. 2 and No. 4. According to the experience of rail transit construction in domestic and foreign cities, the construction is bound to affect the land use along the stations, and even further, the development direction and the spatial structure of the city enormously. Thus, along with the orbit construction, adjustment of the land use, promotion of the land development surrounding the stations is becoming one of the important contents of the construction.

## 25.2 The Necessity of the Construction of Rail Transit in Xuzhou

Traffic, as the artery of city production and city life, is one of the most important infrastructures to maintain the vitality of a city, poor traffic with restrict the development of the urban economy greatly. To achieve sustainable urban transport, it is an inevitable trend to develop public transport network in which the track transit is the backbone, metro and light transit with large, medium capacity are actively introduced. The necessity of the rail transit construction in Xuzhou is mainly reflected in the following aspects:

### *25.2.1 Contradiction Between Supply and Demand, Requirement of Track Transit*

At present, the city zone of Xuzhou, especially the central area of the city, traffic conflicts have been quite prominent. Traffic flow is highly concentrated in the city centre area, road sections are high saturated, and in and out traffic is impeded. The city's main roads have high traffic load, the important roads such as Zhongshan Road, Jiefang Road, Fuxing Road and Jianguo Road; all have a peak hour motor vehicle traffic volume exceeding 3,000. City public transportation is in shortage of transport capacity and low level of service, some bus lines on the main roads in the city centre area have daily passenger traffic of more than 80,000 people. The poor traffic order is making urban transport operation struggling, worsening traffic environment.

### *25.2.2 Future City Passenger Delivery Demand, Requirement of Transit Traffic*

Along with the development of the New City District and Bashan District, the average travel distance of city residents will change from 3–4 km to about 6.5 km, meanwhile, residents travel time distribution is becoming increasingly concentrated and the urban transport intensity in peak hours is continually enhanced.

There are obvious limitations in the existing ground transportation in Xuzhou: it is difficult to adapt to the long distance, centralized transportation of future residents, it is also unable to meet the quality requirements of transportation, such as on time, fast, safe and comfort. Rail transit runs faster, with an average running speed of 35–40 km/h, it has obvious advantages compared with conventional bus running speed, which is 15–20 km/h. Rail Transit can well adapt to the long distance travel trends of Xuzhou residents, using rail transit, residents spend less time in a 6.5 km travel than what they spend in a 3–4 km travel in conventional bus transport. Rail transit has powerful transport capacity, one way peak hour transport capacity can reach 40,000–60,000 passengers, which is seven to ten times to the ground transportation. Thus it is possible to provide a more stable and reliable transportation service during peak hours to accommodate the centralized trend of urban trips. In addition, rail transit is safe and comfort, improving the quality of residents travel prominently.

### 25.2.3 In Favor of Shorten the Spatial Distance, Promoting the Formation of City Structure

Xuzhou city structure will develop in accordance with "Double Hearts, Five Areas" layout mode, along with the continuous expansion of the city area, the continued development of conventional public transport will cause a substantial increase in operating costs and will be overwhelmed by the enormous traffic demand. To promote the connections between central area and other areas, it is an important task only for the large volume, high speed rail transit. Thus it is necessary to build rail transit as soon as possible to shorten the spatial distance between city areas and form a reasonable spatial structure.

### 25.2.4 In Favor of Intensive Use of City Land, Increasing Land Use Efficiency

Xuzhou will move from epitaxial extensive development mode toward connotation intensive development mode, through the construction of rail transit and large-scale public facilities, it is possible to adjust land use structure and improve the intensity level of land use gradually, improving the efficiency in the use of land resources [1].

### *25.2.5 In Favor of Maintaining Safe and Harmonious Ecology and Residential Environment*

In order to improve the quality of the city environment, Xuzhou has taken many effective measures such as giving priority to public transportation, increasing investment in public transportation projects and encouraging people to take buses. However, Xuzhou public transport system can still be considered as a routine ground transportation whose level of service is constantly constrained by the state of the roads. Although the municipal government invested a lot, there is no distinct effect. Construction of rail transit lines with national energy strategy and provides efficient, environmentally friendly and safe city transport, is an effective measure to achieve the sustainable development of the environment.

## 25.3 The Overall Development Orientation Along the Track Traffic Line

### *25.3.1 The Overall Function Positioning of the Rail Transit Line No. 1*

Rail Transit No. 1 is the important track connecting the Old City Area and the High Speed Rail Area, the important channel contacting the old town core area and the high-speed railway station. It plays important role in guiding new area development and promoting the transformation of the old town area. The main stations of Track No. 1 are Hanwang New Town Station, Pengcheng Square Station, Railway Station and High-Speed Railway Station.

### *25.3.2 The Function Positioning of the Areas along Rail Transit Line No. 1*

Considering different land us status and panning constraints in the planning area, according to the analysis results of the maturity of existing construction and the intensity of land use, the land long the track is divided into three categories: arrangement coordination area, appropriate coordination area and part coordination area. Corresponding area function positioning and adjustment strategies are proposed according to different categories in order to guide following detailed planning. See Table 25.1.

**Table 25.1** Area function positioning

| Area category | Area characteristics | Area location |
|---|---|---|
| Arrangement coordination area | New Development Area<br>Moderate or Low Intensity Development | HanWang New Town, Xing Shanzi, Financial Street, No. 1 Road, Latitude 7 Road, High Speed Railway Location |
| Appropriate coordination area | Semi-mature Built-up Area and Aging Area<br>Medium to High, Medium or Low Intensity Development | Gongnong North Road, People's Square, Station East Square, Tongshan Road, Lion Mountain, Qingfeng Road |
| Part coordination area | Mature Built-up Area<br>High or Medium to High Intensity Development | Xi'an Road, Pengcheng Square, Cultural Palace, Railway Station |

### 25.3.3 *The Function Positioning of the Stations Along Rail Transit Line No. 1*

#### 25.3.3.1 Classified According to the Function Positioning of the Peripheral Area

Station function positioning and rank are categorized according to the area of different land uses and the traffic volume generated from it within the service area of each station. In the process, land is divided into three categories: Commercial (C), Residential (R) and others.

Commercial-office Category: commercial, office and public facilities are the main land use within the service area of the station, that is, C-class land use area and the traffic volume generated by it accounts for a high proportion.

Residential Category: large scale residential area is the main land use within the service area of the station, that is, R-class land use area and the traffic volume generated by it accounts for a high proportion.

Industries Category: large scale industrial production area is the main land use within the service area of the station, that is, other land use area and the traffic volume generated by it accounts for a high proportion.

Transportation Category: there are train station, high-speed rail station, bus station or other external transportation facilities within the service area of the station, the traffic land use and the traffic volume generated by it accounts for a certain proportion.

#### 25.3.3.2 Classified According to the Traffic Function of the Station

General Station: relatively good traffic conditions, is usually located in community level business centre; using walking, cycling or bus as the main access mode, without considering the same stage transfer between two-line or multi-line track.

**Table 25.2** Classification of the stations along rail transit Line No. 1

| Name | Commercial: residential: other (Area ratio %) | Commercial: residential: other (Travel volume ratio %) | Peripheral area function | Classification | |
|---|---|---|---|---|---|
| | | | | Traffic | Zoning |
| HanWang New Town | 6.3:25.8:67.9 | 19:71:10 | Residential | Ordinary | CLD |
| Xing Shanzi | 5.7:47.7:46.6 | 18:75:7 | Residential | Ordinary | CLD |
| Financial Street | 37.6:57.3:5.1 | 54:43:3 | Transportation | Ordinary | CLD |
| Road of W&P | 12.5:84.9:2.6 | 15:83:2 | Residential | Ordinary | CLD |
| People's Square | 27.2:72.7:0.1 | 40:59:1 | Residential | Hub | CBD |
| Xi'an Road | 30.2:68.8:1.0 | 46:53:1 | Residential | Ordinary | CLD |
| Pengcheng Square | 57.2:31.2:11.6 | 82:10:8 | Commercial-Business | Hub | CBD |
| Cultural Palace | 46.7:40.9:12.4 | 72:20:8 | Commercial-Business | Ordinary | CBD |
| Railway Station | 31.2:36.2:32.6 | 16:14:70 | Transportation | Hub | CBD |
| Station East Square | 19.9:77.0:3.1 | 34:46:20 | Transportation | Ordinary | CBD |
| Tongshan Road | 15:23.1:61.9 | 26:40:34 | Industry | Ordinary | CWD |
| Lion Mountain | 19.7:63.7:16.6 | 35:57:8 | Residential | Ordinary | CLD |
| Qingfeng Road | 27.2:70.7:2.1 | 61:38:1 | Residential | Ordinary | CLD |
| NO. 1 Road | 18.5:77.6:3.9 | 21:77:2 | Residential | Hub | CWD |
| Latitude 7 Road | 31:68.4:0.6 | 42:57:1 | Residential | Ordinary | CLD |
| High Speed Railway Station | 43.8:11.7:44.5 | 40:9:51 | Transportation | Ordinary | CWD |

Hub Station: having high visitor density as the primary node of the city, is required to consider the same stage transfer between two-line or multi-line track.

#### 25.3.3.3 Classified According to the City Function Area It Belongs to

CBD Area: Central Business District, gathering ground a large number of commercial, office and cultural facilities, facility land use accounts for a high proportion; has intensive population and activities, large scale public facilities often serve the whole district or even the city area.

CLD Area: City Living District refers to the large-scale residential areas located in suburb or urban area, the construction of track and supporting facilities mainly considers the residents' travel demands.

CWD Area: Emerging sub-center in the outskirts of the city area, usually has relatively good land use conditions, rail transit station and land use are designed according to the specific features of the corresponding area (Table 25.2).

## 25.4 Development Strategies on Peripheral Areas of Rail Transit Stations

### *25.4.1 Land Use Adjustment of the Land Surrounding Stations*

The development of the land along the track is in accordance with the area function positioning where it is located in, corresponding adjustment strategies are provided according to different types of areas (Fig. 25.1).

Arrangement Coordination Area: pay attention to adjust land use, improve development intensity and give priority to the distribution of urban function facilities. Rationally organize the land layout and density on both sides of the track. Within 500 m on both sides of the track, it is a high-density multi-functional mixed land use area (such as residential, commercial, office, etc.), the density gradually reduces and the land use changes toward the periphery. This type is mostly new Transit- Oriented Development area where various public transports are well linked.

Appropriate Coordination Area: the land development is mainly land integration and comprehensive transformation, therefore should take full account of the current situation and adjust land use and development intensity moderately. The development intensity should maintain appropriate balance with current regulation index. Pay attention to land use replacement to raise the value of land resources. Adjust

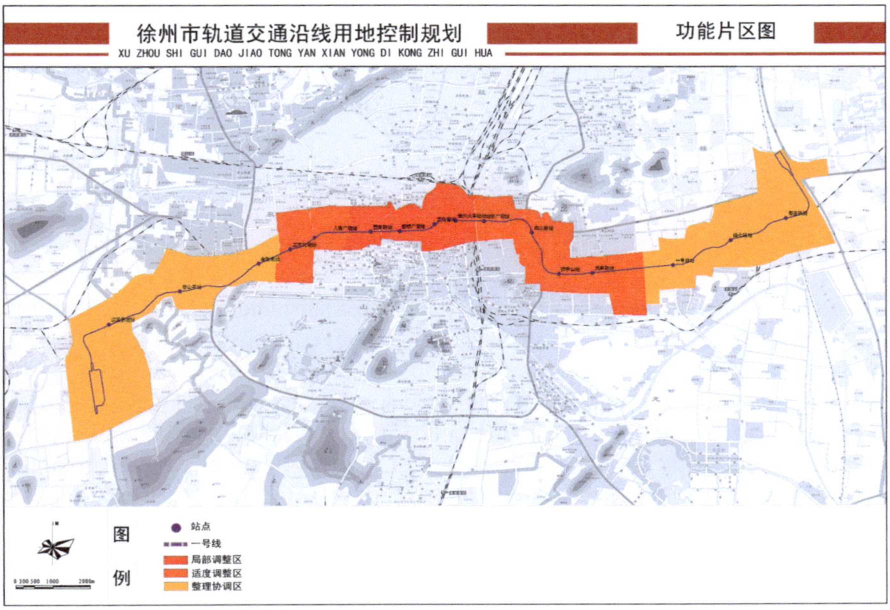


**Fig. 25.1** Area function positioning

land use and development intensity in a range of 500 m on both sides along the track, especially 200 m. Coordinate relations between retention areas and improved areas and improve access facilities connecting public transportation and walking system. Form the TOD mode gradually.

Part Coordination Area: adjust this area prudently and optimize the function moderately. Protect the human spatial scale of the built-up area and focus on the access facilities connecting subway, bus and walking system. It is possible to undertake large-scale underground space development to make up for the lack of development intensity on the ground.

### 25.4.2 Development Function Adjustment of Different Types of Stations

TOD (Transit-Oriented Development) communities assume different functions in different locations of the city, thus have corresponding different organization models and layout patterns. Although, considering the implementation of the plan, the scope of the study is roughly a ribbon area, certain distance away from both sides of the track. But, to take full account of the locations, the line functions, the situation of the surrounding land, to develop more targeted, localized adjustment program, it is reasonable to study the punctuate area surrounding the station [2]. Combined domestic and international experience, the stations are classified into four categories: Commercial-Office Station, Residential Station, Traffic Hub Station and Industry Station, adjustment strategy principles are proposed according to each category, as shown in Table 25.3.

### 25.4.3 Development Intensity Adjustment of Different Types of Stations

Theoretically, the development intensity of the area surrounding rail transit stations can be controlled according to ideal model. However, in fact, stations located in different locations of the city show large differences in development intensity. On the one hand, total amount of development and the overall intensity can be determined according to different types of stations; on the other hand, under the premise of protecting the public facilities, public space and urban culture, aesthetics and ecology, each station should comply with development scale distribution and hierarchical intensity control based on the principle of gradient descent [3].

Residential Station: is suitable for the medium intensity development of living and complementary service facilities. The development intensity control takes gradient decreasing. The community center surrounding the station is appropriate for high density development; the recommended floor area ratio is about 2.0. The

**Table 25.3** Station adjustment strategy

| Categories | Characteristics | Adjustment strategy |
|---|---|---|
| Commercial-Office Station | Located in the area where the commercial, administration, office and integrated transportation hub are distributed. Has high population density, location advantage and agglomeration effects | Improve the land use such as large-scale commercial and financial, administrative office, cultural, entertainment, service and residential. Provide more mixed-use land as well as public open space serving a wide range of the city. Establish comprehensive pedestrian system; ensure the adequate and comfortable space for walking. Ensure the convenient transfer of various modes of transportation and provide a seamless connection with surrounding facilities. Promote high density development on the land surrounding the station |
| Residential Station | Located in large residential area | Deploy the appropriate scale commercial facilities as well as parks and public space serving the community. The land use supports a variety of function to provide more jobs. Create diverse residence types. In the spirit of convenience, the entrance of the rail transit station corresponds to the main travel channel and public facilities of the residential area. Create comprehensive, beautiful walking system. Conduct high-medium density development on the land surrounding the station, mainly residence and supporting service facilities |
| Traffic Hub Station | Has train station, high-speed rail station, coach station or other external transport facilities within the service area of the station | Provide more functional facilities and mixed-use land. Ensure the convenient connection between rail transit station and external transport facilities such as railway station, high-speed railway station and coach station. Create comfortable, comprehensive walking system. In the case of land conditions permit, set the sunken city square, and reasonably organize the links between rail transit stations, public parking space and external traffic station through green open space |

(continued)

**Table 25.3** (continued)

| Categories | Characteristics | Adjustment strategy |
|---|---|---|
| Industry Station | Large-scale industrial production workplace within the service area of the station | Improve service facilities and provide moderate commercial leisure facilities. Pay attention to the creation of beautiful, personalized landscape environment and public walking space surrounding the station. Make sure the road space meet the various moods of transportation, ensure the connections between different levels of public transportation facilities and meet the distribution of large-scale flow of people |

development intensity of residential area outside the centre can refer to the ordinary multi-storey or high-rise residential area, the recommended floor area ratio is more than 1.5. Low-rise, low-density residential development is not recommended.

Commercial-Office Station: is usually city centre or district centre, appropriate to adopt the highest development intensity. According to the level of the centre it relies on, the development intensity varies in size. Based on the relationship between the land use and the maximum floor area ratio in the area surrounding subway station in Japan and Hong Kong, the floor area ratio is related to the level of the centre where the station is located in. The higher the level of the centre, t the floor area ratio is higher.

Traffic Hub Station: there are usually external transport facilities or large transport hubs surrounding the station, supporting commercial service facilities are in the peripheral area. Transportation facilities close to the station have relatively low development intensity; the peripheral area has high development intensity. In actual development, many traffic hubs are arranged in combination with the city centre; the surrounding area, mainly residential, commercial and office, has high development intensity.

## 25.5 Conclusion

Promoting land development surrounding stations in the construction of rail transit is the necessary support and effective extension of rail transit network, it helps to improve the implementation and enhance the stability of the network. Rail transit is an important content of the urban transport system, on the basis of network planning, it is necessary to further stabilize network hierarchy and core framework, determine line alignment and station distribution, develop and control the land along the rail transit effectively and ensure the implementation of the projects, eventually, the transport system would improve constantly and the value of the land would upgrade continuously.

## References

1. Li Wenling, Yan Xiaopei (2002) Research on the development of city rail transit and compound land use – using Guangzhou as an example. Geogr Sci (5)
2. Zhang Yong (2008) The land use pattern of the area along the rail transit in Beijing. Beijing Soc Sci (3)
3. Hui Ying (2002) Research on urban planning and construction of the peripheral area of city rail transit station. Urban Plan Bull (2)

# Chapter 26
# Investigate of Tourism Real Estate Development Model Along Intercity Railways – For Example of Jiangxi Tonggu Spa Tourist Service Center

**Zhenyu Wang and Sicong Huang**

**Abstract** In the past 20 years, with the rapid development of Chinese rail transit, from the first railway, urban rail transit to intercity high-speed rail, rail traffic is looking for breakthrough one after another. As the core issue of rail transportation is funding issues, rail transportation has significant external benefits and give land (real estate) along to add value. How to transit the external benefits of rail into the benefits of rail transportation or converted into government land financial gains become a key research issue. The article elaborates the intercity railway tourism real estate development status as the starting point, analyzes the start-up mechanism of the tourism real estate, explores the diversity of tourism real estate development model with an example of the Tonggu Spa Tourism service center planning and designing in Jiangxi, and excavates the enlightenments for today's tourism real estate development (Jingxiang Zhang, History of western city planning, vol 1. Southeast University Press, pp 15–18, 2005).

**Keywords** Intercity railway • Tourism real estate • Development model

## 26.1 Status of Tourism Real Estate Along Intercity Railways

With the rapid development of China's tourism real estate in recent years, good ecological environment and strong cultural customs become the most valuable wealth and resources of tourism development. Mode of tourism real estate changes

Z. Wang (✉)
Architecture and Urban Planning Institute, Huazhong University of Science and Technology, Luoyu Road 1037 South2-308, Wuhan City, 430074 Hubei Province, China
e-mail: 850505718@qq.com

S. Huang
School of architecture and design, Beijing Jiaotong University, Beijing, China

H. Xia and Y. Zhang (eds.), *The 2nd International Symposium on Rail Transit Comprehensive Development (ISRTCD) Proceedings*,
DOI 10.1007/978-3-642-37589-7_26, 

from sightseeing real estate to leisure real estate to the vacation real estate and continues to heat up.

However, there still exits some failure examples of tourism real estate development that these real estate companies hold areas along intercity railways quickly when the intercity railways project comes. Phenomenon such as lack of popularity of tourism real estate and long investment payback period are due to blindly real estate development, lack of project guide, unreasonable space development mode and inadequate facilities.

## 26.2 Multivariate Start-up Mechanism of Tourism Real Estate

As most of the urban rail transit promotes commercial real estate and residential real estate, mostly intercity railway places develop tourist resort real estate projects. Compared with hospitality tourism, leisure travel, which regard recreation and relax as its basic characteristics, is an advanced tour and product type.

The intercity railway tourism real estate is the result of the combined effects of the policy and market factors. Tourism real estate projects along rail traffic usually develop by the government and rail transportation subordinates Group co-led. The rail transit subordinates Group companies and local governments in consultation, in order to "open entrances intercity railway specific location or set aside under wear corridor" conditions for comprehensive development. By the different characteristics of tourism resources and tourism industry, tourism real estate development has a variety of oriented theme, and therefore the diverse tourism real estate development mechanism will be activated (Table 26.1).

### *26.2.1 The Scenic Spot Business Service Oriented*

As the gathering place of tourism resources, most scenic seize the opportunity of the rapid development of tourism. As the basis of tourism resources, the scenic travel service oriented real estate project operational born.

Natural resource is one of the most important guide factor for the tourism real estate. The scenic rely on natural resources like mountain landscape resources, water landscape resources, and other resources to develop tourism real estate. Human resources are the tourism real estate's another inducible factor. With landscaping, buildings, monuments, folk customs and cultural tourism resources, tourism real estate projects provide people with the experience as fun and participation including culture and entertainment, folk customs experience.

Ocean Spring, Foothills international, Huangshan scenic Qiandao Lake scenic area and other places of the real estate projects are well known for the natural landscape, relatively, the Yangtze River Delta, the Liangzhu Culture and Arts

**Table 26.1** Diversified development model and features of tourism real estate

| Real estate type | Feature | Development location | Typical project case |
|---|---|---|---|
| The scenic spot business service oriented | The better scenic gathering place of tourism resources, natural or human environment, development of the health resort industry, promote scenic gradually transition to an integrated travel services | Landscape resources favorable nature reserves, scenic spots | Zhuhai Ocean Spring, Qiandao Lake scenic area real estate<br>Yangshuo, Lijiang real estate |
| The characteristic resources and theme project oriented | Development of "Spa" special tourism resources, the development of a well-known international conference center, golf theme project, development of resort hotels, the theme of the real estate business meetings | Near the scenic | Pearl River Hot Springs |
| | | Good traffic location | Xianning Hot Springs |
| | | Characteristic resource | Hainan Boao Qianzhou Bay |
| The recreation and leisure oriented | Cultural resources as the basis, to attract the urban tourist crowd, to create "meals, lodging, travel options entertainment platform" | Near a big city | Shenzhen OCT |
| | | Suburban areas<br>Intercity railway | Shanghai Xintiandi |
| The ecology and habitation oriented | Create Features lifestyle, livable city for tourism and leisure and tourism and leisure theme-driven | Cities' Outskirts | Shenzhen Vanke |
| | | Small towns | East Coast |
| | | Intercity rail along exports | Holiday Peninsula |

Village real estate projects to win the cultural landscape. Tourism real estate is not necessarily for the traffic pilot-type real estate, but nice traffic location must make the estate further appreciation.

### 26.2.2 The Characteristic Resources and Theme Project Oriented

In recent years, the government accelerates the construction of rail transit and other municipal infrastructure that has injected new vitality into the development of tourism real estate resort project. Many characteristics of tourism resources have an effective development and utilization [1].

Spa, as one of the important features of the development of tourism resources and their health functions have a very popular gathering function, is the best choice for the winter tourism. Tourism real estates with spa theme along rail traffic such as Pearl River Hot Springs are in the hot development where people can enjoy the gifts from nature.

Golf is a healthy and elegant leisure sports. Jiangxi Lushan International Golf Club relying on tourism real estate projects has been widely recognized. Tourism real estate usually combined with business meetings and business meetings enhance the visibility of the real estate. Matic meeting of the Boao Forum for Asia in Hainan have led to the development of the surrounding real estate.

This type of tourism real estate usually needs a good traffic location guide. Tourism real estate will be developed in areas with a good of landscape along intercity railway, with a development of special theme projects to improve the attractiveness of the resort real estate.

### *26.2.3 The Recreation and Leisure Oriented*

As the emerging model of complex real estate, tourism and leisure real estate exists in different forms and broadly successful in the country. The Overseas Chinese Town in Shenzhen and The New World in Shanghai are two typical tourist complex real estate projects. OCT Happy Valley questions park, combining art, leisure, shopping and other functions, become regional characteristics. Shanghai Xintiandi combined the Shanghai Shikumen residential historic district protection function replacement by cultural entertainment, and integrate into the social and cultural life of the high-quality, in order to attract tourists to visit.

Multivariate development models of the tourism real estate along intercity railway can be fully developed recreation projects, to create a meals, lodging, travel options entertainment platform. The architectural sightseeing, town sightseeing, star hotel, amazing themed inn teahouse, themed dining, street snacks, snack bar, KTV, foot bath, theater, folk performances can satisfy the demand for local matching guide tourism consumption, drive around the estate development.

### *26.2.4 The Ecology and Habitation Oriented*

City tourism and leisure real estate generally choose good places at the edge of the city with natural resources and transportation zone area. Suburban real estate usually rely on the holiday economy and tourism and leisure is driven to create specialty lifestyle livable real estate projects [3].

Shenzhen Vanke East Coast, Country Garden Holiday Islands are in such projects. Vanke East Coast as an opportunity of the Shenzhen metro east line development make full use of its advantages in resources, gold locational advantages, premium advantages of active stations, in order to obtain the outer suburbs of leisure development opportunities. Wuhan Country Garden Holiday Islands located in Wuhan new station near the entrance, to reflect the "small town life" community planning, highlighting the area of low-density intention to focus on the business district, clubs and other exchanges and living spaces, creating a very exotic resort town of demonstration zone. The entrance of intercity railway trends to the development of this kind of tourism real estate projects.

## 26.3 Multivariate Development Mode of Jiangxi Tonggu Spa Tourist Service Center (Fig. 26.1)

Tonggu is located in the northwestern region and the junction of Hunan and Jiangxi Hubei provinces. According to Jiangxi railway network development plan (2008–2020), the planning and construction of the "Nine Great Wall Railway" and "suitable for all ages Kyrgyzstan railway" pass through Tonggu County that will bring new opportunities for Tonggu County development. With its hot spring resources and

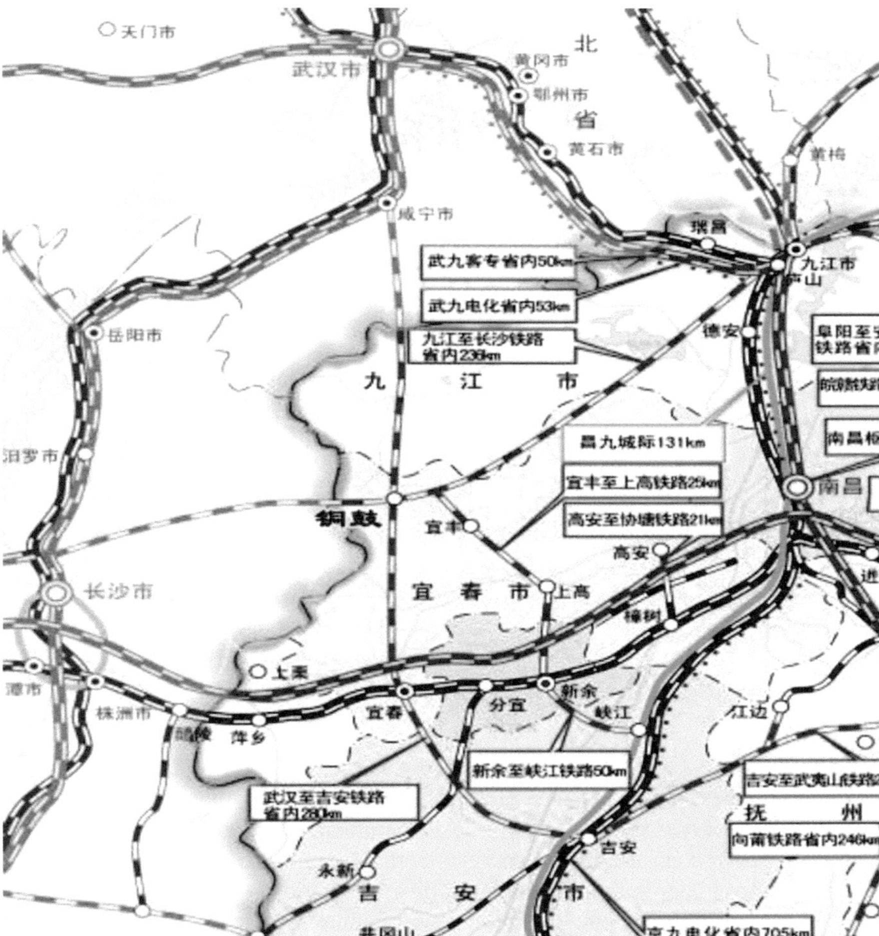


**Fig. 26.1** Intercity rail diagram around Tonggu

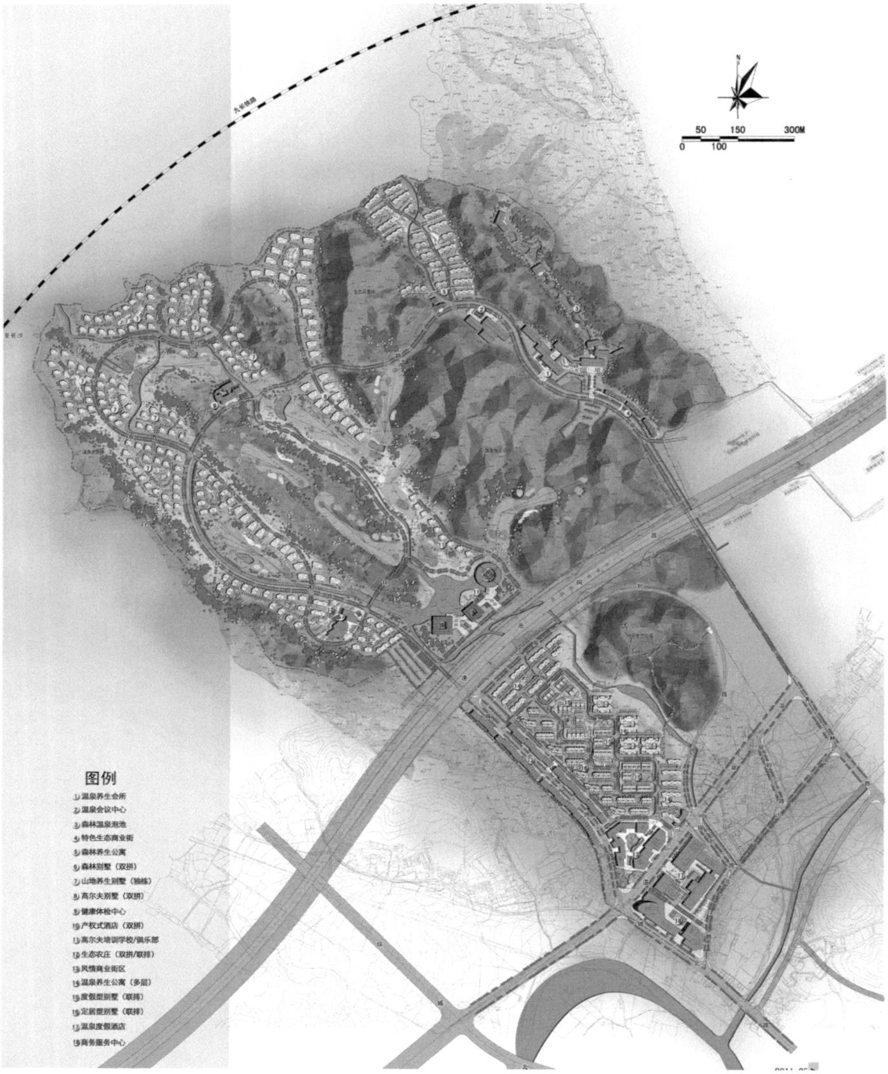


**Fig. 26.2** Tonggu Spa Tourist real estate program plans

traffic advantages, Tonggu County Hot Spring tour is actively promote the construction of the spa tourism service center. The development of spa tourism resort industry as a leader, prominent health spa, tourist resort and conference business, sports and leisure, eco-livable core functionality, building international ecological low-carbon, energy-saving environmental characteristics of tourism service center to achieve expected feasible purpose of travel, and livable (Fig. 26.2).

### 26.3.1 To Plan Characteristic Experience Areas with the Theme of Spa and Business

Spa tourism service center featuring hot spring resources, differentiated spa at the core, the main business and high-end leisure function to enhance the project grade, promote the comprehensive development of tourism real estate.

As mountainous regions and some narrow valley land, Tonggu is not suitable for all types of construction but its natural valley characteristics which is hidden secluded, quiet, eco, making it the best sites for health spa bubble pool. To considering the hot springs relying on the business functions, it's a good choice to establish a health spa clubs. Combine with forest ecological environment and Spa features, to form an ecological health spa with the feature of highly experience.

### 26.3.2 To Design Leisure Resort with the Theme of Sports and Health

Mountain sports, entertainment and enjoy the gift of nature brought to mankind. The Mountain and gentle Strip both create an open and full of rustic charm sports environment. It is suitable for configure mountain golf, tennis, softball, mountain biking, mountains runway project groups for high-end outdoor sports.

Mountain golf is the most important characteristics of spa services center sports. Planning sports park, the stadium satisfied mountains of essence is the venue for the King, to penetrate the greenery, cloud stroll, beautiful mountain feelings. Golf schools, golf clubs and other leisure and configuration functions are built to create a leisure center with the function of resort health, entertainment, conference centers and so on.

### 26.3.3 To Construct Traditional Residential Architecture Demonstration Garden with Cultural Resources Relied

Tonggu County is the main habitat of the northwestern Hakka. Its unique Hakka delicious food, popular and lively Hakka dialect, the sweet sounds of Hakka folk songs, simply and elegant Hakka residential areas has created a unique Hakka culture. The Tonggu is not only the main source of the Autumn Harvest Uprising but also the location of the center of the revolutionary base in Hunan, Hubei, Jiangxi that started the brand Tonggu red tourism.

In suburban areas close to the entrance of intercity rail, create a traditional residential buildings demonstration garden. Inherited the tradition of the excellent red cultural and Hakka culture, the waterfront villa is build in the form of Chinese-style villa, to create living space streets of Chinese traditional characteristics. Through Guzhen commercial space mechanism and explore the ancient town commercial settlement patterns, style commercial street to create a form of "centralized within the Street". The architecture demonstration garden is the first startup items, other projects have been started, and played a major role in promoting the development of Tonggu.

## 26.4 Summary

The Intercity rail development and improving quality of people's life has brought great opportunities for the development of intercity railway travel vacation industry.

Full analysis of tourism resources, multi-start mechanism to find tourism real estate, rational allocation of land resources, to avoid blind construction intercity rail development process, and international perspective starting point plan make sure that the intercity railway sustainable tourism, real estate development, so as to promote the all-round progress of the inter-city rail transport undertakings.

## References

1. Thomas fee Reed (2001) The world is flat, vol 2. China Building Industry Press, pp 36–40
2. Jingxiang Zhang (2005) History of Western city planning, vol 1. Southeast University Press, pp 15–18
3. Jianhong Dong (2004) History of urban construction in China, vol 13. China Architecture & Building Press, pp 356–360

# Chapter 27
# Research on the Control Planning of Land Use Along Rail Transportation – Taking Xuzhou Rail Transportation as an Example

**Qiang Sun and Defang Deng**

**Abstract** In recent years, the complex interaction between the urban transport system and land-use become one of the hot issues. Many cities tried to dominate effectively the land along urban transport system by the land use control planning. Xuzhou City is planning to construct rail traffic, then how to pre-contract the impact of the surrounding land of track traffic from the planning level and how to use the land is a key issue. This paper analyzes what is the key content of land use control planning e along the rail line.

**Keywords** Rail transportation • Land use • Control planning

## 27.1 Introduction

Rail transportation plays significant role to limit urban sprawl, save valuable land resources and improve the efficiency of urban land use. It is necessary to control the land use along rail transportation as early as possible to ease the problem that the construction of rail transportation has a slow investment income period but a long construction cycle.

Q. Sun • D. Deng (✉)
Xuzhou Planning and Design Institute, Xin'an Road, New City, 221000 Xuzhou, China
e-mail: dw0510@163.com

H. Xia and Y. Zhang (eds.), *The 2nd International Symposium on Rail Transit Comprehensive Development (ISRTCD) Proceedings*,
DOI 10.1007/978-3-642-37589-7_27, 

## 27.2 Outstanding Significance of Land Use Control Planning Along Rail Transportation

### *27.2.1 Experience of Domestic and Foreign Cities*

#### 27.2.1.1 Foreign Cities

Tokyo has more than 70 year's history of subway construction. The benefits of the subway in Tokyo are not only to ease the traffic pressure, but also that it is driven the residential and commercial areas along the subway. In addition, the underground space development and utilization in Tokyo started relatively early and covered wide areas, including the subway, underground street, underground pipe network and so on. Thus, it can provide a good example for the other city.

Montreal is famous to the world for the success on its use of urban underground space, and it has a lot of experience and advantages of spatial development patterns, comfort, ground convergence and investment mode in the utilization of underground space.

#### 27.2.1.2 Domestic Cities

Hong Kong has two development models on subway stations: one is the land development taking the subway hub station as the key of planning; another is the new town regional development taking the subway site as the planning core node. This model successfully solves two problems of the construction funds and passenger.

Shenzhen rail transportation network planning achieved the interaction between the rail construction and land use. The site tries to be located in the densely built-up areas or new district. By the adjustment of land use, a lot of living and employed population gathered around the rail site to enhance passenger flow benefits and the land value.

#### 27.2.1.3 Summary

Construction of rail transportation is not only an important means to ease traffic pressure; but also the impact of the land-use surrounding rail transportation change and real estate activities cannot be ignored. With the experience of other cities at home and abroad, the construction of rail transportation must change the nature, use intensity and development mode of the surrounding land [1].

### *27.2.2 Advance Guiding Affect of Control Planning of the Land Use in Guiding*

#### 27.2.2.1 Giving Advance Response to the Irreversibility of Urban Construction

Land use type and intensity will change dramatically along the rail transportation. Taking into account the irreversibility of the urban construction, planning proposal that it should try to begin the research work on control planning as early as impossible, in order to response the change and guide urban construction.

#### 27.2.2.2 Urban Land Was Effectively Reserved for Rail Transportation Construction

Control planning of land use along the rail transportation can effectively guide the planning department to arrange land for construction of rail transportation, avoid the contradictions, and reduce unnecessary land acquisition and secondary demolition. Meanwhile, it can provide the basis for the urban facilities and rail transportation and guide urban infrastructure construction and rail transportation coordinated developing.

#### 27.2.2.3 Strong Guarantee for Changing the Investing and Financing Mode of Rail Transportation Construction

The lack of funds is the key to restrict rail transportation development of our country in current stage. Raise funds from the land-transferring fees in the process of urban construction and renewal has increasingly become an important mode of financing. Therefore Control planning of the surrounding land is a revolution of investing and financing mode, a stable source of investment annually and an important guarantee for the sustainable and healthy development of rail transportation [2].

## 27.3 Control Planning Analysis for Land Use Along Trail Transportation

### *27.3.1 Planning Drawing Experience in China Cities*

#### 27.3.1.1 Guangzhou Control Planning for Land Use Along Trail Transportation

Under the direction of city Overall planning and on the base of trail transportation network planning, the planning area covered 300 m distance from trail transportation and 500 m around station and mainly control from dimension rate and construction rate index, make comprehensive strengthen analysis for the circumjacent land of trail transportation station spots.

#### 27.3.1.2 Planning Adjustment for Land Around Suzhou Trail Transportation Stations

Planning scope covered the lands 500 m distance around stations. Firstly, it confirmed the service scope for every station, provided adjustment strategy for difference of every piece of land and every station; finally make beneficial analysis for the adjusted planning lands. It also advanced control requirement, confirm landing using character, developing density index.

#### 27.3.1.3 Land Control Planning Along Suzhou Trail Transportation Line-3-4, Branch Line of Line, Expansion Part of Line-2

Planning area covers the landing 500 m around trail transportation stations. According to Suzhou City overall planning, every region control planning and land using status evaluation, Optimize and integrate the target planning area function and space structure, to fully advance the land value of the core land around the station.

#### 27.3.1.4 Land Control Detail Planning Along Zhengzhou Trail Transportation Line-1 Phase-1

Planning scope covered lands 400~800 m distance along line-1 and stations. The planning mainly adjust land use layout, construction size, urban design, exit and entrance location, parking construction and transfer system, especially adjust the lands use around the transfer station, develop high- density and high-mixing degree lands.

### *27.3.2 Establishing Key Points Research of Land Using Control Planning [3]*

#### 27.3.2.1 Planning Scope Definition

Based on other cities planning experience, although there are many differences on planning scope standards, more projects use the scope around 500 m (no tired on foot), while some cities adopt GIS analysis methods. Xuzhou High-speed rail transportation land using control planning considers the definition of control scope need more objective.

Take station as a core, the covered scope for every station mentions walking-expend 10~15 min around. In this planning, it is predicted the walking speed is 4.8 km/h, analysis the best walking path coverage in 5, 10 and 15 min separately, and then refer to 10 min best walking path coverage, cooperate with natural lands board similar as roads and water, to designate station control coverage. Finally integrate the control area for every station to combine the whole control coverage for the whole planning.

#### 27.3.2.2 Diversity of Land Using Control Content

Land using control planning of other cities mainly take every station as research unit, adjust and optimize the land using character and develop density of lands around every station. The extent is decided by station type, control planning establishment and present status, and principally respects original control planning. Meantime, consider land character and develop density change leading by public transportation construction.

Xuzhou High-speed rail transportation land using control planning considers that it is in favor of guarantee trail transportation construction and safety running on one side, and it can lead land using efficiently, promote lands develop efficient, complete urban space structure on the other side.

##### Land Required the Demolition and Acquisition

According to the actual needs of the rail transportation construction, there are three circumstances following: (1) within the depot and parking location and 15 m away from centerline of the both sides along the contact line; (2) scope within the 10 m away from the vertical line direction of the station exterior outline and 30 m away from the parallel lines direction, but except the multi-storey buildings and high-rise buildings outside the parallel lines direction above 6 m; (3) scope within the 10 m away from the centerline of the interval line, including the underground line, elevated line and ground line. If the land fits the above-mentioned content, all the building on the land should be removed, if the land still use for the agricultural, land nature replacement work will need to carry on as soon as possible.

##### Land Required Security Controls [4]

According to requirements of the "urban rail transportation construction projects standards (Trial)", "urban rail transportation operators managements", "metro design specification (GB50157-2003)", security requirements of the control distance on both sides of rail is different because of the distinct focus of project construction and business management. In the rail transportation project construction standards, control requirements of planning rail line are higher than the existing ones. Because the existing rail lines focus on orbiting security, but the planning rail lines should also consider the impact of t rail construction. Meanwhile, underground lines have more strict control requirements than the ground lines due to the different rail operations management provisions. Second line of Xuzhou is the underground line, and not construct yet. In this control planning, the security control distance was being confirmed to 50 m away the outside of rail line by the thinking of construction, working security and environmental protection requirements including the noise and vibration. All the construction activities within the scope of the security controls should be monitored by the special rail transportation competent department.

Land Required Be Storage

Most of the area passed through by the plan rail lines in Xuzhou has prepared regulatory planning. On the basis of adjustment and optimization of the nature and intensity of the land surrounding the various sites, the planning has surveyed, arrange the exploitable land within the scope of 10 min walking. Surrounding area alone the line-one in Xuzhou can be sort out approximately 16,472.55 mu, among them residential land has approximately 12,215.12 mu and commercial land has 4,122.61 mu. This practice has reflected the changes in the nature and intensity of the land after the large-scale public transport pass through, and the land benefits can be used by supplement funds.

## 27.4 Conclusions

Establishing the control strategy of land use comply with the laws of the urban rail transit construction has a strategic significance to cities for the development of rail transportation. Effective co-ordination of urban rail transit and land use control programs can guide the rational distribution of urban space, promote the reasonable adjustments of urban form and land use and achieve sustained and coordinated development of the urban economy, society and environment.

## References

1. Zhou Jun, Xu Jiangang (2002) Rail transit corridor effect and urban land use analysis. Urban Rail Transit Study (1):77–81
2. Bian Jingwei (2005) Research on urban rail transit and land use control programs. Planners 21(2):87–90
3. Xuzhou City Planning Bureau, Xuzhou Urban Planning and Design Institute (2010) Xuzhou control planning of the land use along rail transportation
4. Wu Shuang (2005) Land controlled detailed planning of urban rail transportation method. Urban Traffic 3:43–46

# Chapter 28
# Studies on the Design of Rail Transit Complex of Guangda Station in Changsha Metro Line 2

**Wei Sun and Yagang Liu**

**Abstract** This paper analyses the influence the restrictive and decisive factors the construction of rail transit complex of Guangda Station on Changsha Metro Line 2, combining with the city planning and the TOD theories. The construction mode of the rail transport complex in the commercial space is discussed. The design methods are summarized from the aspects of concept definition, function layout, streamline design, space forms and so on. The technology strategies of organizing stream of people with transit hall as the core and developing commercial space with the indoor and outdoor pedestrians are presented.

**Keywords** Changsha • Metro • Complex • Commercial

With the development of urbanization, China's urban rail transit enters a rapid development stage, and the planning of rail transit stations has a huge influence on the land development of surrounding area. The construction of urban rail transit has become a strategy of the development of China's large and medium-sized cities. Subway station can spur the business planning and integration of the region, and enlarge the scope of the business irradiation district. Therefore, through subway station design, the city design can be gradually optimized and the urban functions and the quality of urban areas can be improved.

W. Sun (✉)
School of Architecture and Design, Beijing Jiaotong University, Beijing, China
e-mail: sunwei@bjtu.edu.cncom

Y. Liu
City Track Traffic Design Institute, Fifth Railway Survey and Design Institute, Beijing, China

H. Xia and Y. Zhang (eds.), *The 2nd International Symposium on Rail Transit Comprehensive Development (ISRTCD) Proceedings*,
DOI 10.1007/978-3-642-37589-7_28, 

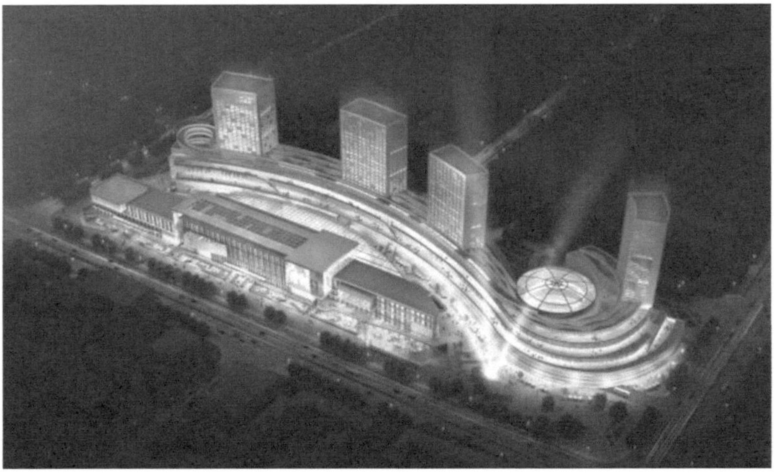

**Fig. 28.1** Rendering

## 28.1 The Project Background

The Guangda Metro Station, which is the terminal of Changsha Metro Line 2 to the southeast, is located in Changsha City, Huang Xing town that has a total area of 84 $km^2$ and a population of 50,000, is located in the eastern suburbs of Changsha City, near the Liuyang River. The town has an area of 5 $km^2$, with a residential population of 14,400. Guangda Station and the surrounding land are in the planning areas of Wuhan-Guangzhou new town near the intersection of Beijing-Guangzhou and Wuhan-Guangzhou high-speed railways. In accordance with the conception of Changsha Municipal Government, a central business district is planned on both sides of the high-speed rail transportation hub and the Liuyang River, and a commercial office cultural center is planned on the east bank of the Liuyang river in Wuhan-Guangzhou new town. The overall planning extents along two perpendicular axis and the east-west axis is the main one in which Guangda station is located. To sum up, Guangda Station is very suitable for commercial development because of its advantaged position (Fig. 28.1).

## 28.2 Concept Orientation

### *28.2.1 Basis of the Design Concepts Orientation*

As an important part of the city public transportation, rail transit station has catalytic action to the development of urban areas and brings convenience transportation connecting work, shopping and leisure for the urban areas [1]. Following the TOD traffic dominant theory, commercial, residential, service and other related functional space are gathered around the transportation hub and a comprehensive and intensive land use pattern is formed. The compact development of land use can make the employment, living and recreational functions close to each other, to share infrastructure and to provide more public transportation opportunities such that energy consumption can be reduced [2].

### *28.2.2 Factor Analysis of the Design Concept Orientation*

The case comprehensively analyzes the upper structure of the urban development and the planning system of the urban traffic. Influencing factors of the design concept is summarized as the following three aspects, first, the positioning of Guangda Station a city transportation hub in Changsha City public transport system, and it is also a transfer station for intercity traffic. The station has a great catalytic action on the regional development; second, according to the conception of the regional planning, Wuhan-Guangzhou new town where Guangda Station is located will become the city sub-center in Changsha. In the future, whether the public services, transportation, business, tourism culture, or commerciality, residentiality will be perfect facilities and city spaces radiating surroundings. The connection between Guangda station and the main city of Changsha is obvious; third, Guangda Station domain, sitting on the landscape belt of Liuyang River riparian with the unique ecological landscape superiority, will form a popular interaction. It can be said that the construction of the Guangda station possesses the advantages of "Climate, geography and human harmoniousness".

### *28.2.3 The Target of the Design Concept Orientation*

Based on the above analysis, Guangda Station is overall orientated as a modern commercial city complex combining the functions of transportation hub, shopping, leisure, entertainment, dining, living etc. in one. The infrastructure, including boutique business, retail business, culture shows, business office, hotel, conference etc. in the region can be further improved and the function of radiation to the surrounding areas can be played through the construction of the station, to enhance the value of the surrounding land especially the residential land.

## 28.3 Functional Layout

The land area of Guangda Station is 8.9 ha and the total construction area of Guangda Station complex is 290,269 m$^2$, with the volume rate of 3.5. The building takes business as the main industry and integrates various functions of transportation hub, parking, apartment and office etc. The commercial space area accounts for

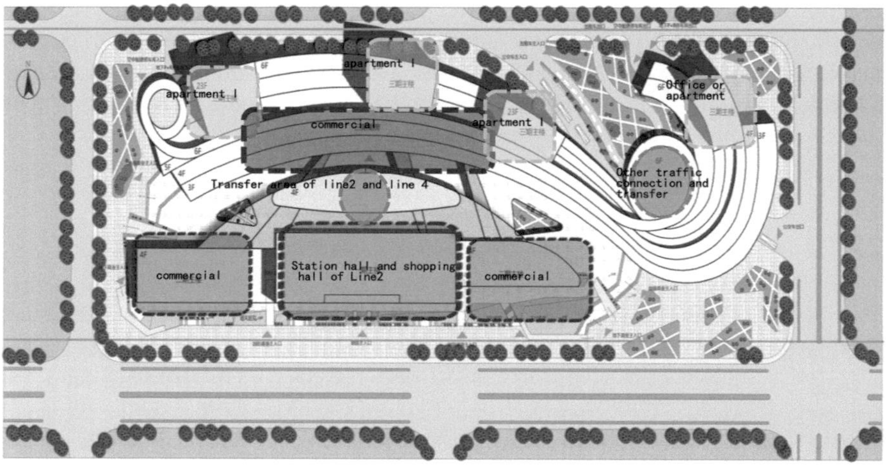


**Fig. 28.2** General plan and functional layout

**Table 28.1** Area and proportion of the complex function space

| Item | | Construction area (m$^2$) | | Amount of parking (unit) | Proportion (%) |
|---|---|---|---|---|---|
| Commercial | Over ground commercial | 174,550 | 147,091 | | 60.14 |
| | Underground commercial | | 27,459 | | |
| Commercial apartment | | 43,632 | | | 15.03 |
| Park for buses | | 4,150 | | 20 | 1.43 |
| Park for taxies | | 5,998 | | 160 | 2.07 |
| Park of social cars | | 7,327 | | 210 | 2.52 |
| P+R garage underground | | 25,573 | | 583 | 8.81 |
| Garage in the air | | 29,039 | | 730 | 10.00 |
| Total construction area | | 290,269 | | | 100 |

about 60 % of the complex, and the amount of parking is up to 1,313 units (Fig. 28.2, Table 28.1).

Functional layout takes "clear partition, convenient links" as the principle with the horizontal and vertical partitions combined. Commercial space is located on the podium and the first floor underground, while the commercial apartment is located in the high-rise building on the north side. The station hall of line 2 is located beneath the southern part of the commercial space, while the central hall and the transfer space of line 4 is located beneath the mid commercial space. The transfer hub of the traffic connection and transfer is located in the circular building which is on the east of the complex, while parking garage is located on the second underground floor, the fifth and sixth floors. The constitution of the commercial space takes diversification and composition as features, and has integrated functions such as supermarket, the main stores, boutiques, retail stores, food court, movie city and so on. Integrating shopping, dining, entertainment, culture, art and other consumption functions in one, the commercial space can even accommodate "flagship store", the themed entertainment, leisure facilities, diversified commercial pedestrian and other various formats.

## 28.4 Flow Line Design

### *28.4.1 Design of the Internal Traffic Flow Lines*

The traffic flow line can be divided into three types: the flow line for accessing the subway stations, subway and other transportation transfer flow line, and business flow lines. Metro stations can be accessed through the sunken plaza directly, or through the ground floor by staircase. The transfer between P+R and other transits is achieved in subway station hall and transfer hall through commercial pedestrian. The design of these two flow lines strives to achieve a seamless connection. Vehicle is separated from pedestrian, and the motor vehicle can enter down into the underground garage from the northern side of the entrance, or up into it in the high-rise park or entry into the Park-and-Ride parking garage on the ground. Both apartments and shopping mall have vertical elevators to the high-rise park (garage in the air) and it provides maximum convenience for business and commercial activities; business flow is flexible to enter into the underground business from either the subway station hall directly, or the sunken plaza, or other multiple entrances of the ground floor. The diversity of the commercial lines provides a good guarantee for commercial passenger flow (Fig. 28.3).

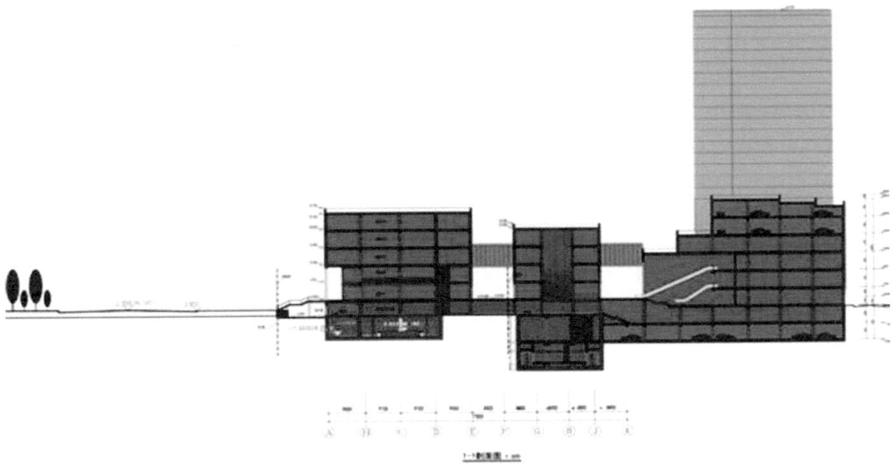

**Fig. 28.3** Organization of the internal traffic flow

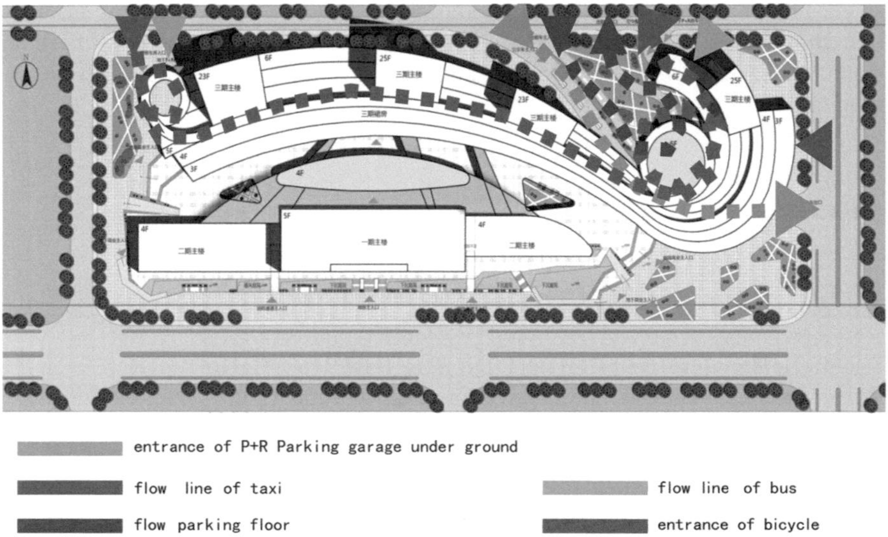


**Fig. 28.4** Organization of the external traffic flow

### *28.4.2 Design of the External Traffic Flow Line*

The design of external traffic fully considers the internal flow and the reasonable connection in the complex, and connects the internal and external traffics. Main flow line, including crowd flow, vehicles flow of social vehicle (P+R or other), taxi, bus lines and so on. The crowd flow is arranged in the south, while the vehicles flow is on the north side, to separate people and vehicles. Social cars, taxies access the station from the north side of base, while buses, P+R society cars enter the station from the northern side and leave from the eastern side. The high-rise park is referenced from Tsing Yi Station in Hong Kong, to solve the traffic flow, as well as other parking problems by utilizing the ramp (Fig. 28.4).

## 28.5 Space Form

The complex building has rich functions and complex flow lines. This case uses a variety of spatial organization forms, to solve the problem of transit, shopping, business, leisure and other functions, of which spaces are mixed and interspersed [3], so as to expand the advantageous design elements to enlarge commercial value, and to form a rich, interesting, convenient and efficient urban public environment.

### 28.5.1 Sunken Commercial Plaza

The main design purpose of the sunken plaza is to expand business interface and to make an inventory to the underground business. In this case, the sunken plaza is designed on the south side of the complex, where the station hall of line 2 is located. The difference in level between the roof and the ground of the station hall is more than 2 m. The sunken square makes proper use of this level difference and save the construction cost. The plaza not only solves many functional problems, but also increases the interest of commercial space, and raises its commercial value.

### 28.5.2 Indoor Commercial Pedestrian

The interior commercial space of the complex uses commercial pedestrian to organize the stream of shoppers. The spaces both on top and bottom of the commercial street are linked up. Glass sunroof is applied making the interior space bright and open. Pedestrian integrates the commercial spaces together, making the space with proper density and with improved comfort. At the same time, the open space is conducive to aggregate more passengers, to increase the balance and quality of the store value.

### 28.5.3 Outdoor Commercial Pedestrian

The land of Guangda Station is deep in the north and south. It is not enough to take the interfaces adjacent to the urban roads in north, south, east and west directions as business interfaces. To solve the problem, an east to west outdoor commercial pedestrian in the centre of the complex is introduced to expand the interface of urban business. At the same time, the pedestrian echoes the axis of Wuhan-Guangzhou Metro commercial space, maintaining the integrity of the overall plan. The introduction of the pedestrian also improves the spatial environment and reduces energy consumption, so that the quality and commercial value of the whole complex space are greatly improved.

### 28.5.4 "Glass" Island

The transfer hall of the complex, which is the transfer space of line 2 and line 4, is located in the centre of the pedestrian where crowd flow concentrates in. In this design, a "glass" island is applied to flow the space and to make the space with clear accessibility and full openness. The centre of the pedestrian is widened to divide the commercial street space into two, so that it can connect different architectural spaces in both horizontal and vertical directions. Above the boarding hall, a large indoor space with a curved glass roof is formed. The "Glass" Island introduces natural light and ventilation for the underground metro station space, making the design more humanized.

### 28.5.5 *Half Interior Plaza*

In order to make the commercial space has better openness and connectivity, the podium of the high-rise apartment retreats floor by floor towards the pedestrian street. Green plants are grown on the platform to form a garden in the air. The centre of the building is the main entrance which is designed as a giant form with three stories, and directly connected with the outdoor pedestrian. The half interior space extends the pedestrian street and forms a semi-indoor square. The open space of the square can well attract people's attention, and can be used as a place for business shows. Two outdoor escalators are set in the square to introduce passengers into the second or third floor directly, expanding the utilization value of the commercial space indirectly.

## 28.6 Conclusions

In the design of the rail transit complex, the land intensive use concept in development of city space [4], is carried out. Composite functions and comprehensive use of the space are the basic strategies that have reached a consensus. The methods to make use of the passenger flow brought by rail transit station, to organize reasonable traffic flow, and to construct positive architectural space have decisive effects on achieving the maximizing functioning of value of the land. The case studies the format configuration of the complex commercial space, arrangement of the flow lines and functional layout deeply, and focuses on the strategy of the integration of indoor and outdoor commercial spaces, and attempts on creating commercial space suitable for the experiential business pattern.

**Acknowledgment** This research was supported by: National Natural Science Foundation of China (51078022).

The Fundamental Research Funds for the Central Universities (2009JBZ023).

Doctoral Fund of Ministry of Education of China (20090009120013).

## References

1. Rubenstein JM (2003) The cultural landscape: an introduction to human geography, 7nd edn [M]. Person Education, Inc, Upper Saddle River, p 97
2. Gregg Doyle D, Chen Zhen (2002) Densification and gentrification in the USA: linking the agendas of "Smart Growth" and Inner-city advocates. Urban Plan Overseas 3:2–9
3. Lu Jiwei, Han Jing (2007) Systemization of mass transit station area and urban design. Urban Plan Forum 2:32–36
4. Li Chi (2006) Probed into city spacial development pattern based on compact land-use. Urban Plan Forum 1:55

# Chapter 29
# The Optimization Design Strategy of Xizhimen Underground Commercial Space Based on Post-Occupancy Evaluation

**Linlin Guo, Haishan Xia, and Linlin Qian**

**Abstract** The construction of underground commercial streets and underground business complexes has become a trend for development and utilization of underground space as the development of China's urbanization process and the construction of the underground rail transit accelerated. Due to the relative limitations of the design methods or design concept when architectures and the environments were built, the introduction of architectural and environmental post-occupancy evaluation (POE) becomes the trend of the underground space construction. This paper explore the development status of the underground commercial space as well as the qualitative and quantitative research indicators analysis by applying the POE method in the current situation of Xizhimen station's underground commercial space. Thus Xizhimen underground commercial space of optimization strategies was proposed in order to promote the development of underground space and utilization.

**Keywords** Underground business complex • Xizhimen underground commercial space • POE • Optimization strategies

## 29.1 Introduction

Along with the rapid economic and social development, the development and utilization of urban underground space, has entered a stage of rapid speed period. Until now, the Beijing underground space run into an area of growth per year 300 $hm^2$ speed, accounting for about 10 % of the total building area. Meanwhile along with the peak period of Beijing's underground rail transit construction, According to the data of the rail transportation planning, the number of Beijing

L. Guo • H. Xia (✉) • L. Qian
School of Architecture and Design, Beijing Jiaotong University, 100044 Beijing, China
e-mail: guo.linlin2007@163.com; hshxia@bjtu.edu.cn; 339246488@qq.com

H. Xia and Y. Zhang (eds.), *The 2nd International Symposium on Rail Transit Comprehensive Development (ISRTCD) Proceedings*,
DOI 10.1007/978-3-642-37589-7_29, 

**Fig. 29.1** The open distribution space of Xizhimen station

**Fig. 29.2** The half outdoor traffic distribution space of Xizhimen station

rail transit operations will improve 50 % (2011 mileage 372 km) to reach 561 km in 2015, and to 2050, the total operating mileage will exceed 1,000 km for the huge leap in the development of rail transportation [1]. This trend, will provide a good opportunity for the development of underground space in the rail transportation.

Currently, the status quo of Beijing city center underground space is mainly on punctate distribution, shallow forms (underground 0~10 m), and is still in the simple form of single stage of development. The main function of the underground space is for parking, commercial, civil defense, transportation, distribution and equipment, including various components. Subway, transportation hubs, and other land use of underground space, which are among the Beijing underground space development in key areas and important nodes, interconnected to form the function more integrated, and more closely linked. Traditional underground single commercial, parking space and etc., all these are gradually developed into the functionality of "underground commercial + underground parking + the traffic distribution space + other public channel network" [2] (Figs. 29.1, 29.2, 29.3, and 29.4).

**Fig. 29.3** The exit of the shopping mall and the subway

**Fig. 29.4** The aerial view of Cade Mall

## 29.2 Problem Analysis

### *29.2.1 The Current Situation of Underground Commercial Space in Xizhimen*

In recent years, combined with rail transit construction, Beijing has built some large-scale transportation hub complexes. Xizhimen is one of the most famous rail transit complex for business and transportation.

#### 29.2.1.1 The Geographical Position of Xizhimen

The famous Xizhimen Xihuan Plaza is in the northwest corner of Xizhimen overpass, and it is in key position of Xizhimen traffic, leading in all directions, even it has obvious commercial atmosphere and traffic advantage. Xizhimen underground commercial space which rely on Beijing North Railway Station, subway, light rail and the public transport development on ground, really has a obvious business advantage compared with other traffic node. As a result, the rail transit station has the commercial advantage because of the highest flow strength, density and turnover frequency that other traffic nodes do not have.

#### 29.2.1.2 Underground Commercial Space of Xizhimen

The total building area of Xizhimen Xihuan Plaza is 260,000 $m^2$, the total investment of the project is 2.9 billion yuan, and it is consists of 300-m high international grade 'a' office building, large commercial and the Xizhimen traffic hub to form a large city synthesis. Xihuan plaza covers an area of 5.99 ha, and it has a use rate of 70 %, rental office building area of 53,000 $m^2$. The current large-scale commercial shopping center – Cade Mall which has the 3 underground layers, 23 floors on the ground buildings is very famous. In the building, underground 2, 3 are for the civil air defense, equipment and the garage, from underground 1 layer to the ground 6 are for the commercial business, the 7th floor above are office area. Therefore, as a successful underground space in Xizhimen, it has big advantages of three-dimensional traffic development and business environment, meanwhile it has greater value for business development and utilization.

## 29.3 The Introduction of Post-Occupancy Evaluation (POE)

### *29.3.1 The Necessity of the Introduction of Post-Occupancy Evaluation*

Post-Occupancy Evaluation [3] is a method to have some comment or opinion when the underground commercial space were used for some time .environment Post-Occupancy evaluation.

POE refers to the buildings that have been built for some time, it collects the user on the environmental evaluation data information standardized and systematic programs, through scientific analysis, then they show their understanding of the target environment evaluation; And with the original design goals for comparison, they will get to know the extent of comprehensive identification of design environment to meet

**Table 29.1** Likert scale structure (self-painted)

| Satisfactions | Strongly agree | Agree | Neither agree nor disagree | Disagree | Strongly disagree |
|---|---|---|---|---|---|
| Scores | 5 | 4 | 3 | 2 | 1 |

**Table 29.2** Membership functions (self-painted)

| Evaluation values (Z) | Five-level | Evaluations |
|---|---|---|
| $X \leq 1.5$ | Strongly disagree | E1 |
| $1.5 < X \leq 2.5$ | Disagree | E2 |
| $2.5 < X \leq 3.5$ | Neither agree nor disagree | E3 |
| $3.5 < X \leq 4.5$ | Agree | E4 |
| $X > 4.5$ | Strongly agree | E5 |

the use of people. Through the similar reliable information to provide the scientific reference, in order to maximize the overall efficiency and the quality of design.

### *29.3.2 Summary of Index on POE*

This paper will use the Post-Occupancy Evaluation of the satisfaction survey for main analysis. And it will establish the basic index set for underground commercial space. Evaluation index set is defined as follows:

1. In this study, using Likert scale structure, each of the indicators of psychological reaction volume is divided into five-level (Table 29.1).
2. 200 copies of the questionnaire, received 179, 170 valid questionnaires, effective rate of 85 %.
3. The composition of consumers: 100 males and 100 females. Including ages, occupations, annual income, the purpose of coming here will be recorded.
4. To determine the membership functions (Tables 29.2 and 29.3).

## 29.4 Research and Analysis for Xizhimen Underground Commercial Space

### *29.4.1 Functional Layout for Underground Commercial Space in Cade Mall*

Through investigation and data analysis, Xizhimen subway unpaid area which connects with the shopping mall, and the basement, the first floor and the second floor of the shopping mall form various interfaces. As a result, this design forms a

Table 29.3 Index set of Post-Occupancy Evaluation for subway commercial space (self-painted)

| Index set of POE for subway commercial space | | | | Average score |
|---|---|---|---|---|
| Target | Criterion | Factor | Index | 4.166 |
| Evaluation for satisfaction of Xizhimen underground commercial space in Beijing | Visual image | Environment of overall space | Overall style | 4.154 |
| | | | Functional layout for Space | 4.045 |
| | | | Spatial scale | 4.136 |
| | | Image for details | Decoration materials | 4.144 |
| | | | Patterns and colors | 3.760 |
| | Function for use | Physical environment | Light environment | 3.687 |
| | | | Acoustic environment | 3.820 |
| | | | Air quality | 3.475 |
| | | Commercial space design | Organization for commercial space design | 3.698 |
| | | | Commercial space size scale | 3.540 |
| | | | Degree of openness of the commercial space | 3.810 |
| | | Commercial location | Commercial districts | 3.760 |
| | | | Types of goods | 3.700 |
| | | | Quality of goods | 3.560 |
| | | Design for traffic space | Entrances marked (size, color, shape) | 3.620 |
| | | | Convenience or accessibility for Entrance | 3.595 |
| | | | Traffic congestion | 3.620 |
| | | | Location for Elevator or staircase | 3.560 |
| | | Organization for traffic flow | Organization for traffic flow | 3.455 |
| | | | Path design of rapid traverse | 3.490 |
| | | | Degree of convenience to subway | 3.290 |
| | | | Guiding definition to subway | 3.130 |
| | | | Transport interchange with other vehicles | 3.560 |

(continued)

**Table 29.3** (continued)

| Index set of POE for subway commercial space | | | Average score |
|---|---|---|---|
| | Supporting facilities | Catering and entertainment types | 3.400 |
| | | Facilities for toilets and drinking water | 3.290 |
| | | Virescence for environment | 3.320 |
| Operation management | Operation management | Hygienic condition of environment | 3.200 |
| | | Public Security Management | 3.650 |

**Fig. 29.5** The current situation of the traffic space in B1 Cade Mall

three-dimensional integrated development pattern. Is a typical Metro business development case; Moreover, Xizhimen station synchronously has the subway and light rail [4], so the traffic composition is more complex which leads to the present complex integration.

Commercial stores locate from the basement to the sixth floor of Cade Mall, without standard floor. The area of each floor is not identical which is from 600 to 2,000. The rental fee of Cade Mall is very high. According to our investigation and research, the current average price is 25 yuan/square meter/day which could change on the basis of store position and different brand; the property costs 80 yuan/square meter/month, high rental fee and property costs make lots of living brand, such as sporting goods, bedding and other living products to step back. According to a survey, Sports 100, lovely lace and Yisiaila brands have been quietly leaven. Especially, the whole Sports 100 store completely disappeared (Figs. 29.5, 29.6, and 29.7).

**Fig. 29.6** The current situation of commercial space in the ground floor Cade Mall

**Fig. 29.7** The connected area of the commercial space and the subway

**Fig. 29.8** The subway entrance to Cade Mall

### 29.4.2 The Commercial Format of Cade Mall

The basement of Cade Mall provides restaurants, supermarket, clothing and hair & beauty services. The six floor of Cade Mall distributed characteristics food, clothing, cosmetics, jewelry, game room, gym, training institutions and other integrated business. The composition of Cade mall underground commercial business has the following characteristics:

1. In this study, using Likert scale structure, each of the indicators of psychological reaction volume is divided into five-level (Table 29.1).
2. 200 copies of the questionnaire, received 179, 170 valid questionnaires, effective rate of 85 %.
3. through introducing catering, such as KFC, Starbucks, Tai Hing wing Fu tea restaurants, this could effectively delay the shopping time of consumers in order to bring more consumers.
4. Introducing the major shops, enhancing brand appeal and visibility of the project as a whole strategy, which can further ensures that the project's overall image.
5. Selection of high rental rate of return business, to ensure that the underground commercial space overall revenue maximization [5].

### 29.4.3 The Underground Commercial Space Environment of Cade Mall

#### 29.4.3.1 The Business Model of Underground Commercial Space Settled by the Well-Known Brand

With the developing of Cade Mall underground commercial business, Starbucks coffee, KFC, Aesthetic design, Vero Moda and Jack & Jones brands are introduced, greatly activating commercial atmosphere and attracting a large number of people to shopping.

#### 29.4.3.2 The Use-Pattern of the Space

At underground business hall, the utilization of the shopping space mainly represent at consumers' wandering, shopping, dining, resting, passing and stopping. Due to seamless docking between the shopping center and the subway, many passengers are evacuated through the commercial part; in addition, the east-west distance of shopping center is very short, clear route, comfortable space scale, and convenient traffic [6]. On work days, the chance which metro passenger transfer into a commercial passenger is small, but will increase on weekend days. Cade mall has less break space, so the most of waiting people stand in the hall or sit in KFC.

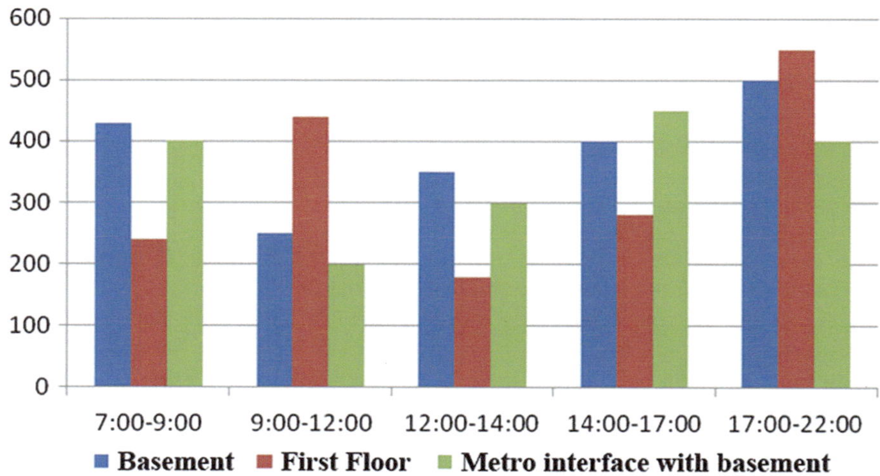


**Fig. 29.9** Flow analysis diagram of one general work day

#### 29.4.3.3 The Diversity of Design Style

The decoration of the internal space of underground Cade Mall emerged diversified styles. Abundant design language and various promotions increasingly improve the shopping interesting and shopping desire of consumers. Although each store has its own selling style, the store in the underground business hall has uniform pavement style. People can clearly recognize their current location and direction. At the entrance to the ground traffic, large posters achieve a concise guidance which not only effectively advertise products but also provide the corresponding guidance to consumers. The research results show, the visual image design of underground space locate into the range of $3.5 < X \leq 4.5$. As a result, the design meets the needs of the users.

### *29.4.4 Analysis of Flow Difference for Cade Mall Underground Commercial Space*

The author do the research on the flow of people, the data statistics is analyzed, flow of the general work days and weekend days. Here we get the following analysis as follows (Fig. 29.9).

#### 29.4.4.1 The Flow Difference of Generally Work Days in Different Time

According to the research, the flow of people is mainly centralized in the subway and underground interface in Xizhimen. The enormous passenger flow of the shopping mall makes Xizhimen an important part of stream of people in commercial space. Stream of people concentrated in the basement in the morning rush hours, especially the dynamic flow of people is greater. Passenger flow mainly concentrated in the storefronts in the morning rush hours, such as catering stores and transit hall. It is intensive period for the passenger flow at noon, There are a large crowd of people on the underground floor and the first floor, as well as the subway and underground interface, then after lunch time there will be a lot of leisure customers flow. In the afternoon, especially after 16:00, metro passenger flow increases, but the flow in the shopping mall would be slightly decreased. At Night, subway passenger travel across, and people enjoy the city life in underground commercial space, as well as the first commercial passenger flow will be increased to a large extent. As the data is shown in Fig. 29.8.

#### 29.4.4.2 The Condition of Space Use on General Weekdays and Weekends

According to the investigation and observation, we found that the guest flow is larger from Monday to Thursday, compared to a average statistics in the shopping mall. From Friday to Sunday, Cade Mall's passenger flow turns out to be of an unprecedented increasing, however, the subway passenger flow peak did not occurs, even the statistics is less than the weekdays. Therefore, for enhancing the purchasing power of people, how to changing the passing flow to becoming static customers flow in Xizhimen Underground commercial space will be important goal of perfect development.

## 29.5 The Optimization Design Strategy of Xizhimen Underground Commercial Space

According to the analysis above, related space optimization policies for Cade Mall located in XIZHIMEN are proposed below:

### *29.5.1 To Promote the Level of Commerce of Underground Commercial Space and to Increase the Actual Number of Well Known Brands Stores*

The whole environment of business has great influence on underground commercial space. The level of Cade mall is located around medium level, however, the overall comprehensive level cannot achieve to the requirement of middle level. One reason is that high priced rent limits the possibility of renting by those traders who have interests. Another reason is because of the distribution of the stores is in a mess, this negative influence may affect the impacts of some major brands in a long run. All in all, it is better to adjust the overall arrangement of this mall and have a long term plans for stable development to meet the requirement of people's expectation for shopping.

### 29.5.2 To Increase the Number of Consumers and Further Increase the Efficiency of Utility of Underground Commercial Space

Current situation of underground commercial space is to induce the cheaper brands and meanwhile construct food court, hair and cosmetology saloon stores, underground cinema etc. However, all the brands are still needed to propagate to gain good reputation. The Mall could retain some consumers from subway. All of these attractions may turn these consumers to loyal and long time consumers.

### 29.5.3 The Optimization Design Strategy of Underground Space

The underground space design of Cade Mall has a certain comprehensive nature and its falling square which forms the semi-closed and half-open architectural environment, all end type space utilization, and the Hall-style Conversion hall, all of these nature provides valuable reference template for the design of the underground space. After investigation, the features of Cade Mall's underground commercial building space are as follows.

#### 29.5.3.1 The Design of Semi-closed and Half-Open Square

The semi-closed and half-open environment which are formed by the falling square has effectively connected the stream of people who are coming from the subway or the underground commercial space. So we suggest that some publicity activities should be set here or we can organize some unique theatrical performances or publicity, and both of these activities will largely promote the gathering of the crowd which can transform the passing crowd to the watching crowd, thus to promote the flow quantity of the underground space.

#### 29.5.3.2 The Space for Corridor

The western part of the underground space is the end of the aisle, so the sequence of the guide is not enough. The stores have just open soon, and they have not yet accumulated much popularity. So it is necessary to introduce more well-known stores to settle and organize some special promotional activities to increase the flow of people. On the other hand, it is also necessary to increase the propaganda here, such as putting up some guide signs of the toilet and stores.

#### 29.5.3.3 The Design of the Transfer Hall

Symmetric space (tetrahedral symmetric or bilateral symmetry), except for centripetalism and centrifugation there is almost no other tendency and it is finally come to a sort of static balance. The subway entrance of Xizhimen and the design of Conversion hall for property of subway upper cover of Cade Mall forms a multidirectional divergence of transport space which is contributes greatly to people's multiple path choice. We proposed that, in order to enhance people's desires to stop and shopping, the logo design of the conjuncture of the underground layer of the main building of Cade Mall and the subway should be more eye-catching and stronger leading.

#### 29.5.3.4 The Design of Oriented Identification System

The proportion and scale of the space and display are coordinated. According to the result of research, it is relatively appropriate of the proportion and scale of Xizhimen underground space and design. But the scores of the exchange streamline organization and the supporting facilities are lower, so it is necessary to strengthen the traffic guidance of underground commercial space. For example, setting up eye-catching signs, striking path prompt on the ground and vivid brand logo and so on, we can form the active guidance for people's behavior in underground space.

#### 29.5.3.5 The Physical Environment of Underground Space

The color is quietly elegant and harmonious, the light is downy and adornment is concise. The underground space of the hall, channels and underground commercial stores, all through the redesign of the decorate and use of physical, optical, electrical environment, further improve the comfort of the underground commercial space and meet the human needs of the people in underground space. We should exclude the use of the explosive discount promotions method which is the major factor to make customers resentment and sense of alienation.

#### 29.5.3.6 Multi Functional Requirements

To realize calm conversion and avoid forced guide line of sight. We can see from the above chart, $3.5 < X \leq 4.5$, which is represent that it is basically satisfied. Therefore, the establishments of appropriate underground commercial space function, not only can improve the utilization of underground space, but also contribute to the formation extensive customer groups. For example, in the area which is close to the North Station, we can follow the setting of the airport's duty-free shops to introduce the souvenirs-gift-typed brand stores and convenience stores in, in that way, not only can we make the function multicultural, but also form a variety of customer groups, and lead the customers of the mall come to shopping in the underground space, at the same time make up for the lack of some commercial function of Cade Mall.

## 29.6 Conclusions

With the great development of Tito and the Metro Rail Transit construction, as well as the surrounding business district livingness improving, a large number of commercial complexes constantly emerge around the subway. However, after an investigation we can find that these commercial development projects are not as good as expected, the actual utilization rate of the rail transportation is not high, the efficiency that subway passenger flow translated into commercial passenger flow is relatively low. On the basis of the research, the author makes an study on Post-Occupancy Evaluation, and then explore the problems and the causes existing in the commercial space, and finally puts forward the underground commercial space development and architectural space optimization strategy.

**Acknowledgment** This research was supported by: National Natural Science Foundation of China (51078022); Doctoral Fund of Ministry of Education of China (20090009120013); Beijing Urban Engineering Design & Research Institute Co., Ltd, Underground Public Space Optimized Simulation Research – Based on Digital Technology (CJ2012-47).

## References

1. Linlin Qian (2011) Research on Post-Occupancy Evaluation of Beijing's metro subway commercial space [D]. Beijing Jiaotong University, Beijing
2. Hongtao Du (2006) Planning and design study of urban traffic hub [J]. City Plan 30(07):85–88
3. Xiaolei Zhu, Shuoxian Wu (2002) The influences of Post-occupancy evaluation on architecture design and its significance to China [J]. Arch J (05):42–44
4. Lin JJ, Gau CC (2006) A TOD planning model to review the regulation of allowable development densities around subway stations [J]. Land Use Policy 23(3):353–360
5. Du Hongbo, Mulley C (2007) The short-term land value impacts of urban rail transit: quantitative evidence from Sunderland, UK [J]. Land Use Policy 24(1):223–233
6. Zhongwei Shen (2009) The core issues of the design of rail transportation hub complex [J]. Time Archit 26(5):27–29

# Chapter 30
# The Spatial Effects of High Speed Rail on Intermediate Cities in China – A Case Study of Beijing-Shanghai Express Railway

**Han Yao and Ze Liu**

**Abstract** Based on case studies of Beijing-Shanghai express railway, this study tries to reveal the spatial effect of high speed rail on urban internal space at the local level. At first, development of high speed rail in China and the Beijing-Shanghai express railway are introduced briefly, and then the general mechanism and special challenges are discussed in detail as to high speed rail's spatial impacts on intermediate cities. Furthermore, potential spatial developments are analyzed with the local backgrounds. At last, This study concluded: (1) accessibility and land rent are bases for high speed rail station's spatial impacts, while locations, connections, station capacity and functions being driving factors for spatial interactions; (2) it should be necessary to balance node-place functions of high speed rail stations based on an evolutionary mode; (3) many typical characteristics for station's development could be observed, especially in terms of urban spatial expansion and reconstruction.

**Keywords** High Speed Rail • Station area • Urban spatial structure • Beijing-Shanghai express railway • China

## 30.1 Introduction

The Beijing-Tianjin intercity rail has been operating from Jul. 2008, which symbols China's entering into the age of high speed rail (HSR), while the launching of the Wuhan-Guangzhou high speed rail in 2009 leads the development of Long-Distance HSR Mainlines. According to the 'Intermediate and Long-Term Plan for

H. Yao (✉)
School of Architecture, Tsinghua University, 100084 Beijing, China
e-mail: yaohan5dd4@hotmail.com

Z. Liu
Administration Center of Urban-Rural Planning, Ministry of Housing and Urban-Rural Development, 100835 Beijing, China

H. Xia and Y. Zhang (eds.), *The 2nd International Symposium on Rail Transit Comprehensive Development (ISRTCD) Proceedings*, DOI 10.1007/978-3-642-37589-7_30, © Springer-Verlag Berlin Heidelberg 2014 

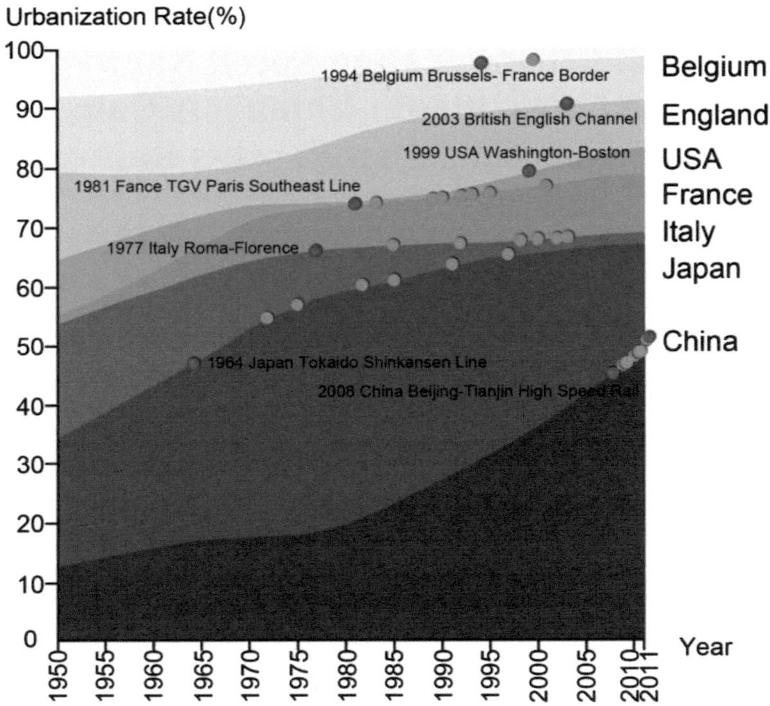


**Fig. 30.1** Comparisons of HSR development an urbanization among China and other countries (Source: (1) World Urbanization Prospects, the 2011 Revision: http://esa.un.org/unup/, (2) China Railway Yearbook 2012)

National Railway Network (revised in 2008)' published by the Ministry of Railway (MOR), in 2020 the total length of HSR will reach to 18,000 km in China, forming a Passenger Dedicated Lines (PDL) network, which will connect every provincial capital cities as well as cities with a populations of more than 500 thousands except Lhasa, and then the HSR population will be more than 90 %.

With comparing HSR development and urbanization among China, Japan and other West European countries (Fig. 30.1), HSR development in China is in a rapid urbanization which would be outstandingly different from other countries. Besides of that, HSR development in China also hold special characteristics as to the economic level, HSR development rate and management system. It can be deduced that those differences would impose significantly different influences to regional and urban development. In a latest study, Chen suggested that the claimed benefit and the real spatial implications of Chinese HSR development should be discussed with more relevant empirical evidences; especially the special economic and institutional contexts should not be ignored [1].

Multilevel analysis is necessary to exploring the spatial influences for urban system and urban space regarding to the HSR [2], while it should base on deeply discussions of each level separately, including international level, national level, regional level, and local level. The existing studies on HSR either emphasize on the macro level of international [3, 4], national [5–6] and regional space [7–9], or emphasize on the development of HSR station and its surrounding areas [10, 11], while other studies on urban level are mainly case analyses. There is few studies focusing on urban spatial development and spatial structure in developing countries, in the background of HSR.

This study is to explore the spatial impacts of the HSR at the local city level, meanwhile, characteristics of urban spatial structure in the context of the HSR. Especially it focuses on the following topics with references to the newly-built Beijing-Shanghai Express Railway:

1. the mechanism through which the HSR impact intermediate cities' spatial development, as well as the special challenges for China.
2. the balance of node-place functions for HSR Terminus Area (HTA).
3. the typical characteristics of urban spatial structures with influences from the HSR and HTA development.

## 30.2 The Beijing-Shanghai Express Railway

In Jun. 2011, the Beijing-Shanghai Express Railway (BSER) was launched, connecting two major coastal economic zones of China: the Bohai Sea Rim and the Yangtze River Delta, with the total length of 1,318 km, the maximum speed of 350 km/h, and reduces the traffic time from 10 to 5 h (Fig. 30.2). Crossing 4 provinces and 3 municipalities under the central government directly, the BESER links 24 stations in 23 cities (one county among them). Figure 30.3 demonstrate the spatial relationship between those 24 stations and their corresponding cites.

Since the spatial impact of HSR at local urban space is recognized to be related to the social, economic and institutional circumstances comprehensively, including factors of population, economic scale, superior industrial, resource advantages and divisions in the urban networks in a larger geographic space [12]. To take into account of these factors, 23 cities along the BSER are classified into four types in this study (Table 30.1, Fig. 30.2), which will be used for to further the discussions of HSR's different spatial impacts.

Moreover, different from HSR stations in European countries, 22 stations along the BSER are newly-built on greenfield at the peripheral or rural area, while only 2 stations are upgraded from old rail station sites with completely reconstructions.

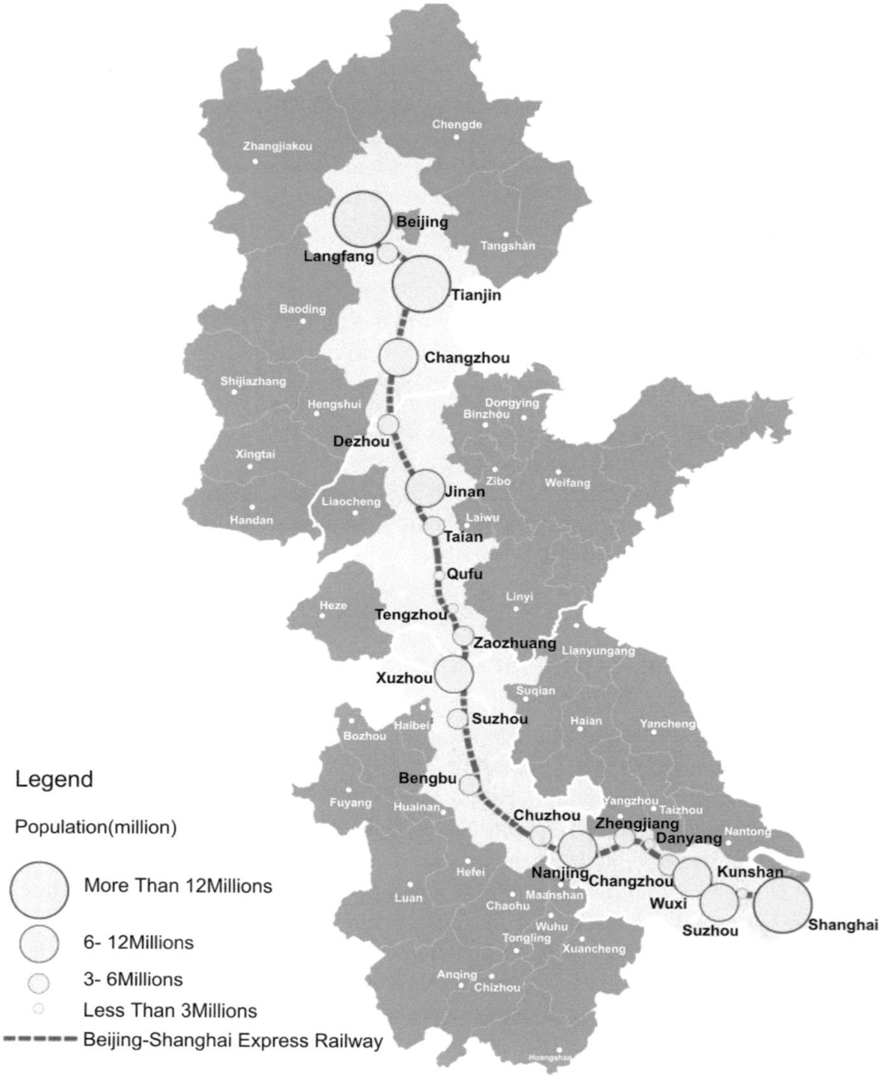


**Fig. 30.2** Beijing-Shanghai express railway in the national rail network and the cites being connected

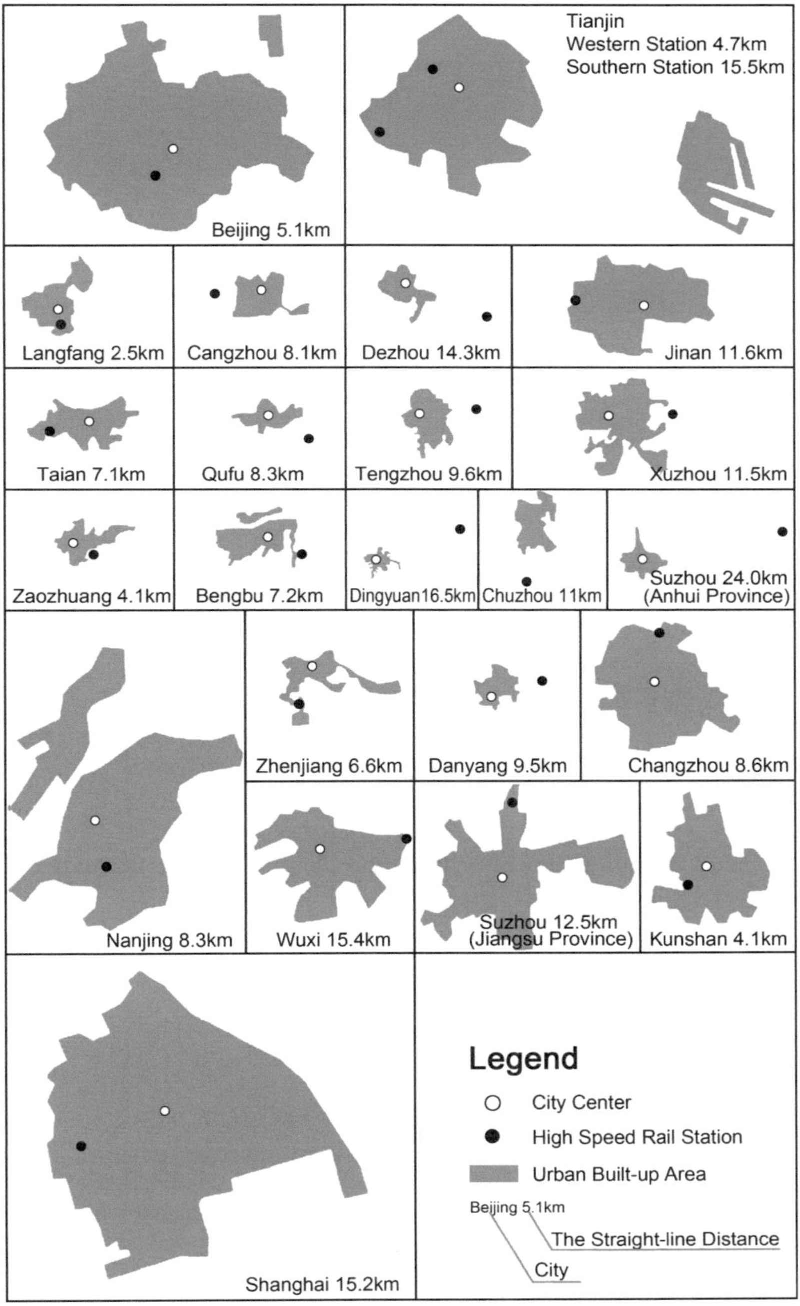


**Fig. 30.3** Spatial relationship between stations and cites along the BSER (Source: revised from: Wang Lan [14])

**Table 30.1** Classifications of connected cities along the BSER[a]

| Type | Cities | Main city characters |
|---|---|---|
| World Cities | Beijing, Shanghai, Tianjin | Population: more than 12 millions<br>GDP: more than 100,000,000 millions |
| Regional Cities | Jinan, Nanjing, Wuxi, Suzhou (Jiangsu Province) | Population: 6–12 millions<br>GDP: less than 100,000,000 millions |
| Local Cites I | Taian, Zaozhuang, Xuzhou, Suzhou (Anhui Province), Bengbu, Zhengjiang, Changzhou | Population: 3–6 millions<br>GDP: less than 50,000,000 millions |
| Local Cites II | Langfang, Cangzhou, Dezhou, Qufu, Tengzhou, Dingyuan, Chuzhou, Danyang, Kunshan | Population: less than 3 millions<br>GDP: less than 10,000,000 millions |

Source: Residential Populations of each city from the Six National Population Census (All the data of cities in this study is collected from the China City Statical Yearbook (2011) except special notes)

[a]The classification is a achieved through qualitative based on many factors comprehensively, so a city of certain type is not strictly in the corresponding value scope of population and GDP

## 30.3 Mechanism of the Spatial Impact of HSR

Being the important catalyst, HSR delivers its spatial impacts at the city level through the development of station and HTA. In following part, the interaction mechanism and influencing factors of interactions between high speed rail and urban spatial development will be analyzed deliberately, together with special challenges for China.

### *30.3.1 Interactions Between the HSR and Urban Spatial Development*

As shown in the outer loop of Fig. 30.4, accessibility improvements from the HSR will reinforce city location, lead to the savings of travelling time and cost, thus promote urban economic development and increase urban traffic demand. But the study on city level should base on development of HTA to realize the interaction between the HSR stations and urban spatial development.

As shown in the inner loop of Fig. 30.4, based on accessibility and land rent, HSR promote the evolution and reconstruction of urban space through the mutual interactions between urban transport and spatial development [13]. The entire interaction process including following components: on one side, the construction

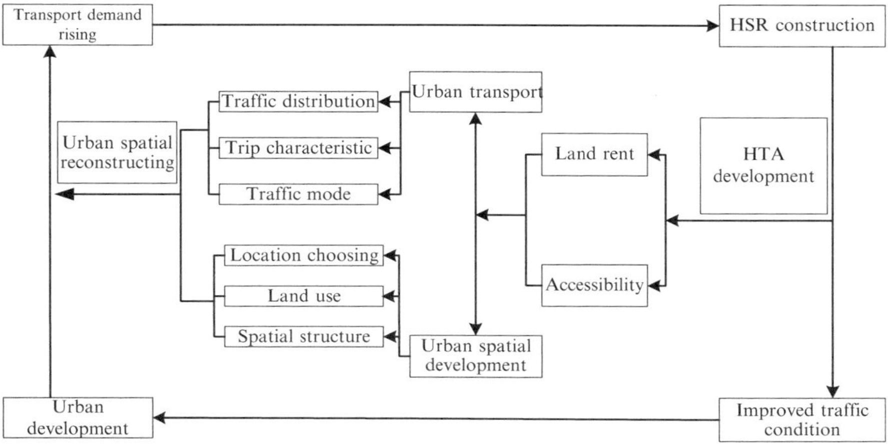


**Fig. 30.4** The interactions between high speed rail and urban spatial structure

and development of HSR stations and surrounding areas promotes the land value and improves the accessibility of the station, which lead to concentrations of populations, offices, commercials and services. Then further urban spatial developments are achieved through location behaviour of individuals and companies, land use, and spatial evolution. On the other side, based on land use, the evolution of urban space in turn propels the development of urban transport through traffic distribution, trip characteristic and traffic patterns. With settings of investments, planning polices and other external conditions, those interactions continue to intensify the land use, increase the traffic demand and improve traffic conditions.

The above analysis reveals the general process of the interactions between urban space and HSR stations, which is based on the accessibility and the land rent. However, more studies focusing on the HSR's potential impacts for Chinese cities will be carried out in Sect. 30.3.2, within Chinese economic and institutional circumstances.

### *30.3.2 Driving Factors and Special Challenges for the Interaction*

The driving factors for interactions between HTA and urban space are discussed in this part, including locations, transportation connections, station capacity, and station functions. Those factors reflect the locational, traffic, scale and functional relationship between station and urban space [14]. General driving factors for the interaction and special challenges in Chinese economic and institutional contexts are discussed in following study; especially 24 stations along the BSER are used as analysis cases.

**Table 30.2** The Location priorities of BSER stations according to city classifications

| Type of city | Central area | Edge area | Rural area | Average distance (km) | Maximum distance (km) | Minimum distance (km) |
|---|---|---|---|---|---|---|
| World cities | ●●● | ●● | | 10.1 | 15.5 | 4.7 |
| Regional cities | ●● | ●●● | | 12 | 15.4 | 8.3 |
| Local cities I | ● | ●●● | ●● | 8.8 | 16.5 | 4.1 |
| Local cities II | | ●● | ●●● | 10.2 | 24 | 2.5 |

#### 30.3.2.1 Station Location in the City

Generally the location relationship refers to place the station in the city to extending access to more populations, jobs and businesses nearby, which will further economic benefits from HSR development. It is concluded that for stations located in areas with limited business activities, very little development has been stimulated [15]. Obviously the city center locations hold priority than those in peripheral and rural sites, for they already have high concentrations of people and economic activities.

Among 24 stations along the BSER, 4 stations are located in city center, while 9 in peripheral or suburban areas, and 11 in rural areas. A striking case is the Dingyuan station which is newly built on the farmland 16 km away from the city center. Table 30.2 shows the location priority of BSER stations based on the city classifications in Sect. 30.2, and a larger number of the black dot represents the higher priority.

There are three major challenges which would be useful to understand those different location priorities, concerning the urban expansion, rail land polices, and assessing methods of local government in China.

The first challenge is the continuing rapid urbanization which is bringing strong growth of populations and incomes for Chinese cities. Not only top cities in forming metropolis belt, such as Beijing and Shanghai, but also the medium and small cites in under-developed regions are experiencing high speed spatial expansions.

The second one is the land policies and related laws. Those laws and policies in use empower the local government to expropriate land from its users with limited compensation comparing to its potential revenues. Besides that, it is banded to use the rail yard over the HSR station for comprehensive development. With the limited unused land around exiting station and city center, sites on suburban and rural parts become the reasonable and profitable choices.

The third challenge is the exiting assessment indicators based on GDP in Chinese bureaucratic system. While transport infrastructures, such as the rail station are recognized as the important impulse for city development, most ambitious mayors prefer to locate the HSR station in a site far away from the downtown, hoping to attract more invest and business to stimulate local economy with cheap land.

#### 30.3.2.2 Transportation Connections to the Station

As has been stressed by many researchers, the intermodal connection with HSR station and giving the traveler the most seamless trip are necessary to deprive the maximum potential of the HSR. Transform the HSR station into an intermodal hub needs good connections and integrations of various transport modal, including automobile, public transport and bicycle/pedestrian networks [16].

Briefly there are two reasons for this necessary integration. At first, as far as the feeding and distribution system, local and regional transport connections extend the ridership catchment areas of HSR, play roles of complementary, instead of competitive modes [4]. Furthermore the high level integration will reduce the transfer penalties and improve transfer experience to the travelers. In a word, better connections offer transport network of high efficiency and high capacity to access and egress to and from the HSR station.

Table 30.3 reveals the transportation connections of stations along the BSER. The 'STT' stands for the Shortest Traveling Time from the city center to the station, which are calculated from the online querying of public transport via Baidu (http://map.baidu.com). Firstly it is a time-consuming trip from the city center to most of those 24 stations, with an average value of near 50 min. Similarly, poor transport connections could be found for many BSER stations, because most cities are lacking of good mass transit, as well as other transfer mode. Those poor connections could be attributed to the current economic trajectory in local governments, especially for medium and small cites.

#### 30.3.2.3 The Station Capacity and the City Size

Both the station capacity and the city size have considerable impacts on the spatial interaction between HSR stations and those cities being connected.

Frequency is the key indicator to measure the station capacity, together with number of dock and rail line, building area, which is usually determined by potential ridership, and size of the corresponding city. Regarding to the average waiting time, frequency is seemed as more important than travel time, and have the potential to provide service of good quality with high values [17]. In addition, besides the scale economy and the agglomeration effect, city size also influences station development through the station location and transportation connections.

Table 30.4 demonstrates service frequencies of stations and city sizes in terms of population, GDP and built area. Although it is not a linear relationship between the city size and the HSR service frequency, large city are always the popular traveling destinations with high frequencies. This relationship exposes an urban system of central place pattern, which is the typical picture in modern China.

**Table 30.3** The transportation connections of BSERHSR stations in intermediate cities

| Station | STT (Minute) | Interchange mode[a] | Num. of subway line |
|---|---|---|---|
| Beijing Southern Station | 30 | City Bus, Subway, Regional Bus, Conventional Rail, EMU, HSR | 1 in running<br>1 in planning |
| Langfang Station | 30 | City Bus, EMU, HSR | – |
| Tianjin Western Station | 40 | City Bus, Subway, Conventional Rail, EMU | 1 |
| Tianjin Southern Station | 80 | City Bus, EMU, HSR | 1 in planning |
| Cangzhou Western Station | 50 | City Bus, EMU, HSR | – |
| Dezhou Eastern Station | 100 | City Bus, EMU, HSR | – |
| Jinan Western Station | 80 | City Bus, EMU, HSR, Airport | – |
| Taian Station | 60 | City Bus, Regional Bus, EMU, HSR | – |
| Qufu Eastern Station | 65 | City Bus, EMU, HSR | – |
| Tengzhou Eastern Station | 60 | City Bus, EMU, HSR | – |
| Zaozhuang Station | 35 | City Bus, EMU, HSR | – |
| Xuzhou Eastern Station | 70 | City Bus, EMU, HSR | – |
| Suzhou Eastern Station (Anhui Province) | 170 | City Bus, EMU, HSR | – |
| Bengbu Southern Station | 60 | City Bus, EMU, HSR | – |
| Dingyuan Station | 140 | City Bus, EMU | – |
| Chuzhou Station | 75 | City Bus, EMU | – |
| Nanjing Southern Station | 30 | City Bus, Subway, EMU, HSR | 1 |
| Zhenjiang Southern Station | 60 | City Bus, EMU, HSR | – |
| Danyang Northern Station | 70 | City Bus, EMU, HSR | – |
| Changzhou Northern Station | 70 | City Bus, Regional Bus, EMU, HSR | – |
| Wuxi Eastern Station | 90 | City Bus, EMU, HSR | – |
| Suzhou Northern Station (Jiangsu Province) | 90 | City Bus, EMU, HSR | 1 in planning |
| Kunshan Southern Station | 50 | City Bus, EMU, HSR | – |
| Shanghai Hongqiao Station | 40 | City Bus, Regional Bus, Subway, EMU, HSR, Airport | 2 |

[a]According to official classifications of MOR, passages dedicated rails in running are classified into three types, namely HSR (with a speed of more than 250 km/h), EMU (with a speed between 200 and 250 km/h) and Conventional Rail (other trains)

Furthermore to compare with the Table 30.1, it can be concluded that HSR station in those World Cities and Regional Cities tend to location in the inner urban area or a site proximity to the existing rail station or airport, like Beijing and Shanghai; to the contrast, those Local Cities II with smallest population and GDP, like Qufu and Danyang, prefer to locate the station in rural areas far away from the city center. For cities with a size amid of those three types, consequently most stations situate in fringe and near suburban areas, such as HSR stations in Jinan, Xuzhou, and Suzhou (Fig. 30.5). Similar conditions could be found with transportation connections concerning the station capacity and city size.

**Table 30.4** Service frequency of the stations and size of corresponding cities along the BSER

| Station | Service frequency | Population (million) | GDP (billion) | Built -up area ($km^2$) |
|---|---|---|---|---|
| Beijing Southern Station | 96 | 1,961.24 | 1,390.44 | 1,186 |
| Langfang Station | 11 | 435.88 | 33.43 | 59 |
| Tianjin Western Station | 10 | 1,293.82 | 856.15 | 687 |
| Tianjin Southern Station | 17 | | | |
| Cangzhou Western Station | 26 | 1,293.82 | 42.93 | 46 |
| Dezhou Eastern Station | 29 | 713.41 | 36.67 | 89 |
| Jinan Western Station | 108 | 556.82 | 295.98 | 347 |
| Taian Station | 23 | 681.40 | 61.14 | 107 |
| Qufu Eastern Station | 19 | 549.42 | 26.58 | 21.2 |
| Tengzhou Eastern Station | 12 | 64.05 | 63.39 | 46 |
| Zaozhuang Station | 14 | 160.37 | 72.81 | 119 |
| Xuzhou Eastern Station | 35 | 372.93 | 177.95 | 239 |
| Suzhou Eastern Station (Anhui Province) | 11 | 858.05 | 24.65 | 53 |
| Bengbu Southern Station | 16 | 535.30 | 31.52 | 105 |
| Dingyuan Station | 2 | 316.40 | 8.65 | 37 |
| Chuzhou Station | 13 | 77.92 | 17.42 | 60 |
| Nanjing Southern Station | 114 | 393.80 | 451.52 | 619 |
| Zhenjiang Southern Station | 17 | 800.47 | 84.49 | 109 |
| Danyang Northern Station | 10 | 311.34 | 60.77 | 33.2 |
| Changzhou Northern Station | 24 | 96.04 | 231.63 | 153 |
| Wuxi Eastern Station | 28 | 459.20 | 298.66 | 231 |
| Suzhou Northern Station (Jiangsu Province) | 38 | 637.26 | 357.28 | 329 |
| Kunshan Southern Station | 19 | 1,046.60 | 210.03 | 150 |
| Shanghai Hongqiao Station | 100 | 164.63 | 1,697.16 | 866 |

#### 30.3.2.4 Functions of HTA and City

As the 'node-place' pattern proclaimed, significant transport node should ideally be the vibrant place for the city [18]. With regarding to the development of technologies innovations, information and service economic in current globalization world, HSR terminus area's has also been perceived to correlate both functions of node and place in a time-space shrinking space, which is viewed as 'second rail age' [19]. Based on deeply examinations of HSR station development of Euralille, Amsterdam and Rotterdam, Trip concluded the keys for planning 'quality of place', including diversity of people and functions, integration into the city and public space creation [20]. As to the economic activities and land use at HTA, many literatures have highlighted the economic development in service and information sectors generated by HSR, including business, public administration, leisure, commerce, tourism, exhibition and so on [21].

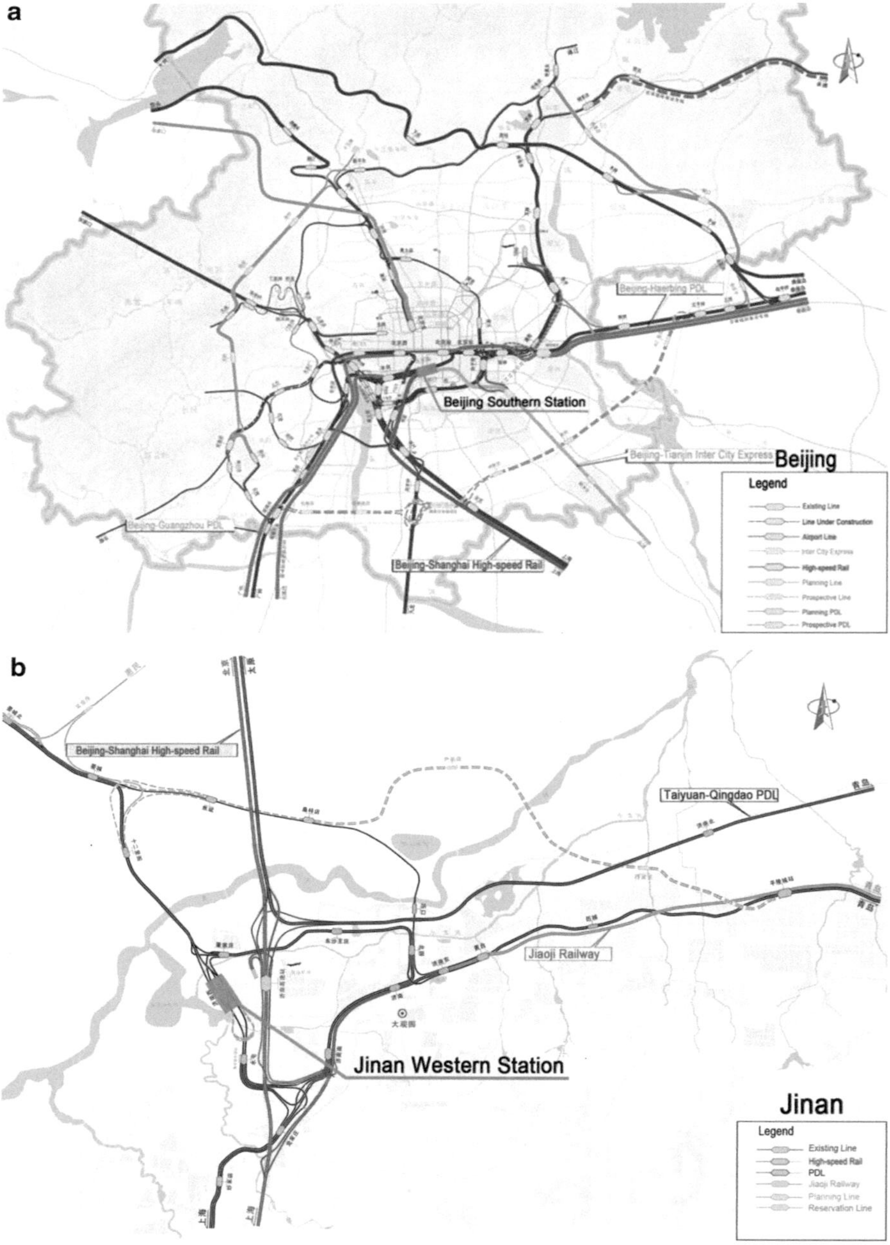


**Fig. 30.5** Location characters in cities of different sizes (**a**) Location characteristics in World/ Regional Cities (Beijing) (**b**) Location characteristics in Local Cites I (Jinan) (**c**) Location characteristics in Local Cites II (Qufu)

c

**Fig. 30.5** (continued)

However the situations are quite different for cities along the BSER. Firstly, most connected cities are in a industrialization process with slow development of information and service industries. Figure 30.6 displays the frequency distributions of second and tertiary industrial proportion of GDP in 23 cities along the BSES, with the mean value of 49.3 % and 44.8 % respectively, comparing with national values of 46.8 % and 43.1 % in 2010.

Secondly, existing functions of terminus areas indicates it is a hard work for a HTA to achieve the ideal functions. For those of World Cities and Regional Cities, the obstacles always lies in too many node or place functions (Fig. 30.7c); while for those of Local Cities I and Local II, a newly-built station means a long way for its integration into the external and internal structures, to fulfill both node and place functions (Fig. 30.7a, b).

Thirdly, the established experiences have proved that only limited types of services industries have succeed in Chinese railway termini area, which would not help to attract multiple activities and shape a desirable place. A survey of land use for the place functions of seven railway terminus area found that, real estate, especially housing, had become the main benefits for HTA development (Table 30.5) [22]. Yet it should be noted that those 7 are selected from 17 terminuses in case studies, for their better economic performances than others.

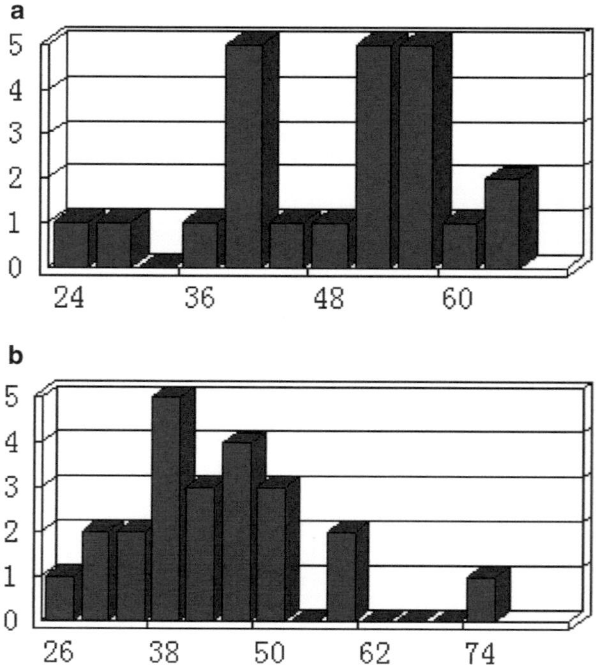


**Fig. 30.6** frequency distributions of the second and tertiary industrial proportion (**a**) Second industrial Proportion (%) (**b**) Tertiary industrial Proportion (%)

## 30.4 Prospections on Urban Spatial Development

According to the preceding study, some special challenges is arising for interactions between HSR stations and urban space, because of the economic and institutional circumstances in China, especially in terms of location strategies, transportation connections, services frequencies and station functions. Base on those discussions, the possible urban spatial development will be revealed in the following part.

### *30.4.1 How to Achieving the "Node-Place" Balance?*

#### 30.4.1.1 An Evolutionary View of the Terminus Area's Development

It seems urgent to check the whole development process of a station, in order to incorporate above challenges and discern the changing roles of the HSR terminus, because the 'Node-Place' pattern can only be used as an indicator reporting the existing states of station' functions. To locate the station's position in its evolutionary process, then the possible spatial development could be figured out considering different urban background.

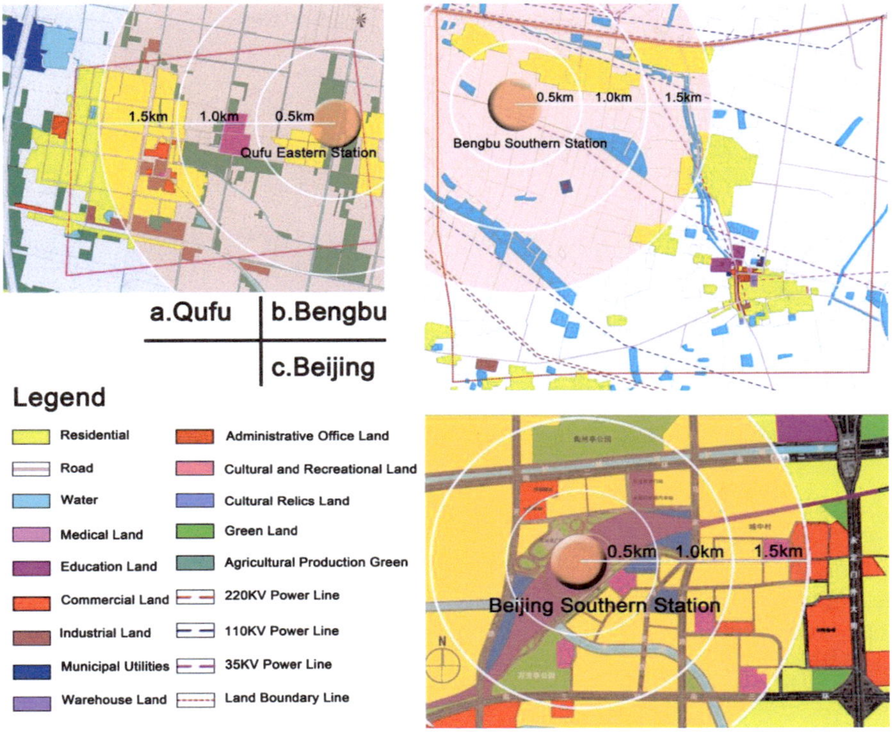


**Fig. 30.7** Exiting functions of terminus areas as the construction of station

**Table 30.5** Area composing of main functions of seven railway terminus area in China (%)

| Land use | Tianjin Station | Hefei Station | Kunshan Station | Shenyang Northern Station | Guangzhou Eastern Station | Wuxi Station | Shanghai Station | Average |
|---|---|---|---|---|---|---|---|---|
| Market | 2.41 | 18.79 | 0.00 | 10.33 | 6.82 | 15.61 | 0.00 | 7.71 |
| Hotel | 6.26 | 3.87 | 1.08 | 11.45 | 11.06 | 2.85 | 12.05 | 6.95 |
| Office | 26.59 | 16.20 | 0.00 | 58.86 | 36.22 | 14.34 | 24.69 | 25.27 |
| Residence | 57.78 | 59.20 | 94.76 | 17.44 | 37.86 | 64.55 | 61.61 | 56.17 |
| Commercial | 4.49 | 0.24 | 0.21 | 0.47 | 0.00 | 0.85 | 0.00 | 0.89 |
| Others | 2.47 | 1.70 | 3.95 | 1.46 | 8.03 | 1.81 | 1.64 | 3.01 |

Source: Lin Chenhui and Ma Xuan [22]

Diagrammatically Fig. 30.8 depicts the evolutionary process of a railway station from the introduction to its running stably, considering locations, transport connections, land rent, 'node-place' functions and influenced areas of the station and its surroundings.

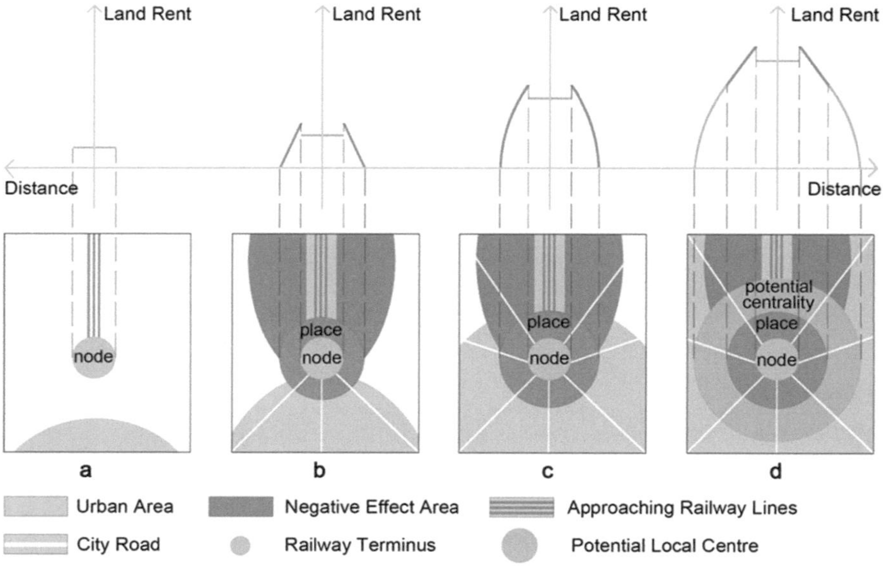


**Fig. 30.8** The evolutionary process of railway terminus. (**a**) Island (**b**) Polarizing (**c**) Integration (**d**) Maturity (Source: revised from Paksukcharern Thammaruangsri [24])

Figure 30.8a -'Island' stage represents the immediate introduction of a station on a barren land out of the inner city. The construction input raise the value of that site, which would be lower than city center. To be the main transport infrastructure, apparently the terminus's chief function is node-oriented in limited building spaces.

Figure 30.8b -'Polarizing' stage shows a high-speed growth of the terminus area. On one side, node functions are strengthened with continually improved transportation connections and services level. On the other side, With the aggregation and diffusion effects, extensive flow of economic, labor, information and technology are gathering and spreading from here; the terminus area now are performing its place functions to attract many urban activities. The land value has being increased with a fast spatial growth, as the influenced area has extended to the urban edge.

Figure 30.8c- 'Integration' stage defines terminus area's integrating into urban settings. With the expansions of both terminus area and urban space; gradually the terminus area merges with urban areas, leading to an inner location. Terminus' place functions are strengthened to accommodate more urban activities, and become one important component of urban public space system. The land rent keeps growing, but with a slower rate than Polarizing stage. It should be emphasized a negative effect area caused apparently by the HSR, and related infrastructures and buildings, would emerge around the HSR line and HTA areas.

Figure 30.8d- 'maturity' stage depicts the stable development of a station. No matter the node-place balance has been fulfilled or not, the type and component of its function are too solid to change because of the threshold effect [23]. While the location has been moved to the inner city area with a larger negative environment, Moreover this stage also stands for what European countries are experiencing currently, where redevelopment strategies and stimulating programs are employed to revitalize the area.

Last but not least, it can be concluded the above four stages will be turned into a new circle with large-scale investment or important transportation innovations, such as regenerations of an old conventional railway station with a HSR station.

#### 30.4.1.2 Ways Ahead for Different Terminus Area

Considering above evolutionary process, newly-built or reconstructed stations along the BSER can be categorized into two types, those at the Maturity stage and the others at the Island stage.

For stations at the Maturity stage, they are renewed from old stations with completely newly-built building at the city center, while the surrounding area is all the same as before. These stations would usually be found in the World Cities and Regional Cities along the BSER, such as the Beijing Southern Station, Tianjin Western Station. What those Maturity stations facing are problems resulting from existing functions and lacking of usable land. As it can be observed in Fig. 30.5c, there isn't any empty space around the Beijing Southern Station, while most land has been used for housing and commercial industry. So it should be the groundwork for those stations to build a robust cooperation mechanism between public and private investment, central government and local government through varied financial modes. While local governments and MOR had taken the main costs for station constructions to across the unavoidable threshold, an institutional framework should be useful to import other place functions into the HTA areas, which should be time-consuming and costly.

For stations at the Island stage, they are always in the Regional Cities, Local Cities I and Local Cities II as the new stations in greenfield, while their locations depend on priorities as the Table 30.1 has revealed, generally at the city edges or rural areas. Being at the first stage of a station's evolution process, they are in the same urgency to improve their transportation connections, build public space, foster place functions to speed into the next Polarizing stage, while the situations are varied between different cities. Generally the situations for Regional Cites and Local Cities I usually seems optimistic. Such cities like Nanjing, Suzhou (Jiangsu Province), Wuxi, play key roles in regional or inter-regional development with superior to other cities, and they may get best benefits from the HSR and HTA development. But for cities of Local Cities II and some of Local Cities I, there should be a long time for them to meet the substantial influences before being prepared for them, especial for cities like Cangzhou, Chuzhou and Dingyuan.

## 30.4.2 Urban Spatial Expansions and Reconstruction

With the rapid urbanization in China, all HTA developments are carried out in the urban background of spatial expansions and reconstructions, so it is essential to analyze HTA's spatial impacts with perspectives of expansion. Further discussions will firstly focus on which type of spatial expansions will be generated, and to what extent should they impact the urban space; then to calculate the potential urban spatial forms resulting from the HTA development.

### 30.4.2.1 Differences of Spatial Impacts for Intermediate Cities

Based on classifications of intermediate cities in Sect. 30.2, different urban spatial expansions derived from the HSR and HTA development could be discussed separately.

Firstly, for HSR stations in the World Cities, the spatial expansion caused by HTA development advances spatial expansions in regional area, while reconstructions of urban space and functions in city urban areas. With the ruling role in national urban system, those cities, such as Beijing, Shanghai, tend to get the best benefits from HSR, with superiorities than other cities in terms of station locations, transport connections, HSR frequencies, functions of station area and city, and other influenced factors. From the perspective of metropolitan development, the spatial expansions would promote integrations of cities in the metropolitan area to form polycentric mega-city regions. While the spatial expansions caused by HTA development in the city center areas should not be so obvious, the development promoting the HTA to be a subcenter or a secondary functional node of the city. And this could be found in Shanghai at to the spatial development influenced by Hongqiao Station (Fig. 30.9a).

Secondly, as to stations in Regional Cities, a new HSR functional region should accelerate urban spatial sprawl and reconstruct the urban space into a polycentric metropolitan. Figure 30.9b shows the urban spatial development of Nanjing city, considering the influences from Nanjing Southern Station.

Thirdly, for stations in Local Cities I, HTA has been the catalyst for local governments' ambitious urban spatial planning and development. Taking the HTA as the scarce chance to strengthen their comprehensive competitiveness, those cities try to achieve an enterprising jump in spatial and economic development through HSR. For example, according to local spatial plan, it could be found many cities of Local Cities I have the goal to build new towns around HTA, together with other ambitious development plan. Although the spatial influences of those development planning should be evaluated in the future, but it could be deduced that HTA development in such a city would impose a remarkable influences to urban space.

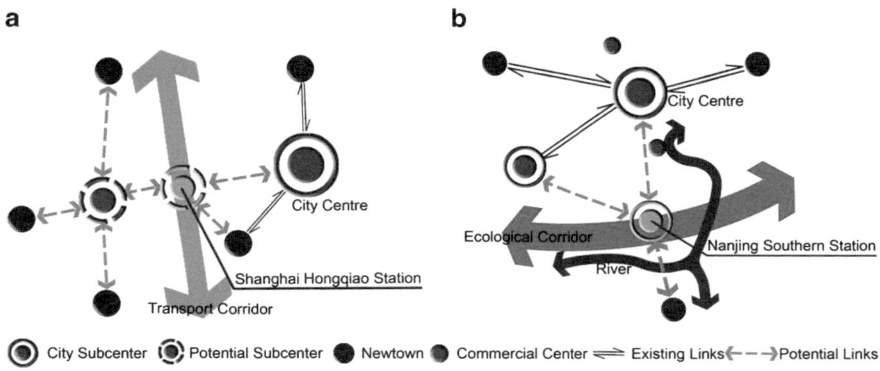


**Fig. 30.9** Reconstruction of urban space in (**a**) Shanghai and (**b**) Nanjing

At Last, for those stations in Local Cities II, special functional district should be the main development goal in near and medium future, while the spatial influence will not be so striking for those cities.

According to above differences, it could be summarized that the HSR and HTA developments bring outstanding influences to urban functions and its functional space for mega cities in metropolitan or regional area, especially for those of World Cities and Regional Cities, but as to the urban form, the impacts are not so distinct; while for medium and small cities of Local Cities II, the situation is just contrary to that for mega cites.

### 30.4.2.2 HTA and Urban Spatial Reconstruction

Based on above discussions of potential spatial developments, some typical characteristics of the urban form, brought out by the HSR, could be concluded, which are mainly depending on regional conditions and the urban scale.

The first one is integrations of urban space, in terms of polycentric mega-city regions and metropolitan areas, and urban integrations for city pairs, and the second character refers to the formation of subcenter in large cities. Moreover, special functional district resulting from HTA developments could be found in medium and small cities.

## 30.5 Conclusion and Discussion

To be concluded, based on a short introduction of HSR development in China, this study firstly argues that the spatial effects of high speed rail should be taken into account with special considering of social, economic and institutional factors. Secondly, referencing the interactions between urban transport and land use, the spatial relationships between the HSR and a city are revealed, which shows the importance of accessibilities and land rent. Furthermore, station locations, transportation connections, station capacity, functions of stations and cities are concluded as key driving factors for those spatial relationships. Thirdly, based on the revolutionary process of railway terminus, this study proposes a conceptual framework to understand the node-place balance of the HTA. At Last, the different spatial expansions and reconstructions of intermediate cities are investigated considering their special settings.

## References

1. Chen C (2012) Reshaping Chinese space-economy through high-speed trains: opportunities and challenges. J Transp Geogr 22(0):312–316
2. Ureña JM, Menerault P, Garmendia M (2009) The high-speed rail challenge for big intermediate cities: a national, regional and local perspective. Cities 26(5):266–279
3. Gutiérrez J, González R, Gómez G (1996) The European high-speed train network: predicted effects on accessibility patterns. J Transp Geogr 4(4):227–238
4. Vickerman R (1997) High-speed rail in Europe: experience and issues for future development. Ann Reg Sci 31(1):21–38
5. Chen C, Hall P (2011) The impacts of high-speed trains on British economic geography: a study of the UK's InterCity 125/225 and its effects. J Transp Geogr 19(4):689–704
6. Inglada V, de Rus G (1997) Cost-benefit analysis of the high-speed train in Spain. Ann Reg Sci 2(2):1–20
7. Blum U, Haynes KE, Karlsson C (1997) The regional and urban effects of high-speed trains. Ann Reg Sci 31(1):1–20
8. Bonnafous A (1987) The regional impact of the TGV. Transportation 14(2):127–137
9. Sasaki K, Ohashi T, Ando A (1997) High-speed rail transit impact on regional systems: does the Shinkansen contribute to dispersion? Ann Reg Sci 31(1):77–98
10. Bertolini L (1996) Nodes and places: complexities of railway station redevelopment. Eur Plan Stud 4(3):331–345
11. Pol PMJ (2003) The economic impact of the high-speed train on urban regions. In: Conference paper from European Regional Science Association, pp 1–20
12. Kim KS (2000) High-speed rail developments and spatial restructuring: a case study of the capital region in South Korea. Cities 17(4):251–262
13. Wegener M, Fuerst F (2004) Land-use transport interaction: state of the art (January 1, 2004). Available at SSRN: http://ssrn.com/abstract=1434678, doi:10.2139/ssrn.1434678
14. Wang Lan (2011) Research framework of high-speed railway impact on urban space. Urban Planner 27(7):13–19 (in Chinese)
15. Greengauge 21 (2006) High speed trains and the development and regeneration of cities. London: 15. http://www.greengauge21.net/assets/European_Regeneration_Experience.pdf. Accessed 15 Mar 2012
16. Preston J, Wall G (2008) The ex-ante and ex-post economic and social impacts of the introduction of high-speed trains in South East England. Plann Pract Res 23(3):403–422
17. Lu L (2003) The vital role of metropolitan access in commuter, regional, intercity, and overnight rail passenger transportation. Massachusetts Institute of Technology, Cambridge
18. Bertolini L (1999) Spatial development patterns and public transport: the application of an analytical model in the Netherlands. Plann Pract Res 14(2):199–210
19. Hall P, Banister D (1994) The second railway age. Built Environ 3/4(19):157–162

20. Trip JJ (2007) What makes a city?: Planning for 'quality of place': the case of high-speed train station area redevelopment. Delft University Press, Delft, pp 67–81
21. Haynes KE (1997) Labor markets and regional transportation improvements: the case of high-speed trains an introduction and review. Ann Reg Sci 31(1):57–76
22. Lin Chenhui, Ma Xuan (2012) Function type and development model of HSR station in China. Urban Transp China 10(5):41–49 (in Chinese)
23. Hurlin C (2006) Network effects of the productivity of infrastructure in developing countries. Policy Research Working Paper Series No 3808, The World Bank, pp 5–9
24. Paksukcharern Thammaruangsri K (2003) Node and place, a study on the spatial process of railway terminus area redevelopment in central London [D]. Doctoral thesis, University College London, pp 89–90. http://eprints.ucl.ac.uk/4384/

# Chapter 31
# Regional Studies on Form Design of Metro Station Entrance – Taking Changsha Metro Line 2 Entrance as an Example

**Guangyao Zhao, Wei Sun, and Xiaojie Wang**

**Abstract** This paper studies on the building form design of the entrance of Changsha Metro Line 2. The method of how to form the local characteristics in metro entrance design is discussed. For the form design of the entrance building of Changsha Metro Line 2, regional and cultural characteristics are taken as the main design elements and the creative on natural environment, humanistic environment and city space order are presented.

**Keywords** Metro station entrance • Form design • Regional culture

Since 1863 when the world's first subway has been constructed in London, subway has changed the pattern of city public transportation due to its large transportation volume, high efficiency and it greatly promoted the development of the city traffic. Metro entrance is a significant part of a Metro station. Therefore, entrance design influences the rationality of the metro traffic organization, and restricts the function of the city space structure. Most metro entrances are specially located in the city center or downtown area on both sides of the street sidewalk or intersection corner [1], thus their architectural forms are directly related to the road landscape and city image. With the rapid development of the city culture construction, Metro station entrance is often considered as the important element in the city landscape endowed with rich cultural connotation.

G. Zhao • W. Sun (✉)
School of Architecture and Design, Beijing Jiaotong University, Beijing, China
e-mail: sunwei@bjtu.edu.cn

X. Wang
Fifth Railway Survey and Design Institute, Beijing, China

H. Xia and Y. Zhang (eds.), *The 2nd International Symposium on Rail Transit Comprehensive Development (ISRTCD) Proceedings*,
DOI 10.1007/978-3-642-37589-7_31, © Springer-Verlag Berlin Heidelberg 2014

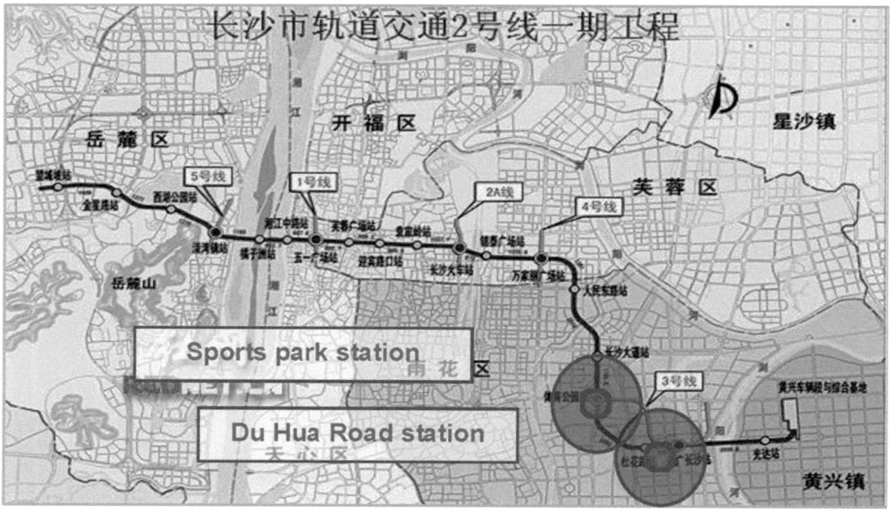


**Fig. 31.1** General plan (provided by the Fifth Railway Survey and Design Institute)

## 31.1 Introduction of the Project

The first-stage project of Changsha Metro Line 2 project which is crossing Xiangjiang from west to east, going through the city centre and up to the new Changsha Railway Station is 21.925 km in length. There are totally 19 stations on line, from Wangchengpo station in the west and Guangda Station in the east. The entrance buildings of the Sports park station and Du Hua Road station locate in Yuhua District in Changsha City and the Changsha County. On the north side of the sports park station is the planning new sports town, while on the southeast corner of the station is the Changsha Wan Guo commercial square, and on the south side is the planning sports park. Du Hua Road Station is on the west side of the intersection of Li Tuo Road and Du Hua Road and on the east side of the Beijing-Zhuhai highway. Fish ponds and warehouses are near the station. The station is surrounded by a large area of land to be developed. Figure 31.1. This paper will take the entrance designs of Sports park station and Du Hua Road Station as examples to show the regional culture characteristics in the design of Changsha Metro Line 2 entrance buildings.

## 31.2 Idea of the Regional Metro Station Entrance Form Design

The metro station entrance can be classified into three forms: independent, integrated and sinking type. Fully enclosed entrance has a roof and enclosed maintenance and its architectural form has significant independence. With the development of society, the form design work is of rich mental functions, and after the law beyond the general sense, it finally return to the personality and to the local geographical mood. The metro station entrance is a space with universal significance though the single function. In the form design, it is particularly important to highlight the image characteristics and the integration into the regional culture.

### 31.2.1 Focus on Natural Environment and Keep Harmonious with Regional Environment

Geographical environment is the basic characteristic of a region. Different region has different natural environment such as climate, topography, landforms, vegetation and so on. With the distribution of the subway lines over the world, the metro station entrance is easy to show industrialization and stylized phenomenon because of its small volume and less constitution components. Full consideration should be given to the natural factors in the form design of entrance building to select the appropriate wall thickness, material and constructivism that are in line with the local conditions.

### 31.2.2 Combing Cultural Environment Elements, Manifesting the Regional Cultural Landscape

In the process of urban development, city culture is generating and continuing the historical context, forming the social environment of the city with its physical facility and intangible forms. Locating in city, metro station entrances clearly identify the environment coordinates. To emphasize the humanities elements such as local customs and historical context, the design of public facilities should be combined with the city culture, to stimulate the building city catalyst function, get rid of surface design and give cultural connotation to the architecture.

### 31.2.3 Follow the Integrity of the Urban Environment, to Integrate into the Urban Space Order

Although the metro station entrance has very strong identity, visibility and explicitness [2], it is an important part of the overall urban environment. Human behavior is a continuous flow in the traffic facility, such that the overall design is particularly important. It is very necessary to coordinate with the surrounding landscape, and to echo with the surrounding architecture forms, and to create space sequence on base of the integrity.

### *31.2.4 Using Modern Technology, Promoting Regional Cultural Vitality*

The characteristic of the times of Metro station entrance is a kind of creation of urban vitality. In the design, it is need to select a reasonable new building material, to pay attention to the choice of the structure form, and to make full use of aesthetic feeling of structure to embody characteristics of modern building [3]. It is also need to explore constructive innovation and architectural modern science and technology to make the protection of regional culture and make the heritage more vivid. Therefore, the city public facilities can give better service to the residents in a more comfortable way.

## 31.3 Form Design Conception of Metro Entrance

Form design of the metro station entrance shape takes the regional culture as breakthrough point, and pays attention to the combination of architectural form and culture. Cultural elements are extracted from geographical and cultural characteristics. In general, Changsha City cultural style can be summarized into two aspects, one is about the Ancient Changsha – unvarnished and elegance, with Chu and Han legacy, Yuelu Academy, Mawangdui Han dynasty tombs, the Three Kingdoms bamboo slips, Baishajing cultural relics and historic sites and so on; the other is about Fitured Changsha – creative vitality of landscape, known as the capital of television and media. The regional culture characteristics as shown in Table 31.1 provide a large number of creative sources for the design conceptions as shown in Table 31.2.

### *31.3.1 Orange Island in Water*

Design idea comes from one of the unique natural features in Changsha – Orange Island. Orange Island in water is a famous landscape in Changsha. The bright color of orange reflects a kind of active innovation spirit. Design scheme selected orange as the basic color, implicating the enterprising city culture of Changsha. Besides, identification guidance of the metro station entrance can be achieved by using this color. Stereoscopic techniques are used on building construction. The solid wall is in the color of orange yellow, and the glass wall is with blue transparent glass. Blue and orange colors abstractly express the unique geographical landscape in Changsha – the orange island in water.

**Table 31.1** Regional culture characteristics Changsha

| Orange Island | Tangerine | Volume in Yuelu | Bamboo slip | Ever-changing |
|---|---|---|---|---|
|  |  |  | 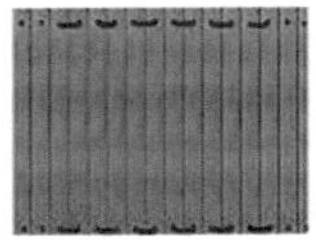 |  |

**Table 31.2** Building form design of the entrance

| Orange island in water | Boat sailing in the Star city | Xiang Boat to Sail |
|---|---|---|
|  |  |  |
| Petal of Fallen Flowers | New Rhyme of Bamboo Slip | City Seal |
|  |  |  |
| Water of Xiangjiang 1st | Water of Xiangjiang 2nd | Yuelu Style |
|  |  |  |

Zhang Yunan, Jia Xiujuan, Qian Linlin, Yang Da, Zhang Wenchao, Liu Yi, and Liu Xing are taking part of the design

### *31.3.2 Boat Sailing in the Star City*

Star city is another name of Changsha. Rail transit represents the development of transportation technology, thus the design tries to create a sense of speed and modern temperament, making the entrance building become a "milestone" landmark of the city. This design selects the form of "boat arrow" to express the meaning of "pilot symbol", with the horizontal skeleton. The arch in entrance stretches out, forming a strong momentum. Glass veneer based on steel skeleton and aluminum alloy skin are chosen as the building materials. The texture of "bamboo slip" combines the new material and the traditional culture, making the viewer experience monthly development speed in Star City and taste the charm of history and culture.

### *31.3.3 Xiang Boat to Sail*

The design focuses on showing the bright future development of Changsha City, highlighting the metropolitan style full of vigor and vitality. Building entrance is upward lifted, while roof curve decreases rapidly with the downlink from entrance. The junction line of the upper and lower part in facade parallels to the roof curve, and the glass interface in lower part highlights the complete form of the upper solid wall. The overall shape is full of strong dynamic sense. The column in entrance increases the stability of the structure, despite the dynamic sense of the building.

The slash on the facade increases the details of the building. The slanting column is inspired by bamboo forest in Hunan, not only makes the building facade more lively, but also creates rich lighting effects for the interior.

### 31.3.4 Petal of Fallen Flowers

The colorful fallen flower petal is beautiful scenery in Changsha. This design uses simple shape, with light color frosted glass as maintenance of the structure, and the hollow thin steel skin at the outer layer. When sunlight goes through, a indistinct projection effect appears. Pattern on epidermis like rain of petal falling on the building skin forms beautiful imprint. The texture is abstracted from the window lattice pattern in Changsha ancient town, reinterpreting the traditional components form with modern materials by screen printing and other modern technology.

### 31.3.5 New Rhyme of Bamboo Slip

The design extracted creative elements from bamboo slips as important cultural relics, reflecting the heavy local culture in Changsha. The vertical window bars in the form of "bamboo" is arranged high and low on the façade. Building materials and its color are clean and simple. The side and top of the architecture are aluminum alloy windows. The construction details strive to be simple, and the connection of stone and glass is like connection between bamboos, which is concise and clear.

### 31.3.6 City Seal

The original idea comes from the interpretation of the traffic signs in Changsha. The architectural form is expressed as a loop pattern, like a seal. It draws the Changsha rail transit logo, implicating the modern traffic accessibility. Gray granite and glass are selected as the two kinds of building materials, to maintain stable and transparent at the same time.

### 31.3.7 Water of Xiangjiang

The design uses steel and glass as the main materials, with which curve shape is constructed through sheer contrast and integration. The overall shape is like the flowing water of Xiangjiang which is with firmness and flexibility. The texture of the glass skin is from the fluctuating water surface. The solid walls made of concrete present different gestures in changing shapes of curve. The surface is in color of dark red or white, making the entire subway line full of variances.

### *31.3.8 Yuelu Style*

Old Yuelu Academy is a symbol for Hunan culture and education, and also the cultural base of historical heritage of Changsha. This design transforms three-dimensional form into plane form, and each facade of the building is a page of manuscript with sections of text. Acrylic plate as concave-convex texture and decorative elements made of metal and imitative wood materials are applied on the building construction. "antiquity" and other cultural contents of Changsha are recorded on the decorative components in the form of painting and calligraphy.

## 31.4 Conclusions

Although the metro station entrance is small in scale, it is rather influential and significant for a city. It is of great importance to establish good relationship between architecture and culture based on the study of regional culture, and to seek the breakthrough in the form, material, color, shape and so on, in order to pursue the artistic effect consistent with regional culture. A metro station with unified style, regional characteristics and integrated function, can serve the public better.

**Acknowledgement** Fund: National Natural Science Foundation of China (51078022).

## References

1. Editorial Committee of Series (1994) Architectural design information set, 6th edn [M]. China Building & Building Press, Beijing, p 87
2. Sheng Haitao, Zhang Jie (2008) Case study on station territory development and design of Tianjin Metro Line 1 – metro square design of area B in Xinanjiao Station. J Tianjin Univ (Soc Sci), 158–163
3. City Building Design Institute (2003) Code for design of metro [M]. China Planning Press, Beijing, p 45

# Chapter 32
# Survey of Xizhimen Station of Beiping-Suiyuan Railway in 1928–1949 and the Relationship Between the Station and Urban Space

**Wei Gao, Mei Zhao, and Xuejie Zhou**

**Abstract** Through the interpretation of a particular era and city background, the paper reviews Beiping-Suiyuan Railway Xizhimen station building process. The text is written according to time sequence, which start from relatively active period of 10 years construction time to stagnating period of 8-year war, then momentary recovery period of postwar. The author introduced the building and functioning history of Xizhimen station in the Beiping-Suiyuan Railway period, as well as the political environment and urban decision-making on the important influence of Xizhimen station at that time. And then the author offered analysis of the relationship between Xizhimen Station and the surrounding urban space.

**Keywords** Xizhimen station • Beijing North Railway Station • Beiping-Suiyuan Railway • City

## 32.1 Introduction

Beijing North Railway Station is a well-known one of railway stations in the whole country, and also an important transportation hub in Beijing. As the station of the firstly independently designed and built railway in China—Beijing to Zhangjiakou Line, it had a long history, and became one of the most important railway station in

W. Gao
Architecture and Art Institute, Beijing Jiaotong University, Beijing, China
e-mail: wgao2@bjtu.edu.cn

M. Zhao (✉)
School of Design Art, Beijing Institute of Technology, Beijing, China
e-mail: maezhao@qq.com

X. Zhou
China Wuzhou Engineering Co. Ltd, Beijing, China
e-mail: wwgg00@126.com

H. Xia and Y. Zhang (eds.), *The 2nd International Symposium on Rail Transit Comprehensive Development (ISRTCD) Proceedings*,
DOI 10.1007/978-3-642-37589-7_32, 

China's modern history from the beginning of its establishment in 1905. Beijing North Railway Station was closely linked with various aspects of the urban development in Beijing starting from its completion. Every adjustment and change of Beijing North Railway Station influenced surrounding areas hardly in aspects of urban space, transportation organizations, business development, daily life.

Beiping-Suiyuan Railway Xizhimen station was built in a special time of the history in the development of Beijing city and the North Station. After a rapid development period at the beginning, Xizhimen station experienced the turbulent times of the war with the change of the domestic environment. While In 1928 Beijing was renamed Beiping, no longer as the capital, Jing Zhang (Beijing Sui) railway line was renamed Beiping-Suiyuan line. Beijing North Railway Station and the city of Peking, together experienced a small development of 10 years. And then the war coming, it was charged by Japanese colonial rule for 8-year, and then it experienced the baptism of the 4-year civil war. Everything left was riddled and devastated. During the period of colony, it played a fatal role of translation for the war, so it was totally militarized and lost civilian functions. Gradually, the railway station was isolated from the city's development artificially. Until 1949 after the founding of New China, Xizhimen station gradually returned to the normal process of construction and development.

## 32.2 Relatively Active Period for 10-Year (1928–1937)

In 1928, the Northern Expedition ended the domination of the Northern Government, while the capital moved to Nanjing, Beijing renamed Beiping special city. With the political center moving to south, urban development faced many obstacles. But with the accelerated pace of Western Learning, Western modern urban planning ideology also gradually emerged in Peking city planning and construction. Therefore, the urban planning and construction of this 10-year period is relatively active. From the identity of the capital, Peking took a clear position of the North political, cultural center and tourist places. "After capital has migrated to the south, the political power of Peking was hardly weaken, but the city is still the political center of North China...... nice climate, appropriate geographic location, people driving off to travel or move home to this every year, are like a constant stream" [1], Also they made up "three-year construction plan", issued urban construction plans like *The Beiping Resort Construction Plan*, *Beiping River Maintenance Plans*. In the same time, the government also gradually started to focus on the railway construction, offering the policy and economic support. Due to these two aspects, Xizhimen station had a small-scale development.

In a book named *The Beiping-Suiyuan* (divided into two parts), which was published and edited by Beiping-Suiyuan Railway Administration, there are lots of detailed describes for the construction.

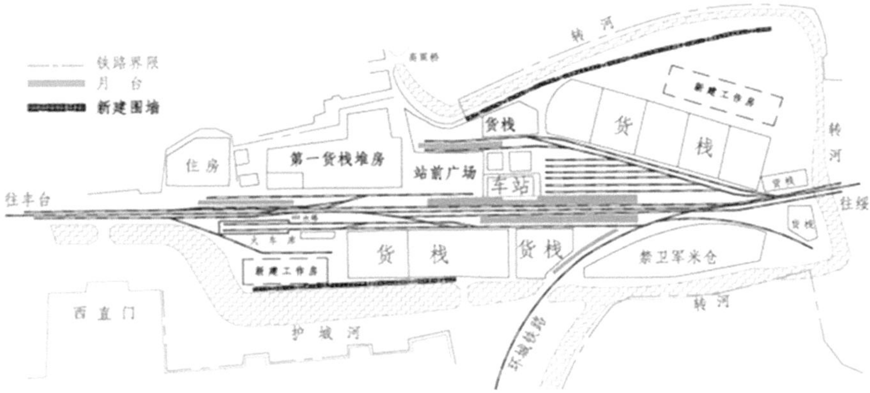


**Fig. 32.1** 1934 the speculated inside plan of Xizhimen Station (painted by the author in accordance with the relevant information)

In volume No.1, 1933 July to 20 June 2003 (1933.7–1934.6), it is written:

"(A) Improvements of Xizhimen Station". "Xizhimen station, is the business center, full of trips, and the primary station, for the reason that building was old, all platforms and waiting office rooms were broken ... itself should be improved ... it should be taken into account, (1) the box office need to be mended (2) the platform need a cement ground (3) A new entering door need to be added and also neighboring walls (4) Fence need be installed (e) Station West need add adobe walls. In total five items almost need 12000yuan", " the Chi Square in south of station and the road from the station gate to the bridge should also be... renovated," In addition to the above, there are some work like lengthening platform, adding waiting room, building sigal plate, recovering and demolishing the houses of military Yonekura need to be carried out in phases adding four rooms for the Xizhimen garage, one big room for furnace pipe repairment, two rooms for administration, one each room for machinery, ironwork, and furniture.

In volume II: In 23 years between July 1935 June (1934.7–1935.6), there are also descriptions about the Xizhimen station construction matters: "Xizhimen is the first station of this line, which freight is extremely busy. In respect that there was one abandoned garage in the station, now we will transform it into a large-scale warehouse in order to facilitate business...", "...while the original electric service house is lack of space, another six rooms for material storing need be built, as well as a wall"

What we can find from the above information is that, although in this period merely a few functional buildings were constructed, and a little repairmen had been done, Xizhimen station still brought considerable improvement to the overall environment (Fig. 32.1).

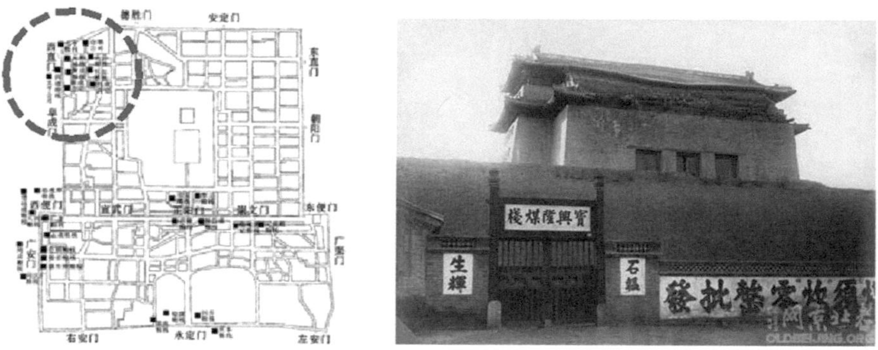

**Fig. 32.2** Food market distribution of Peking (Source: Research of Peking cities Grain Market Distribution and level in Republic time. Tong Meng, Han Maoli. Chinese Historical Geography. July 2008 3 Series, Volume 23) and the Xizhimen watchtower in 1931 (Source: http://www.daoyoupeixun.com)

In this short development time, socio-economic revived and civil supplies traffic increased. The massive growth of civil freight made the Xizhimen station transportation capacity be developed, and surrounding warehouse storage capacity be developed again. In <Beiping-Suiyuan> Volume II property part, it reads: "Xizhimen station, the old train room and adjoining material yards are leased by Peiping Branch of Kincheng bank, where were going to build a warehouse. For the high cost of this construction, we borrow 17000 yuan from Kincheng bank firstly" (Fig. 32.2).

Around the station, the railway transport of the Republic of China in Peking's food supply has replaced the waterway in Ming and Qing, so a typical phenomenon showed up, which large grain markets appeared around the important railway site. At that time, there were three main lines of food supply for Peking, bei-ning, Beiping-Suiyuan, Ping-han. While Xizhimen station is the primary station of Beiping-Suiyuan Railway, a lot of the food arrived here from Beiping-Suiyuan railway. For this reason, lots of grain stack appeared in the Xizhimen surrounding areas which become a important food market in Peking. Due to the large amount of food supply, a big storage space is required, so Xizhimen station surrounding urban space gradually transformed from a disorderly scattered plots to large warehouse spaces. At the same time, as the Xizhimen station led the waterway gradually fading, roads construction as the gateway to the western suburbs had become a necessity. In this period, some main roads (Table 32.1) were built around the Xizhimen Station. The improvement of traffic is conducive to the use of cars and other vehicles, and these roads combined Xizhimen railway formed a better transport system.

**Table 32.1** List of the construction of roads from Xizhimen[a]

| Location | Road name | Start and terminal | Length (m) | Type | Reason for improvement |
|---|---|---|---|---|---|
| Inside of city | West Fourth Street North, Xinjiekou South Street, Xizhimen Avenue | By the West Sipailou to Xizhimen | 2,902 | Asphalt fried oil pavement | From inside and outside the city to the western suburbs of the main road |
| Outside of city | The Xizhi Qinglong Road | Xizhimen to Qinglongqiao | 3,101 | The asphalt spilled oil pavement | Resort roads to the Summer Palace, Fragrant Hill Spa and Tangshan |
| Suburb | The Xizhi Museum Road | From Xizhimen to the Natural Museum and Wutasi | 5,000 | Ditto | From Xizhimen to the Natural Museum and Wutasi highway |

[a]produced by the author in accordance with "Construction Plans of Resort Area in BeiPing City"

## 32.3 Stagnation of the Construction Period (1937–1945)

After the "7.7 Incident" of 1937, Peking was conquered. Japanese invaders developed the "Beijing Urban Planning Outline", and continued their promotion of railway colonial plan, which have had a great impact on the Beijing city and Xizhimen station.

Beijing Urban Planning Outline was programmed in 1938 by Sato Chun long and Yamazaki Gui, which has three purposes: "(1) to cope with the increase in population. (2) City roads are not sound, which are the barriers of economic and military. (3) as far as possible to avoid mixing Japanese and Chinese people living together, in order to avoid friction" [2]. The outline positioned Peking as the center of political, military and special tourist city, and has developed a number of key elements, such as the industrial area of the plan will be concentrated in the eastern suburbs of Tongxian County, an area of about 3 km$^2$; while a "new urban" used for accommodating the new municipal government "and dispersing some Japanese nationals" is located in the western suburbs, the area east to Gongzhufen, west to Babaoshan, south to Fengtai, north to Xiyuan, about 65 km$^2$, etc [3]. Where thoughts of the construction of the new city of the west, and protect of the old city are far-reaching. Beijing Urban Planning Outline adopted some advanced Western urban planning ideas and methods, but the essence is serving its aggression, colonization and plunder. During this period, Peking also showed colonial characteristics: First, the surge of Japanese nationals. Followed by changes in the partition of the city, including the city had a "special area" many are only allowed the Japanese to live and prevailing. Railway colonial plan progressed, the occupation

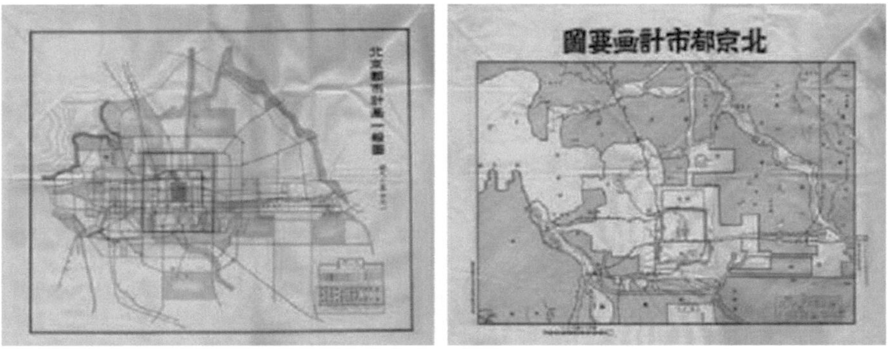


**Fig. 32.3** The general graph and important graph of Beijing Urban Planning Outline (Source: Beijing Archives)

and improvement of the Peiping railway, train station and the surrounding road infrastructure became the primary means of the Japanese invaders to plunder resources and strategic materials (Fig. 32.3).

Japan Railway colonization program began in 1906 in the northeast of China by establishing the South Manchuria Railway Company, referred to as the Manchuria.

Railway, part of the Japanese government, the state monopoly capital colonial enterprise, it is a major economic plunder of Japanese imperialism and the implementation of the mainland policy tools. In 1936, the Japanese had occupied the railway and stations of the Peking& Tianjin region, and managed by establishing the north branch Firm of the South Manchuria Railway, where Xizhimen Railway Station is concluded. After the fall of Peking, the railway and stations had always been under the tight control of the Japanese, being the "lifeline" of the Japanese troops to transport war material to consolidate and expand the encroachment area.

During the occupied period, the city municipal administration was broke down, meanwhile the Xizhimen station only served for military but not opened to citizens. In the "Outline", referring to the partition of the old town, the surroundings of Xizhimen was listed as "ordinary residential area", that non- Japanese-populated, but no detailed solutions were listed about the military area. The important area of city construction even had nothing to do with the surroundings of Xizhimen. Therefore, during this period, there were no improvement in the surrounding cities' construction, Xizhimen was separated totally with the development of the surroundings. For the convenience of the militarily management of Xizhimen, a lot of troops residences were constructed in the area near to railway to the northeast of Xizhimen station, which made the surroundings more difficult to be managed. On the political situation, the Japanese was into a passive soon after entering the North China war area, the tiring struggling Japanese invaders, could hardly afford the normal operation of the station, not even to carry out the construction of the station. Therefore, the development of the Xizhimen Station in this period is basically in the phase of stagnation.

## 32.4 Construction Recovery Period (1945–1949)

Eight years rule of Japanese invaders in Peking left heavy damage to the city and Xizhimen station. After the war, the disagreements between the KMT and CPC again led the development of the city into challenges. Until the foundation of the state, urban construction and development could not be the main task during this short period.

Nevertheless, the two mayors of Peking both tried their efforts on urban construction. Especially Mayor He Siyuan, during his tenure, he implemented some urban construction. He paid most attention on modifying the transport and renovating roads. He also drafted a traffic regulation map of the Xizhimen surroundings, based on the mapping of Peking. The traffic regulations surround Xizhimen once again caught the attention, but it was not implemented with the limitation of the current situation.

At that time, Peking Railway was taken back to be state- owned, and Xizhimen station was also restored to the original organizational system, attributed to the KMT Ping& Jin District Railway Authority. During the War of Resistance against Japanese Aggression, the station was completely militarized which lead to its crude infrastructure. Shortly after the civil war breaking out, the KMT Government had no energy to construct the station. Therefore, Xizhimen station had no big development during this period. In 1946, the KMT Government somehow repaired and expanded the stations along the Pingsui railway, for the restoration of rail transport undertakings. Xizhimen station was included, which was expanded to 18 garages, with an area of 25,780 square feet. In addition, the station inner parts were only functionally recovered, with the surrounding cities basically extending the Puppet state, where there was no further development.

## 32.5 Epilogue

The Xizhimen station of Pingsui Railway in the 1928–1949 year was in slow development stage among the course of Beijing North Station's development. The very history and background of the times lead the Beijing North Railway Station into a period of ill-fated. The construction of the station was stalled, and the relationship between station domain and the neighboring cities were also gradually splitted. But this period is a transitional stage, which followed the period of rapid development in the beginning of the establishment of the Xizhimen station of the Beijing-Zhangjiakou Railway, and is followed by the steady development process after the foundation of China. As can be seen from this period of history, the Xizhimen railway station is combined tightly with every aspects of the Beijing urban development, whose single important modification and change impacts the Xizhimen area profoundly. The chaos of urban development will inevitably lead to the stagnation of the railway construction. After the foundation of China, Xizhimen Station together with Beijing city, ushered in a period of stable development and gradually developed into a major transportation hub of Beijing, while this course will also need us to sort out and analyze.

**Acknowledgment** This work was supported by the Fundamental Research Funds for the Central Universities. (Grant No. 2009JBM131).

## References

1. Zhuangke (1936.6) The particularity of Beijing city. Municipal comment, June in 23 years of Republic of China 1:235–246
2. Shun Donghu, Wangjun (2000) Urban planning and the implementation of Beijing city during the 8 years of Japanese – occupied. Chin Hist Geogr 2000(3)
3. Yue Zheming, Huang Shimeng (Trans). Urban planning of Beijing. Taiwan Univ J Archit Urban Rural Areas 3(1)
4. The Japanese Traders North China Traffic Corporation. Ping-Sui Railway survey report (Operations Relations). Publication date and publisher is unknown, Beijing
5. Wang Yanan (2008) Research on Beijing's urban planning and construction 1900–1949 [M]. Southeast University Press, Nanjing
6. Lin Cuanjia. Jingzhao geographic records of Greater China. Zhonghua Press, Beijing, 1919 (8 years of Republic of China), p 208
7. Compilation Committee of chorography of Beijing (2004) Chorography of Beijing Municipal roll Railway transportation. Beijing publication, Beijing
8. Han Yingkui et al (1997) Chorography of Beijing north station (1909–1996). China Railway Publishing House, Beijing

# Chapter 33
# The Discussion of City Track Traffic and Underground Space Development Problem

**Xia Zhao and Haishan Xia**

**Abstract** The metropolis of China are in orbit during the peak period of construction, change of space function and the rapid development of rail transportation on the city spatial structure plays a huge effect. The development and utilization of the core concern of this paper for the development of rail transit to the underground space which has brought the promotion and problems. In this paper, on the basis of investigation, summed up the impact of rail transit in city underground space and ground space coordination development of city space, put forward the sustainable development of rail transit guiding strategy.

**Keywords** Rail transition • Underground space • Collaborative development

## 33.1 Introduction

At present, the major large cities are entering the peak period of a large number of rail transportation and rapid construction. Along with the development of some of the problems of urban underground space is also very prominent, such as underground space development model is very single, basically just the traffic space and a small portion of commercial space; the ownership of land development is also very confusing, resulting in urban underground space structural imbalance, blindly seeking large scale, underground space underutilized. The development of high strength, the fixed pattern is one of the reasons for these problems. In order to alleviate these problems, we need to explore a new model of development, the use of underground space and rail transit in full contact. In this paper, to the

---

X. Zhao • H. Xia (✉)
School of Architecture and Design, Beijing Jiaotong University, 100044 Beijing, China
e-mail: 603122633@qq.com; hshxia@bjtu.edu.cn

H. Xia and Y. Zhang (eds.), *The 2nd International Symposium on Rail Transit Comprehensive Development (ISRTCD) Proceedings*,
DOI 10.1007/978-3-642-37589-7_33, 

development of underground space in Shanghai, for example, to explore the problems of underground space development and how to solve these problems, from the topic of urban rail transit and underground space:

## 33.2 Rail Transit and the Development of Underground Space

### 33.2.1 *Development of Rail Transportation Status*

At present, the major large cities are entering the peak period of a large number of rail transportation and rapid construction. The first subway of China was built in Beijing in 1969, after a halt for over 20 years, to 1995, built in Shanghai Metro Line 1. At the end of 2005, more than 20 cities including Beijing, Shanghai, Guangzhou, Tianjin, Dalian and other domestic are under construction or ready to construct and planning of new rail transit line, the length of over 4,000 km, by 2050 Chinese urban rail transit line length will be more than 4,500 km.

### 33.2.2 *Inevitability of 1.2 Underground Space Development*

The nineteenth century is the century of the bridge, the twentieth century is the century of high-rise buildings, the twenty-first century is the century of human development and utilization of underground space.

With the social and economic development, the role of the underground space will become increasingly evident. Future society, development and utilization of underground space is a good way to complete urban functions, the efficient use of land and facilitate the daily life. Therefore, a long period of time in the future, comprehensive development and utilization of urban underground space will be an important measure of the intensification of land use and inevitable trend to build-saving harmonious society [1].

## 33.3 The Current Situation and Problems of Underground Space in China

### 33.3.1 *The Development Status of Underground Space in Shanghai*

Compared to other cities in China, Shanghai's Metro underground space development is better. Holds many lessons, has prompted the traditional city business

**Table 33.1** Shanghai subway stations commercial space statistics

| Station name | Belongs line | Station type | Metro business type | Business area ($m^2$) | Surrounding environment |
|---|---|---|---|---|---|
| People's Square Station | 1, 2, 8 | Transfer station | Station mall<br>Channel Business | 48,500 | Urban centers<br>Political and cultural center<br>Business circles |
| South Station | 1, 3 | Transfer station | Station mall<br>Channel Business | 3,200 | Transport hub |
| Xujiahui Station | 1, 9 | Transfer station | Cover the mall<br>Station mall | 15,000 | Urban Vice Centre<br>Business circles |
| Jing'an Temple Station | 2, 7 | Transfer station | Underground shopping mall<br>Cover the mall | 8,215 | Business circles<br>Heritage conservation area |
| Shanghai Railway Station | 1, 3, 4 | Transfer station | Channel Business | – | Transport hub<br>Business circles |
| Science and Technology Museum Station | 1 | Intermediate station | Station mall<br>Channel Business | 30,000 | Political and cultural center<br>Business circles<br>High-end business district |
| South Huangpi Station | 1 | Intermediate station | Underground shopping mall | 2,000 | Business circles |
| Zhongshan Park Station | 2, 3, 4 | Transfer station | Station mall<br>Channel Business<br>Cover the mall | 5,000 | Business circles |

circles upgrade of metro district, the formation of the Shanghai Railway Station, People's Square, South Shanxi Road, Xujiahui four famous underground shopping district, the city is changing business landscape (Table 33.1).

The Shanghai subway station domain underground space development of commercial facilities are mainly concentrated in the dense flow of people of Metro Line 1 and Line 2 stations within Apart from the Science and Technology Museum Station, the other sites are located in the urban district, transportation hubs, business district of the old Town. Commercial scale due to the different type of station and station domain surrounding environment, ranging from 2,000 to 48,500 $m^2$, commercial space type is relatively rich, but few longitudinal development of three-dimensional. The Metro Commercial space grouped into four types in Shanghai: the subway aisle shops, subway station hall business, subway underground business street (Metro Mall) railway superstructure commercial properties [2].

### *33.3.2 The Problem of Underground Space in Jing'an Temple Area*

Shanghai Jing'an Temple area is located in the west end of Nanjing West Road, business, commercial, luxury residential, tourism, shopping, leisure and entertainment in one of the city sub-center. Typical representative, of course, also in the underground space development, there are some very common problem.

#### 33.3.2.1 Underground Space and the Urban Environment Out of Touch, Lack of Effective Urban Design Guide

The combination of Jing'an Park and the sunken plaza is very creative. Jing'an temple with this historical legacy protection combined with the development of the subway station simultaneously. Development the Jing'an temple of underground layer as equipment, reservoir in the Book of space, the ground floor to the east and west sides of the development of commercial space, to make up for the lack of business, to improve the shopping environment and meet the shopping needs of pilgrims. Successful realization of the Jing'an temple modernization. And commercial, business and other general volume of building development of this region, integration with the surrounding environment [3] (Table 33.2).

But not taken into account in the development of the surrounding buildings, the construction period of time, was not able to co-ordinate arrangements for the construction sequence, making the delays have been able to solve the problem is unlimited. For example, the Huashan Road on the west side in the construction of Wheelock Square, from bidding to bid to the final completion of, lasted a long time. This makes Huashan Road underground underpass lack of timely implementation of mixed traffic remain unresolved [4].

#### 33.3.2.2 The Single Functional Form

Jing'an Temple Metro Station underground space is a total of four. The basement and ground floor, are mainly commercial, pedestrian passage, supplemented by the garage, entertainment; Subway station hall is in the underground floor and the ground three; ground three and four underground are garage, equipment room (Fig. 33.1).

**Table 33.2** Jing'an Temple subway station node graph

| | |
|---|---|
|  |  |
| 1 Vents and water | 2 Inquiries Service Center and barrier-free access |
|  |  |
| 3 Sunken plaza and subway entrance | 4 Underground shopping mall |

(continued)

**Table 33.2** (continued)

| | |
|---|---|
|  |  |
| 5 Jing'an Park entrance plaza | 6 Park entrance coffee bar |

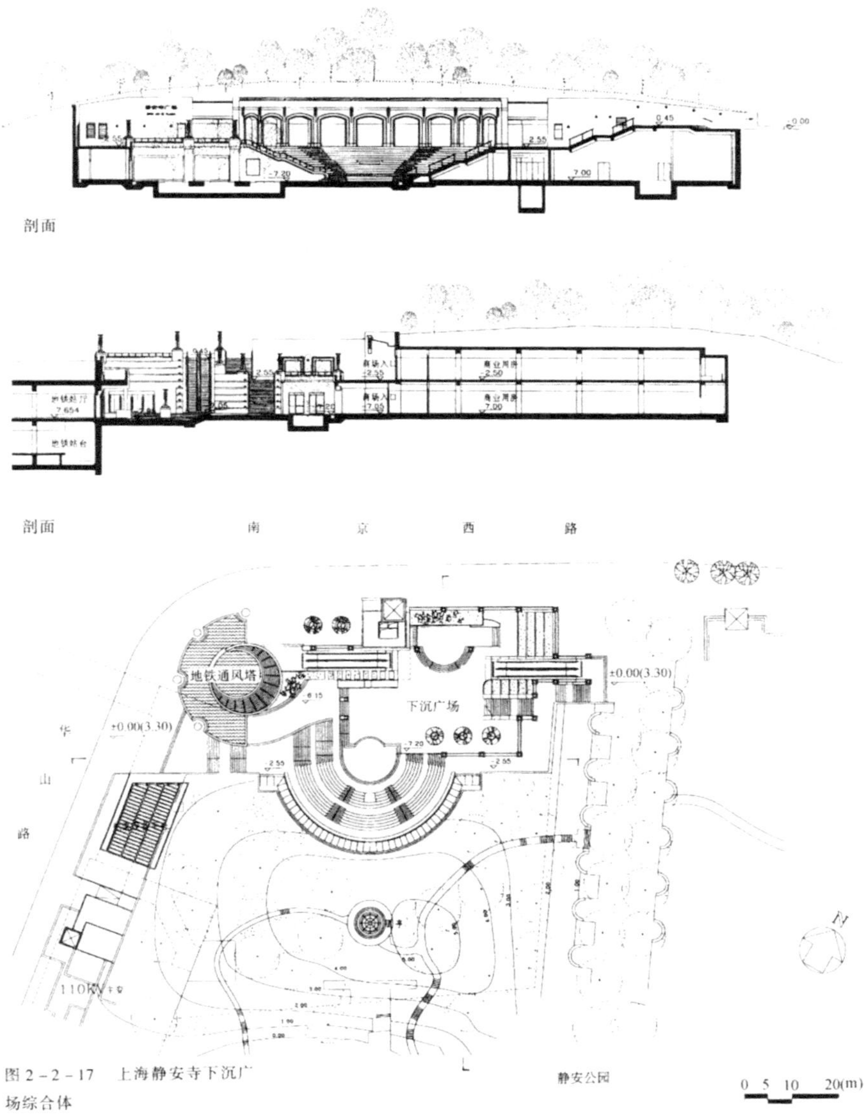


**Fig. 33.1** Jing'an Temple subway station cross-sectional view, plans

Urban Underground Space Development and Utilization are not meet a single one function, and should be based on the overall construction of the city and the feature requests which should be to meet the transportation, commerce, supply and environmental urban complex. At the same time, it is no longer an isolated space form, but a bit line, surface and other forms of space through a flexible combination of organic, space as a whole. Build the node composite facilities of the underground facilities, connected nodes facilities for the network facilities, underground facilities integrated planning and design.

Jing'an Temple area as a whole underground space planning point + line + surface space layout around the walk. "Point" which is important space node example, site entrance sunken plaza and channel business and shopping malls connection point; "line" which is the linear underground passage and the underground business street; "face" which is connected through the commercial network various commercial points, mainly appear in the form of shopping malls. A massive overall space compact, orderly, formed as the leading commercial space underground space system.

#### 33.3.2.3 Underground Space Ownership Confusion, Lack of a Unified Underground Space Planning and Construction Management

Underground space development and utilization of special legislation in China is still blank, without the express approval of construction land and ownership of underground space management, thus exposed some problems in the practice of underground space development and utilization:

1. There is no legal land approval documents, many have been built underground buildings can not handle the real estate registration, it caused can not be transferred, pledged, affecting the enthusiasm of investors in the development and utilization of underground space, seriously affecting the underground space development efforts;
2. the uncertainty of the scope of the rights of the underground space, some construction units to obtain land use rights after the disorderly development of underground space, resulting impact on traffic; development and over-exploitation of underground space like the Jing'an Temple Station subway stations, resulting in serious waste of resources of underground space, making the size of the underground space is too large, impacting unified planning and construction of urban underground space [5].
3. According to author's research interviews, the Shanghai Metro also said, there was no preliminary planning and policy guidance and support when Shanghai built underground commercial space, mainly by the MTRC to plan by itself. So that not only cannot form a unified system of underground space, are also prone to a lot of technical and other aspects of the problem.

## 33.4 Solutions

Japan is a small territory, the land is short. In 1930, Tokyo, Japan Ueno Train Station underground walk on both sides of the channel to open commercial counter the underground street-side. So far, the underground street from purely commercial nature evolved to include a variety of urban functions, transportation, commercial and other facilities composed of interdependent underground complex [6].

### 33.4.1 The Combination of Underground Space Design and Overall Planning

Shinjuku is located in the west of central Tokyo area, about 6 km from the Ginza, the city's main downtown in Tokyo, Shinjuku Fukutoshin built business a total land area of 16.4 ha, commercial, office and office building area is more than 200 million square meters.

The Shinjuku Fukutoshin the economic, administrative, commercial, cultural, information and other departments gathered in the business district, finance and insurance, real estate industry, retail and wholesale industry, the service sector has become the main industry of the Shinjuku population composition of employment has been close to the center of Tokyo Metropolitan area.

The Tokyo Shinjuku CBD core central business district Shinjuku Station, the radius of the range of 7,000 m, gathered more than 160 banks, Shinjuku, Japan has become a microcosm of the "Bank War", is a replica of the "Wall Street".

### 33.4.2 Multi-functional Combination of Underground Space

Tokyo Metro Shinjuku Station link eight rail transit lines, but Shinjuku Station is not only an underground commercial street, bus stop, taxi stand, underground parking, as well as shops, banks, underground commercial street is arranged in the same building.

The Shinjuku underground pedestrian street system, railway stations, shopping malls, cultural, entertainment and other facilities of the entire region as a single entity. Organic link of the bases in the Shinjuku West End is bound to underground and aboveground elevated pedestrian street and set up two automatic pedestrian trails, pedestrian can be easily and quickly from the station to the Tokyo Metropolitan Government, the business office facilities and Central Park [7].

### *33.4.3 Sound Rule Specification*

Generally speaking, Japan has the urban underground space development and utilization of the most perfect legislation. Countries in the development and utilization of the integrated systematic, coordinate the promotion of engineering design and construction, the state administrative management in the world leading level. Japan's progressive approach to the management of urban underground space is gradually shifted to include both sides of the road and the city's business transportation hubs and other facilities intensive municipal facilities by a single road. From the construction of the laws and regulations: *Co-groove construction Special Measures Law on the road, Deep underground public use Special Measures Law*, promulgated reflect Japan to speed up the pace and effectiveness of the legal system construction and implementation.

## 33.5 Conclusion

In this paper, the discussion of rail transit and underground space development, and draw some experience. First of all, with regard to the development of the subway stations underground space, do a good job in the pre-planning and coordination with the urban planning is very important, but also the structure of the city will bring some changes. Be aware that the size of the underground development, both can not ignore the core of the traffic problems of the subway station of underground space disorderly development, resulting in a waste of space. Finally, the construction of rail transit need the relevant departments of our country to the introduction of relevant policies, guidance and support in the planning of urban space, more consideration to the factors of the rail transportation and underground space.

**Acknowledgment** This research was supported by: National Natural Science Foundation of China (51078022); The Fundamental Research Funds for the Central Universities (2009JBZ023); Doctoral Fund of Ministry of Education of China (20090009120013).

## References

1. Yabin Dong (2006) Urban underground space development. Sci Technol Info 24:106
2. Qian Linlin (2011) The Beijing subway commercial space use evaluation[D]. Beijing Jiaotong University, Beijing
3. Ji Cui (2011) The subway commercial development trend and policy recommendations – the case of Shanghai [J]. China Real Estate (04):23–26
4. He Lei, Shu Yu, Wang Xuan (2006) Subway station and the surrounding ground, underground space in urban design. Undergr Space Eng 2(7):37–41
5. Yang Fan (2009) Metro stations underground commercial space design[D], Southeast University

6. Wang Jianhong, Liu Xin-rong (2006) Japan urban underground space development and utilization. Underground Space Eng 2(3):349–353
7. Unkonw (2005) Tokyo Shinjuku _ typical of successful business district[J], Shanghai Econ 8:22–24

# Chapter 34
# Railway Station Integrated Business Development Planning and Architectural Design Integration and Application Research

**Yuan Lu, Ke Qin, and Mu Wang**

**Abstract** Nowadays, railway station business development is serious lag behind in our country, compared with mature development experience of developed countries, station space has a huge and mining business potential in our country. On the basis of seeking integration opportunity of business planning and architectural design, this article structures integration research framework between station building business planning and architectural design and use the collection response of architectural space of station and traffic stream to optimize architectural design of transportation building, excavate the business potential of station building, achieve the proper business value and retribution of station building space and realize sustainable development and utilization of space station.

**Keywords** Railway station business development • Architectural design • Business planning integration

Y. Lu (✉)
School of Architecture and Design, Beijing Jiaotong University, 100044 Beijing, China
e-mail: luyuanup@foxmail.com

K. Qin
CCDI Group, Rail Development Business Sector, Train Transit Comprehensive Development Research Centre, 100013 Beijing, China
e-mail: willqinke@163.com

M. Wang
Train Transit Department, CCDI, 100013 Beijing, China
e-mail: wang.mu@ccdi.com.cn

H. Xia and Y. Zhang (eds.), *The 2nd International Symposium on Rail Transit Comprehensive Development (ISRTCD) Proceedings*,
DOI 10.1007/978-3-642-37589-7_34, 

## 34.1 Status and Issues

### *34.1.1 Macroscopic Background*

In recent years, China's railway network is planning and constructing with an unprecedented speed. In January 2004, The State Council Standing Conference adopted the "Long-term railroad network plans" [1]. And in 2008 National Development and Reform Committee adopted "Long-term Railway Network Adjustment Planning" [2] to newly expanse railway network scale further. According to the plane, the investment of high-speed rail will be more than 20,000 billion yuan in 2020, length of railway lines in service will be transferred from 10,000 to 12,000 km with transferred passenger transport from 1,200 to 1,600 km.

Construction of railway network also brings plenty of planning and construction of railway station building. According to statistics, total 548 railway stations call for construction during the 11th five-year plane: divided by scale, capital level stations are 24; city level stations are 95, county level stations are 428; divided by lines, passenger special lines are 158, inner-city passenger lines are 83 and general main lines are 292.

Promotion of railway transportation passenger function and specialization of station passenger service close the relation of railway station and city function [3], so a high-speed rail metro construction was carried out as the leading of high-speed rail and intercity railway, the core of the central business district of railway station and construction of deputy city center area. The function of station building also changed from the original single external traffic gateway to the transition junction of multiple-urban transportation methods and external transportation.

### *34.1.2 Deficit of Railway System and Station*

However, on the other side of a full blast huge investment and railway construction is a consecutive deficit in railway system. On one hand, it is hard to receive operational effectiveness in a short period; on the other hand, newly increased investment requires a large number of debt which also increased the financial cost, Ministry of railway has already went into a deficit period under the double pressures. As the first half year report shows: up to June 30th, 2011, the debt is up to 2.09 trillion yuan, asset-liability rate is 58.54 %, the profit is only 42.9 billion yuan [4]. As the important asset in railway system, the operation of station is also in a deficit status.

Under the current situation, passenger transport and freight transport income cannot balance large-scale investment and railway operation expense. However, compared with the railway income statistics of developed country, the proportion of our non-railway transport income takes up small scale of the total income of railway (about 33 % in 2005 which is far away from Japan, 60 %), and wholesale and retail industry as the important part of non-transport income (account for 33.76 % of the profit in 2005), the low income rate (in 2001, Beijing Railway Bureau shows: 1.6 %) shows the operation status is in a low efficiency status.

### 34.1.3 Open Question of Station Business Development

Function promotion of railway and station isn't improving the environment of station passenger transport service and business service. At present, station business service is still in a primary stage including lots of newly constructed ones mainly reflected in the following:

Business development scale is small. In station design period, station business matching is always 4 % of the station construction area, which is only, satisfies the passenger necessary consumption needs. Although business development scale has already highly improved in the newly constructed railway station, it still is some distance from the foreign mature station, which 50–70 % business area matching.

Business layout isn't good with incomplete supporting facility. The construction layout of station totally considers the traffic lines, so the business space only can use the fragmentary constructed gray space, it is very inconvenient for causing long distance walking for customers to conduct business consumption; it is hard to form a complete business atmosphere for such a cut-up arrangement of point layout of business; Business layout and traffic lines combines badly, business area development is lopsided with some of them supported by passenger flow with a good economic benefit and some of them are in bad benefit because of poor location; In addition, it is difficult to introduce business service of mature, thorough, but highly-required of electricity and drainage due to the insufficient station foundation construction.

Business service environment is bad. Insufficient supporting facility and deficiency of environment management cause a poor light and healthy environment, and the large portion of self-employed servicer mainly come from laid-off workers or families of workers with a certain employee's welfare, who normally are lack of professional training and service awareness, which is hard to improve the quality of business service.

Low initial consumption desire and customer satisfaction. As a sample survey shows 70 % passengers hardly consume in the station and the consumption mainly are necessities or temporary consumption goods, such as beverage, solid food, instant noodles and newspaper and it is low consumption desire to other goods due to price, environment and quality factors. Except for necessities, customers always prepare by themselves purchasing from other place.

Less type of business service, simple status, high repetition rate, business potential hasn't been fully taped. Customer consumption is only for necessities for causing other business service cannot maintain to quit of the station. Business rent level is 20–30 % higher than others, which is not match with passenger

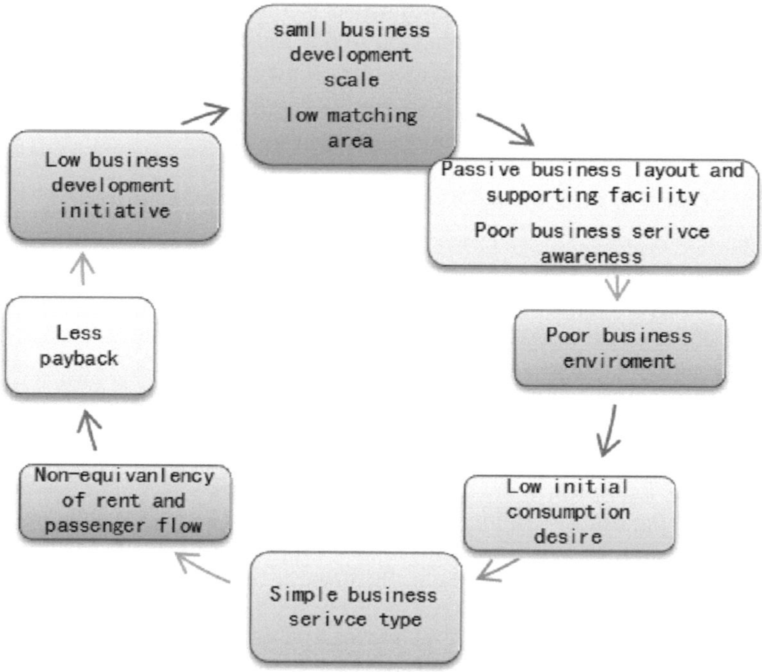


**Chart 34.1** Sketch map of negative circulation of railway station business operation

collection effect, and the business income and rent always be kept with the support of higher price.

It seems to form a negative circulate of railway station business development. As the following chart shows: this kind of development patterns can never be improved by inner spontaneous adjustment, unless there is a certain impetus from external (Chart 34.1).

## 34.2 Development Necessity and Trend Analysis

### *34.2.1 Railway Station Business Development Necessity*

#### 34.2.1.1 Railway and Station Economic Sustainable Construction Development Requirement

Railway station business development necessity reflects on economic sustainable development first. During the long-term debt operation and investment construction this year, the current liability level of Ministry of Railway is from 37.53 % in 2005 up to 57.44 % in 2010, the debt is up to 2 trillion yuan in 2011, and the annual investment of railway foundation construction is more than 600 billion yuan.

From simple estimation, good operation business development can bring 13,400 yuan financing per square meter, if the station configured 40 % business area, then 54 % of station construction fund can realize from the financing of business development [2].[1]

Therefore, fully activate the potential economic benefit of station can relief the construction fund pressure of railway system: on one hand, economic benefit from business development can directly solve the construction fund problem of railway department; on the other hand, stable cash flow and profit can help asset investment and financing and attract bank investment and folk private capital to obtain more construction supporting fund.

#### 34.2.1.2 People Oriented and Humanity Requirement

Railway related service is more like semi-militarize management mode for a long period without much attention on the feedback from customer of service. With changeable railway operation service goal and construction of passenger transport line coming into service, railway system especially station service calls for travel requirement more close to customers' daily life.

### *34.2.2 International Station Business Development Tendency*

In Japan and European countries with well-developed railway system, railway privatization is an important factor to make developers pay much attention on recovers of cost for construct large-scale business service facilities inside of the station to seek the economic payback. Traffic and business synthesis, which combines station construction and multiple-business service and passenger traffic space, is already became a mature development method.

For example, overall floorage of Osaka Railway Passenger Station operated in 1983 is 138,000 m$^2$ including 4 floors underground, 27 floors over ground, total height is 122 m, which combined traffic function, shopping, catering and hotel; overall floorage of Kyoto Station operated in 1997 is 238,000 m$^2$ including 3 floor underground, 16 floors over ground, which includes except traffic, the other multiple-city-function, such as hotel general merchandise, shopping center, antique store, coffee shop, restaurant, movie theatre, museum, exhibition hall, district government office and parking and also overall floorage of Berlin Central Train

---

[1] The calculation is based on 12 % rate of return on investment, commercial profit is 1,800 yuan/year, station comprehensive cost of construction is 10,000 yuan/per square meter.

Station in Germany operated in 2006 is 175,000 $m^2$ also includes over all kinds of business retail and office space over 3,000 $m^2$.

### *34.2.3 Possibility of Railway Station Business Comprehensive Development*

It possesses great possibility on designation, construction and transition of China's railway station to traffic business synthesis in space, demand and policy.

#### 34.2.3.1 Possibility of Architectural Space

At present, floor space of railway station in our country generally is big, but with small covered area, which means plot rate is low. Meanwhile, the atmosphere and grandiose appearance requirement with square, regular spatial arrangement mode on station make a low space use ratio, which always is more non-fully-used constructed gray space except for traffic function. These all shows that it is greatly improved for the current station space usage efficiency.

#### 34.2.3.2 Possibility of Business Demand

For the consideration of uncertainty of train arrived time of city traffic time and station security check line up, domestic passengers usually arrive station waiting room ahead of time. As the statistics shows that waiting time in train station is about for 0.5–1 h, which means the retention period provides possibility of some potential business service in time.

On the other hand, domestic passengers suffer a long-term business consumption demand during travel, which can directly feel from plenty of food, appliance customers taking along. After completion of high-speed train, more and more passengers transfer from short-distance aviation to railway traffic; and compared with the railway passenger's consumption frequency of coastal developed cities, European and Japan, it will be great promotion space for our status of low consumption frequency and expenditure.

Station business, as a gateway commercial centre is not only serves the traffic flow in station area, but also business platform and city new community rely on hub, which conducts important supplementary or business centre newly area development in area function [5].

#### 34.2.3.3 Possibility of Policy Impellent

As "Economic Observer", at the beginning of 2011, Ministry Railway convened an internal meeting to give more autonomy of business operation to local Railway Bureau; railway business will be back to the major work with Ministry Railway extracts partial profit and the most of them still be Railway Bureau.

### *34.2.4 Factor of Traffic Business Synthesis*

The biggest challenge of traffic business synthesis is combine the function of traffic and business. In our current station design process, the traffic stream line comes first and after design and deepens of construction project, then comes business space planning and design. This kind of design with an inadequate deficiency of hardware causes a hard situation of station business development, which is difficult to improve and promote with the follow-up soft treatment operation.

Therefore, first of all, a better traffic business synthesis requires a set of working process different with the general business planning and design, which means combine business planning and traffic stream line and architectural space to the whole process from the beginning of construction to operation, which is the scarcest in the current station design.

Second, User (mainly are passengers) of traffic business synthesis as the business consumer is different from general business synthesis consumer in consumption aim, consumption tendency, consumer psychology and consumption environment expectation, which raise special requirements in station function layout, space design and business configuration, so it needs to solve the following questions before business development of traffic business synthesis:

1. What kind of passengers in this traffic architecture? What are the consuming characteristics? How is the purchase ability? What are the requirements of business service and environment of all kinds of passengers?
2. Wht kind of business type suitable for? What is the design requirement of each architectural space?
3. What kind of scale the traffic architecture require? What kind of scale will be supporting by the passenger's consumption ability?
4. How to arrange space suitably for passenger to get the convenient business service?
5. How to combine the two lines, on the basis of guarantee the traffic line function to get the enough passenger flow support for realizing maximizing the benefit.
6. How to strength usage efficacy of architectural space to arouse the space potential?
7. What kind of management system, supporting facility and financial support strategy call for?

## 34.3 Consumer Type and Demand Characteristics

### 34.3.1 Status and Problem

There are a few attentions on passengers during the previous railway station designation and operation that generates a relation of "manage" and "be managed", which causes institutionalization of station inner function space and modeling of design pattern, so the changing starting point is treating the station users as a "customer" who purchase service.

The foundation of providing good service is getting customers' request, which corresponded to railway station business development is understand what kind of type of the station users? What are their consumption tendencies? How is the purchase ability? What are the requirements of business service and environment of all kinds of passengers?

### 34.3.2 Consumer Type Analysis

Obviously, the major users and potential consumers are the passengers. Passengers can be divided into long-arterial line (above 5 h riding), inter-city train passengers (2–5 h) and suburbs train passengers (within 1 h) according to trip length; also can divided into tourist car, soft sleeper, first-class, second-class and average with ticket level; divided with trip purpose can be business, travel, roundtrip (students or workers), visitor and transportation; with trip stage for arriving, leaving and transfer, etc.

Different kinds of passengers have different kinds of business required type, grade and frequency. According to the survey, consumption level is higher for evection and travel; visiting, commuting and schooling always pay much attention to benefit; long-arterial line focus on diet, entertainment and specialty; high consumption frequency for passengers who are in high-speed train, bullet train or high-level seats. These analyses have already summarized in many studies, such as Miss Zhu Xiaojuan (2008) [6].

### 34.3.3 Research and Design Direction

Railway station is an external door for city and population size, development characteristic, area function and external economic vitality decide composition of passenger style. For the same county-level city, there is clear difference among coastal developed cities, interior under-developed cities and tourist cities. So the diversity of passengers requires various business service and station function space designation.

Zhang Fan (2008) [7] summarizes the comprehensive demands of station passengers in convenience, aggregation of business space, option and identification of space. Different types of passengers call for different station inner space. High-incomers care much on the minimum space between people, inner environment and noise. Arriving passengers expect a rush through function, and fro interchange and waiting passengers are more interested in option and space diversity.

Therefore, station business space planning and design foundation is the angle to inspect the customers' comprehensive requirement of business service in macroscopic view of city function and microcosmic view of individual consumer type requirement.

## 34.4 Type and Characteristic of Business Service

### *34.4.1 Status and Problem*

The business service type and level of current domestic station is simple mainly on beverage and solid goods, which not only reduce the selectivity of business service also it is hard to stimulate and promote the potential consumption desire. So during the business planning and designation, it is important to know what kind of business type suit for traffic architecture interior and the different space design requirements of these kinds of business type.

### *34.4.2 Classification of Station Business Service*

There are many methods to classify the business types as the passenger classification. The article suggests that divide into "optional consumption" and "non-optional consumption" by the initiative and passivity during the business service of consumers.

#### 34.4.2.1 Optional Consumption

Customer can choose consumption service type, time, quantity and payment based on their own preference and actual demand for this kind of customers. This kind of consumption includes traditional retail business, service and financial industry [8], which can be divided detailed into necessity type business and guidance business with passengers' demands. Necessity type business includes catering, daily, health, finance and expresses service; and guidance business includes electronics, Shoppe goods, Movie Theater and other leisure industry.

#### 34.4.2.2 Non-optional Consumption

This kind of consumption cannot be chosen by consumers belonging to passivity consumption, mainly for attention (eyeball) economy which shows on commercial advertisement, product exhibition, showcase and shop signboard inside of traffic architecture. Attention economy obtains economic benefits for attracting public attention which occupies important position in business area with the powerful promotion of media society. Now, we are the fourth biggest advertising market after U.S.A., Japan and England. The amplification is 460 % from 1997 to 2006 outdistance the other business type.

The other characteristic for attention economy is that the users don't need to pay; the economic income is from information a provider who is manufacturer. It is raises a big acceptance of this kind of business service for the traveling passengers for free and no weight increase for the luggage. The most important successful factor is directional delivery for this kind for business service, which means place it at the most related consumers for maximum economic value of attention. The classic case of directional delivery is in internet, advertising giant Google is the first company to issue internet advertisement according to the key words from users visiting website.

### *34.4.3 Research Direction and Design Features*

Different business type requires different architectural designation. KFC chain industry calls for a high healthy condition, electricity and water distribution, drainage facilities and large-scale business area; 3C Shoppe needs better lighting; automobile exhibition requires open public space.

Also with different business development strategies: Necessity type business strength service coverage area to promote convenience; Guidance consumption highlight integrity of business atmosphere to provide diversity to compare and select; Attention economy pay much attention to vision accessibility of architectural space, visual coverage frequency and audience group characteristics to formulate rent price strategy.

Multiple-correspondence of different station passengers' response business service demands formulates the complicated station business demands composition. Meanwhile, changeable composition of passengers causes changeable business demands with the development of city area function, which needs fully consideration this kind of space change during designation.

## 34.5 Estimation of Business Facility Scale

### *34.5.1 Status and Problem*

As mentioned, business development scale of current station is small which is harmful for formulating business atmosphere of selectivity and guidance. The income account for a low proportion in station operation income, the business function needs substantially rise. On the other side, over-large-scale business development without enough consumption ability supporting causes huge waste of investment and space.

The size of business development scale is directly determine the integrity, diversity and hierarchy of business activities, which should be considered in the stage of traffic architectural planning and feasibility study with mainly on what is the passengers call for the business facility? How large that passenger can support? And the business return that station expect.

### *34.5.2 Confirmation of Business Development Scale*

Business development scale is close related to consumers' demands, potential consumption ability, development goal of station and location. With the different goal orientation, summarize the following business development scale methods:

1. Based on passenger service goal direction: cases comparative method

   Mirror mature station business development scale to summarize the suitable proportionate relationships among business supporting, passengers sends and station construction scale. As shows the following statistics, business supporting area of these international stations is up to 40–80 % which are far above China's.

| Station name | Daily passengers sends (million) | Covered area (million square meter) | Percentage of business area in covered area% |
|---|---|---|---|
| Kyoto Station in Japan | **30** | **23.7** | **79.3** |
| Osaka Station in Japan | **20** | **13.8** | **64.4** |
| Waterloo Station in England | **20** | **4.4** | **70.4** |
| Paris North Station in France | **56** | **5.6** | **76.7** |
| Berlin Central Station in Germany | **24** | **16.4** | **45.7** |

2. Based on performance target direction: performance inverse method

   In accordance with international standards of production three business incomes occupies the station operating expenditure proportion relationship confirm business development scale goal. For example, non-transportation income occupies a large proportion in total incomes of JR East Japan Railway Company with average 28 %, and steadily rising.

3. Based on passengers demand direction: consumption ability inverse method

   Analysis consumption possibility with passenger pace time inside the station and inverse their supporting business development retail consumption ability with spending habits and potential consumption ability to estimate the reasonable business development scale.

4. Based on city function goal direction: biz services rocking method

   Different stations have different city function in different location. Under a compact type development mode, development of station also can provide business service to peripheral inhabitancy and city activity. Under this circumstance, the development of station scale needs to consider the city business function.

## 34.6 Spatial Arrangement and Format Distribution Optimization

### *34.6.1 Status and Problem*

Long-term weakness position of station causes a serious restrain on station business layout which is only as the affiliated space form. The general station area business layouts are the following methods in China:

1. Completely separating: completely separation between individual business development and station space is for the consideration of land ownership and asset management, which means customers have to leave the station and walk for a certain distance into another business synthesis for the corresponding business service.
2. Single-point layout: divide a few spaces inside the architectural space to provide comprehensive service. Usually the space is located in the corner of hall. Because of the large-size of the hall space, the average distance between passengers and business service is long.
3. Combined waiting space layout: divide some spaces in the waiting room as the business services area. Because the individuality of waiting rooms, these business spaces are always in small scale with similar products and services.

4. Independent tiering layout: in recent years, there are some improvements of business space layout to use interlayer or divided a bulk of area in transfer area for concentration development, so it can provide complete business services in such a large area, but because of different combination effect between space and traffic streamline, there are obvious differences in different business development value. For example, Shanghai Hongqiao Railway Station, the concentrate business space of 1 floor underground is locates between exit of train station and entrance of metro which makes a good business benefit; but the business value of individual interlayer business development in the upper of waiting room isn't fully expressed because of the inconvenience of waiting rooms.

Because the layout strictly follow geometric figure for the convenience of management in space, which is harmful to use; business space always use individual construction or corner space to arrange which cause too much distance for passengers and it is hard to guide the potential consumption desire.

## 34.6.2 Development Direction

### 34.6.2.1 Analysis of Foreign Business Layout

1. Combination of comprehensive and disperse: in Kyoto Station in Japan, it applies "concentrate larger-scale business, disperse for small scale combines traffic streamline" business layout, which concentrate clear consumption purpose business types at the sides of the architecture such as movie theater, hotel, catering and brand garment; place small-convenient business at the sides of traffic stream and combines large business streamline in the hall such as food, grocery and small articles of daily use to obtain enough traffic and business streamline support.
2. Sandwich style: Berlin Station applies three-dimension layout with 2 floors underground, ground floor and 2 floors over ground for traffic, separately provide different city traffic services such as train, metro, bus and light rail; in the middle of 1 underground floor and 1 over floor are main for business to provide business services covered all aspects of daily life. When passenger flow is flowing in different layers, its business layer also can obtain enough support.

### 34.6.2.2 Research Direction and Design Features

Many factors restrain business layout mode, first of all, complexity of traffic. Due to the integrity among many city traffic function in space, different means of transportation is the direct reason to restrain layout of business space; second, there are business space scale and space distribution of composition of business.

With the development of intensive tendency of architecture space, business layout is transferred from simple plane layout mode to composite three-dimensional layout mode.

In the horizontal level, business space is a combined construction of individual, completed "large space" and small scale, flexible arrange "small space" not simple addition of fragmentary space. For example, for strong purpose of business consumption (movie theater, large-catering, hotel and branded garments), it can be far away from traffic space and arrange at the end of space or gallery; for weak purpose of business consumption (beverage, pot foods and daily) which needs be close to traffic space to gain enough flow supporting.

In the vertical level, business development uses interlayer space of each traffic function to provide complete business services, also have a great promotion of saving the walking distance of passengers, intensive land and integrate various spaces.

In selection of business, different grade and type of station is different for adaptability of different business [9], so it is important to choose different business for different area, which is a significant part for railway station business development.

## 34.7 Business and Traffic Streamline Organization

### *34.7.1 Status and Problem*

The streamline design of station is mainly for traffic streamline because of deficiency of business function in current design, the organization method is waiting going through three stages: Beijing Station as the representative of closed separation waiting mode; Chengdu Station and Hangzhou Station as the representative of concentrated waiting and viaduct waiting; and Shanghai Southern Station as the representative of open-space combined [10].

The main node of station interior can be simply summarizes into "arriving entrance"-"ticket previewing"-"luggage security check"-"waiting"-"ticket checking"-"go on the train"-"go down the train"-"ticket checking departure"-"leaving". In this streamline, each passenger needs 3–4 times line up checking, which greatly reduce passing efficiency with no doubt and also increase nervous mood of passengers.

### *34.7.2 Research Direction and Design Features*

Fang Jing divided passengers into three categories: necessity behavior, spontaneity behavior and social behavior based on Yang. Gail's research [11]. It also can be simplified into necessity behavior (traffic) and social behavior (shopping, asking and mutual class behavior), which is traffic streamline and business streamline for streamline description.

Although the main organization method of domestic station is waiting in a certain time, its streamline factor will transfer from traditional waiting method to modern through mode for the factors of enhance of train punctuality rate, transport operation and proportion promotion of high-speed, short-distance business, less-luggage passengers of traveling. Through mode streamline organization method requires more effective node through efficiency, and business streamline requires free shuttled business space and mainly traffic function area.

Inside business streamline of mature business development should consider physical condition, economic benefit factor, business distribution regulation and customer physiology and psychology of architectural space [12]. Because of its traffic factor, stable and safety traffic function should be considered first on streamline organization, and then seek the opportunity of driving business streamline by traffic streamline: How to design reasonable flow to seek fully traffic streamline supporting to business development.

## 34.8 Strengthen of Space Usage

### *34.8.1 Status and Problem*

The main malpractice of old station design is low inside space usage. Its design is always style, solemn with large covered area as the portal image of city with 20–30 height only includes 1–2 floors. The whole space function apply diamonds, tile pattern, which cause low space usage and imbalance, also cause huge difference between construction, maintenance cost and space value. Some of the areas are overcrowded, and some of them exist plenty of gray space cannot be used reasonably, and also excessive opened architecture can increase the distance between passengers and business services.

### *34.8.2 Research Direction and Design Features*

Under the premise of explicit the next purpose, principle of pedestrian choose path is according to the minimum consumption in space (function of distance and crowd factors). Under the same stream gather density, pedestrian always have straightway from point to point, which means space usage of pedestrian prefers a banding or broken line.

J.J. Fruin (1971) introduced the concept of "service level" [13] to evaluate spatial density of pedestrian, C. Alexander (1977) [14] put forward concept of

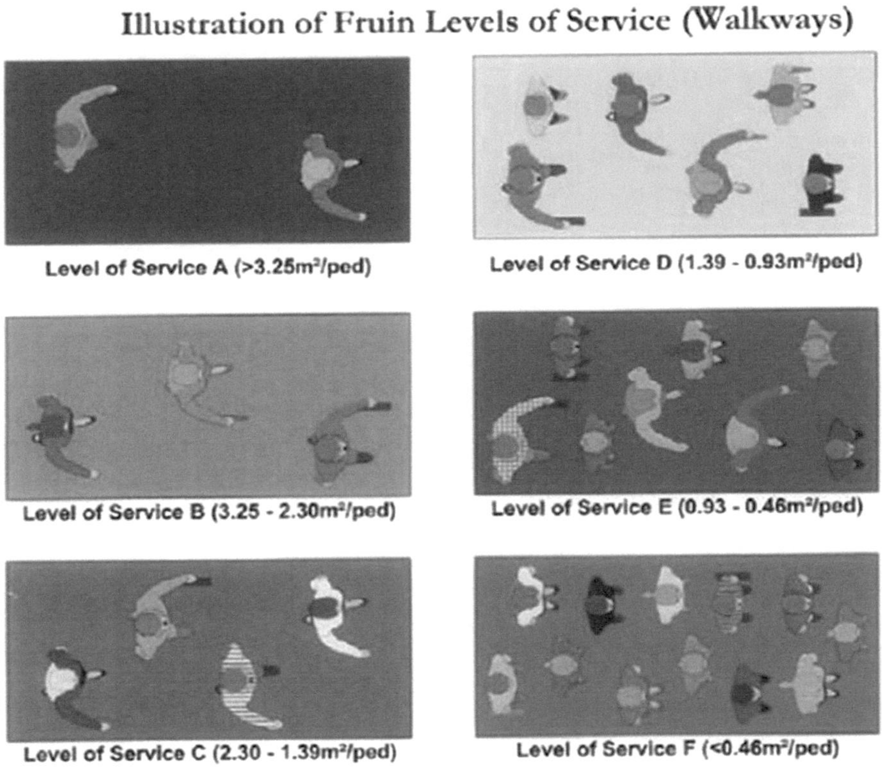


**Chart 34.2** Illustration of Fruin Levels of Service (walkways)

"annoying distance" between function space and users to evaluate the relations between space distance and usage frequency. It is important to satisfy not only a certain traffic service but also provide various services within the annoying distance under the circumstance of the same importance between traffic and business (Charts 34.2 and 34.3).

High usage rate of space causes uncomfortable feeling, crowding feeling and safety problem; Low usage rate shows insufficient space of waste, which isn't suitable for business consumption atmosphere. Therefore, establish comprehensive evaluation system of space usage urgently to control inside space usage into a proper range. Evaluation of space usage rate can guide business value estimation of space to set different rent level and confirm reasonable business return expectation.

Different technical methods evaluate from different view of perspectives on the basis of this evaluation system. Apply usage service level or space usage accumulated number of people to evaluate from individual angle; Analysis different business service covered range with GIS system from business services configuration. Make an accumulated analysis to individual sight range on the basis of people simulation from the vision covered range angle, count received-attention degree of each space to evaluate vision value of space and actual release effect of advertisement (Chart 34.4).

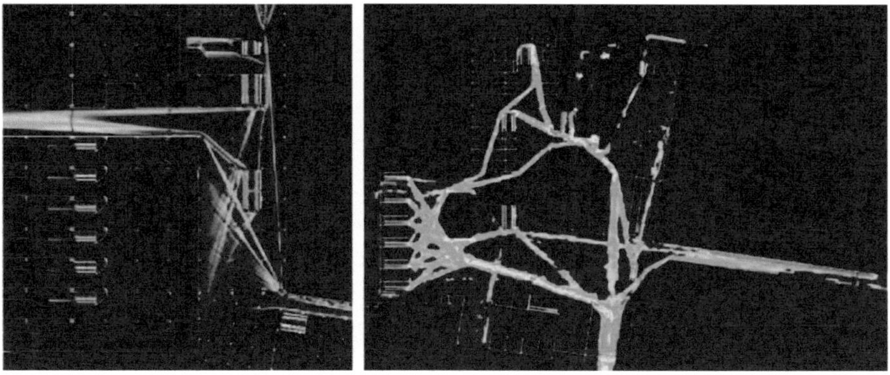

**Chart 34.3** Passenger density map for transfer space in a typical railway station of China

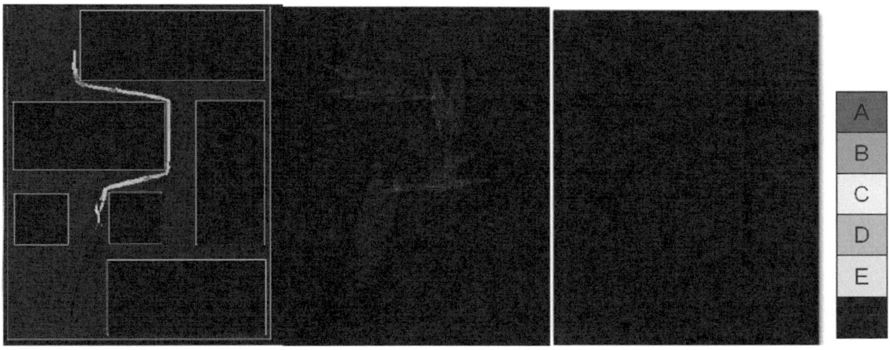


**Chart 34.4** Demonstration analysis based on flow simulation and business value of passenger's view area (passenger flow analysis, View area analysis, business value analysis)

## 34.9 From Node Design and Consultation to Full Services

### *34.9.1 Status and Problem*

Business planning is not a necessary work in China's station designation. Because of small business area, this kind of design always be omitted or completed by architecture designer. Even though the current business development brought to the forefront, business planning always starts after completion of architectural scheme designation. Delay of business planning causes the situation that the planning work distract by the architectural design scheme, and the business function layout, space design and supporting facilities cannot reflect in the architecture design, also it is hard to communicate and understand in design philosophy, working time point and stage objectives in time within such a inattentive working mode between business planning institution and architecture designer.

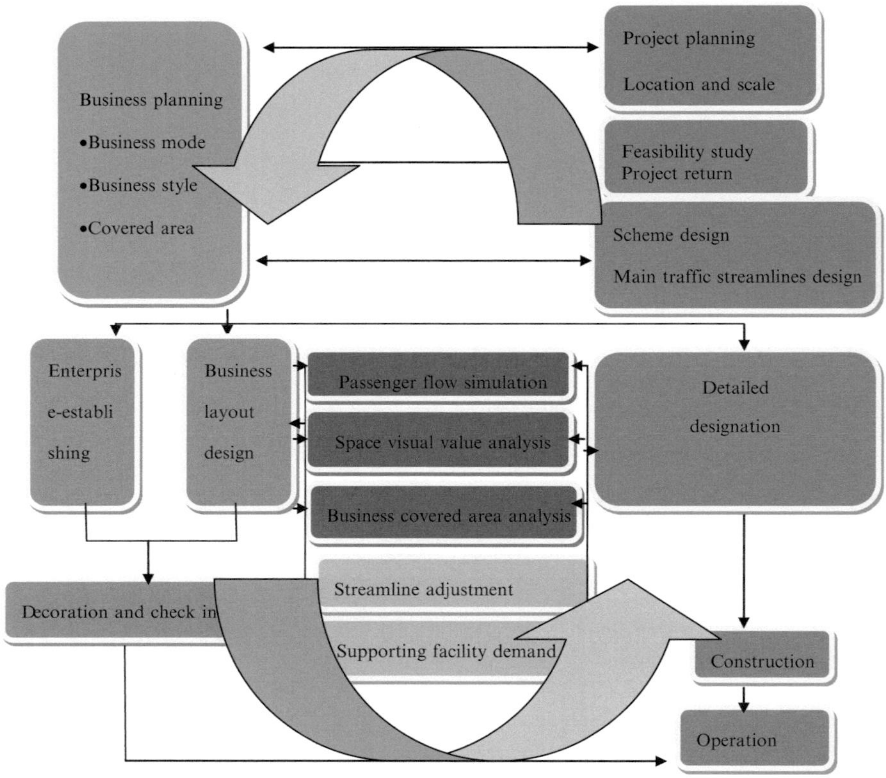


**Chart 34.5** Business planning and architecture design and integrated design flow chart

## *34.9.2 Development Direction*

Operation and earning position of business development affect the operation of traffic architecture more and more when it is in a certain scale, business planning and design calls much more attention as the important part of architecture design and run through from architectural planning, feasibility study, project design, detailed planning, investment an operation stage (Chart 34.5).

In the previous stage, business planning confirm business development scale, grade and primary business compositions with passenger flow mode, train service level and other planning materials, and as the precondition of architecture scheme designation.

Design scheme stage needs to raise primary layout and design pattern to business development according to previous business planning. In the following detailed design stage, business planner cooperate with architecture designer to make sure of consist of business, detailed space layout, space design and supporting construction, and also business operation management and related management modes, practice investment intentions, deepen architecture design with business conditions of specific business brand.

In the operation period, adjust suitable in management, operation mode and business layout according to passenger flow and changeable traffic services to satisfy passengers' demands of each stage.

Meanwhile, business comprehensive development brings a series of new designation requirements to fully consider during the construction design procedure. Plyability tendency of station space function brings demands of large-scale comprehensive space to propose new requirement of architecture structural design; Promotion of space development strength increase the supporting requirements of electricity, water and drainage. Versatility of space and complexity of streamline make it hard to guide the architecture design with the current fire protection design specification, and also it calls dynamic fire protection performance estimation of architecture space normal use.

## 34.10 Management Innovation and Diversification Exploration

### *34.10.1 Status and Problem*

Under the current system, Ministry of Railway mange the local railway department into two lines of income and outcome, which means, the total incomes of railway department hand in Ministry of Railway, and redistributes by Ministry of Railway according to the operation cost and profit norm of each department. This kind of "as a big pot" management achieved makes up deficit of western local railway department by using profit of developed areas, but it also restrict initiative, and this is the policy reason of lagging of station business development.

In human resource, centralized management of railway department cause that the managers and servicers are inner workers or related families, non-professional human resource lead to non-professional service.

### *34.10.2 Development Direction of Management System*

In policy, the reformation of income and outcome and discussion of business together of Ministry of Railway bring the new light of station business development. Transfer of managerial authority and reformation on distribution of business income greatly encourages optimization and strengthens development of station space source. In management, introduce professional traffic business Operation Company to develop, manage and operate according to larger-comprehensive development mode fully use market sources and adjustment ability to keep dynamic balance between provided business service and passenger changeable demand.

### *34.10.3 Diversification Investment Exploration*

High debt rate makes Ministry of Railway considered to transfer the subject of construction investment to local government [15], except a few developed cities, the majority of interior cities' revenues cannot afford the huge and short period construction investment. Introducing bank capital and private capital is the necessary model to solve future railway construction financial matter. Promotion of station function can bring extra initial construction investment, but the stable cash flow and profit of business function will provide stable investment and finance fund foundation and enhance rate of return on investment on railway foundation construction.

In the construction of urban rail transit, local government applied several of investment and financing modes to introduce external capital. The construction of railway system can fully take the experience of urban rail transit an example, such as: Beijing Investment, Hong Kong metro, Beijing capital group uses a public-private partnership in construction of Beijing Metro Line 4 (PPP); Chongqing investment uses the "government procurement, installment of" asset securitization (ABS) in the construction of light rail line 2, line 3; Beijing metro line ten company uses "construction – transfer" (BT) mode in Beijing Subway Olympic extension part of the project (BT); Hong Kong Metro Inc, Japan's Tokyo Corporation used "track + land mode" (R + L); and Hong Kong Rail AC (Shenzhen) Company used "build – operate – transfer" (BOT) mode in the second-phase of line 4 construction of Shenzhen metro project, etc.

Introduce of external source of finance needs to clear the asset relation between passenger line station and lines first during the railway construction, make sure asset attributes of passenger line station, hyalinize the single station clear rules and operation income and distribution under the unity of management of railway.

## 34.11 Conclusion

There are various reasons to cause the lagging status of station business development, it calls multidimensional profiling transfer from planning concept, construction design process and management policy to realize a comprehensive positive development and provide powerful supplementary of follow-up construction fund and better railway travel environment. This article puts forward a preliminary frame for station business planning and development research, it needs further study for detailed integration design and practice application of each module.

**Acknowledgement** Supported by National-sponsored Funding Program for Social Sciences of China (13&ZD026), Supported by National Natural Science Foundation of China (51108018).

## References

1. Mid-long term planning for China's railway network, The State Council (2004) http://jtyss.ndrc.gov.cn/fzgh/t20050720_37500.htm
2. Mid-long term planning for China's railway network (2008 adjustment), National Development and Reform Commission (2008) http://jtyss.ndrc.gov.cn/fzgh/t20090605_284526.htm
3. Chen Lan, Yin Qiong (2011) ) Research of Composition and Optimization of Urban Functions of Large-scale Railway Stations [J]. Huazhong Archit (7):30–35
4. Ministry of Railways servicing 180 billion loss of a foregone conclusion [N]. China Stock Finance http://finance.huagu.com/jd/110902/97792.html
5. Li Lei (2010) From the Transit Synthesis to the City Synthesis: Development of Rail Transit Hub Commercial Center [J]. Modern City Study (7):25–30
6. Zhu Xiaojuan (2008) Passenger flow of large railway station layout research [D]. Beijing, China
7. Zhang Fan (2008) Railway station business space design research [D]. Tianjin, China
8. Fan Zhigao (2006) Large railway station business space design research [D]. Wuhan, China
9. Yu Yan (2010) Railway station business service area industry choice and its influencing factors [D]. Chengdu, China
10. Zhu Xiaojuan (2008) as above
11. Fang Jing (2007) New model of railway station building space of Post-Occupancy Evaluation (POE) study [D]. Beijing, China
12. Yang Bo (2006) Shopping Center (Shopping Mall) space environment design research [D]. Changsha, China
13. Fruin JJ (1971) Pedestrian planning & design[M]. MAUDEP, New York
14. Alexander C et al (1977) A pattern language: towns, buildings, construction [M]. Oxford University Press, New York
15. Yu Bingbing and Zhou Pengfeng (2011) Railway construction brewing major reformation local government to prepare maintaining road with land [N] http://www.tianxiagold.com/html/Comment/ticai/201108309146.html

# Chapter 35
# The Idea and Strategy in the Urban Design of Tangshan North Station

**Honghong Zhang, Chi Zhang, Yingdong Hu, Ming Liu, and Gaoyuan Wang**

**Abstract** With the development of urbanization and intensive land-use, as well as high-speed rail construction, a lot of Chinese cities have built a large-number new train stations, also the station squares area have been reconstructed. This large-scale development and land use intensity, however, cannot find an effective solution to solve the main problem of the urban design of railway station and square. This paper will take the urban design of Tangshan North Station as an example to discuss the idea, strategy and some experience, expecting to provide some valuable information for future train stations design.

**Keywords** Railway station • Station square • Urban design • Comprehensive exploitation • Composite function • Traffic organization • Urban landscape • Architecture design

## 35.1 Project Overview

### *35.1.1 Background*

Tangshan North station as one of two railway stations in Tangshan, is located in Fengrun district. This station not only serves the local public traffic, it is also the most important traffic hinge for high-speed rail to Beijing. In recent years with urbanization

H. Zhang (✉)
Professor of Beijing JiaoTong University, Beijing, China
e-mail: hhzhang@bjtu.edu.cn

C. Zhang • M. Liu • G. Wang
Master Student of Beijing JiaoTong University, Beijing, China

Y. Hu
Associate Professor of Beijing JiaoTong University, Beijing, China
e-mail: ydhu@bjtu.edu.cn

H. Xia and Y. Zhang (eds.), *The 2nd International Symposium on Rail Transit Comprehensive Development (ISRTCD) Proceedings*,
DOI 10.1007/978-3-642-37589-7_35, 

**Fig. 35.1** Location of Tangshan North Station

development, Fengrun district will be replaced by "Gengyang New Town", Tangshan North Station takes the great reconstruction opportunities and challenges.

### *35.1.2 Study Basis*

Tangshan North Station area is positioned by local government to build Tangshan's new card, driving area economic growth. As the railway departments have taken charge of the new station building design, the local government focuses on the urban design of station square and traffic organization. In 2011, we take the research of Tangshan North station on the basis of combining analysis of control detailed planning, early planning design of North Station square and surrounding area and etc. We research the necessary traffic function and focuses on the scale and function-orientation of the square.

### *35.1.3 Location Analysis*

"Gengyang New Town" is located on the north side of Tangshan city, connected to the downtown area by the Tangfeng Road. Tangshan north station is located in the southern end of the main road axis in "Gengyang New Town" (Fig. 35.1).

This area is high-quality trade and commercial region with high development intensity and small plot scale division, which increase the land value. Meanwhile there is Tianning Temple Palace which is in the cultural axis. The station square is located in extension of commercial axis and cultural axis.

## 35.2 Problem Research of Last Urban Planning

Tangshan North Station square, covering an area of about 300 m*300 m (D1 + D2 block) (Fig. 35.2), divided up to two parts by the city main road, In order to solve the problem of walking connection between D-1 and D-2 plots, the Last Urban Planning sets the elevated platform, dividing the station square into double layer and forms three-dimensional layout with the platform as main square, ground floor in D-1 block mainly sets a large supermarket. This layout also brought problems as follows.

### *35.2.1 Connection with Local Traffic Network*

In the last urban planning, Xiangyun Road goes through the square and joins up with Guanghua-West Road at the north-east corner, so the pedestrian system is cut into two pieces. Moreover, after connecting with Guanghua-East Road, Xiangyun Road has become the vital communications line in this area. It is easy to think about how difficult for Xiangyun Road to carry both of the urban traffic and station traffic (Fig. 35.3). The probability of traffic jam and accident will be higher than before. Also it is very dangerous to the people who are passing the road.

### *35.2.2 Square Scale*

Now empty boring large square platforms have appeared in some city stations, make citizen feel nonhuman and cause land and money waste. So confirming square scale is key issue. The elevated platform in the planning of Tangshan North Station has formed 300 m*300 m double space. In contrast to other stations in its scale, we

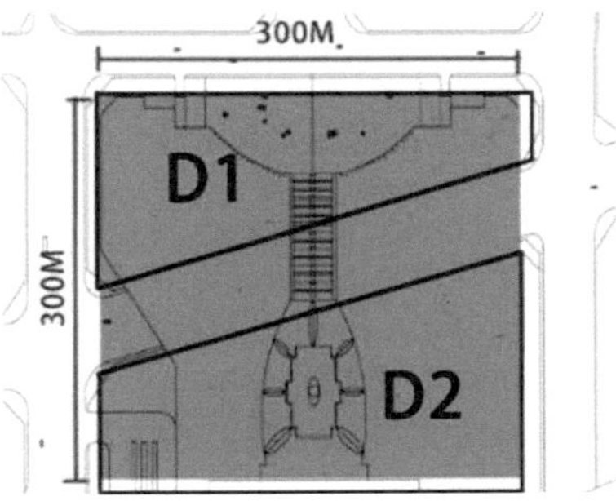


**Fig. 35.2** Dimension of the station square

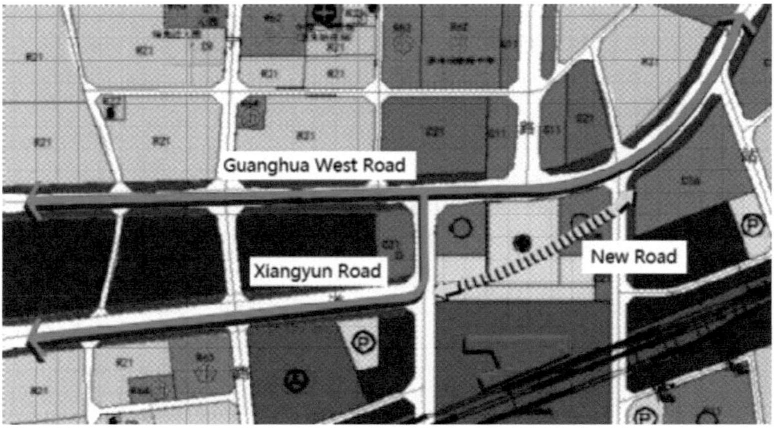


**Fig. 35.3** Relationship of Guanghua West Road, Xiangyun Road and the new road in front of Tangshan North Station

can see its area is too large. Such as Tangshan west station, the station square size is about 120 m*200 m, even Beijing Railway Station square is about 70 m*280 m.

Traffic distribution based on the city development situation, if not explicitly specified, according to city traffic "bus priority" principle, in the Plaza no rail access, passenger travel mode is 50 % by bus, 30 % by taxi, 20 % by social car. According to the following Tangshan North Station passengers-data:

1. Average number of people send every year: 1,060,000
2. Average number of people send everyday: 2,920
3. Average number of people arrival every year: 1,180,000
4. Average number of people arrival everyday: 3,250
5. Maximum gathering number of people: 3,000
6. Average number of people send in the peak hour: 2,000. (from the Tangshan Railway Bureau)

We calculated the required site area of each mode of transportation, and with the existing data for comparison, found the existing actual Area was clearly larger (Table 35.1).

### *35.2.3 Land-Use Value and Development Intensity*

D-1 plot from the station 300 m, only set up a supermarket. It is not making full use of the good location next to the station square, and not associated with the business district along Station Road. Also the square platform is without forming a good transition and articulation with the surrounding architecture. Apparently, the land-use value is not developed enough.

Table 35.1 Required site area and actual area of each mode of transportation

| | Required area ($m^2$) | Covering rate (%) | Actual area ($m^2$) | Covering rate (%) |
|---|---|---|---|---|
| Bus station | 6,000 | 11.8 | 12,704.9 | 24.9 |
| Parking area | 7,418.4 | 14.5 | 10,729.9 | 21.0 |
| Pedestrian | 6,000 | 11.8 | 22,129.5 | 43.4 |

### 35.2.4 Urban Landscape

On the big platform of Tangshan North Station, a height difference exists between the ground and the platform. If people on ground near the platform a certain range, it is unable to feel the full picture of the square and railway station (Fig. 35.4). The Arriving layer is completely covered in underground, the passengers in this layer are unable to feel the city characteristics and landscape charm. The exit design should not only take care of traffic route is short and smooth, consideration should also be given to passengers on the appearance of the city's first impression.

## 35.3 Core Concepts and Strategies

### 35.3.1 Core Concepts

Combining functional orientation and master planning, make full use of passengers flow at the same time to strengthen the core position of the railway station. Aim at forming a full of vitality and charm transportation center, city living room, commercial center, shaping a new image of the city. Reflecting the characteristics of times, meeting public various demand, we tend to research about energy instead of shape, variety instead of single, personality instead of materialization, local culture instead of universality.

### 35.3.2 Design Strategy of Composite Function

Combined the conditions of site and size, Station square has comprehensive functions, such as solving traffic function, to increase the commercial, office, hotel, cultural function, at the same time to increase necessary elements the superior city design, such as square, distinctive ladder, cafe and shop, office and other contents, emphasize land intensive and functional composite (Fig. 35.6).

Traffic Hub: As traffic hub, to integrate a series of traffic relationships: walking and car lines of city and square,, walking line of the square and car line of city. Give the layout of the basic traffic facilities.

Commercial Development: A train station has high potential commercial value. The region adjacent to Station Road is business center, and square scale is too large,

**Fig. 35.4** View in front of the big platform

therefore to break the existing platform layout, development the D-1 plots within limits, raise land utilization rate and form a three-dimensional, complex spatial pattern. At the same time, create high-quality, open public space of business service area.

Continuity 3D Landscape-square: An elevated platform was reexamined, as terminal square. It needs to connect city road, thus liberated ground space, setting a ground square in front of an elevated platform as transition and between city streets and the platform. Specially, we design a continuous flow folded plate to link different height square is, just like a magic carpet with social character Platform as a fabric, connected to the ground and roof, city and architecture.

## 35.4 Urban Design

### *35.4.1 Function Adjustment of Plots and Layout*

D-2 plot: Traffic function, mainly as the transportation facilities to railway station and organizing traffic flow, such as bus station, public parking, taxi station etc.

D-1 land: Commercial Office and cultural landscape facilities as the basic principles, to create high-quality business area and the city landscape; D-1 plot is located in the station exit and the main urban axis. The business district is only 500 m from the railway station, in order to make full use of the advantage, the plan of D-1 plot has a dual meaning -not only has the function of commercial services, but also make passengers arriving business district on foot more convenient, to form 800 m business circle and the open walking system. In summary, D-1 eastern block positioning for commercial pedestrian street, including retail, catering and leisure shops mainly; D-1 western block for absorbing large-scale supermarkets and business office (Fig. 35.5).

### *35.4.2 Traffic Organization*

#### 35.4.2.1 The Station Square and City Traffic

Analysis of intercity arrival and departure passenger characteristics, through the integrated traffic hub, focused on solving the city traffic problem inside and outside connection, build efficient passenger distributing system.

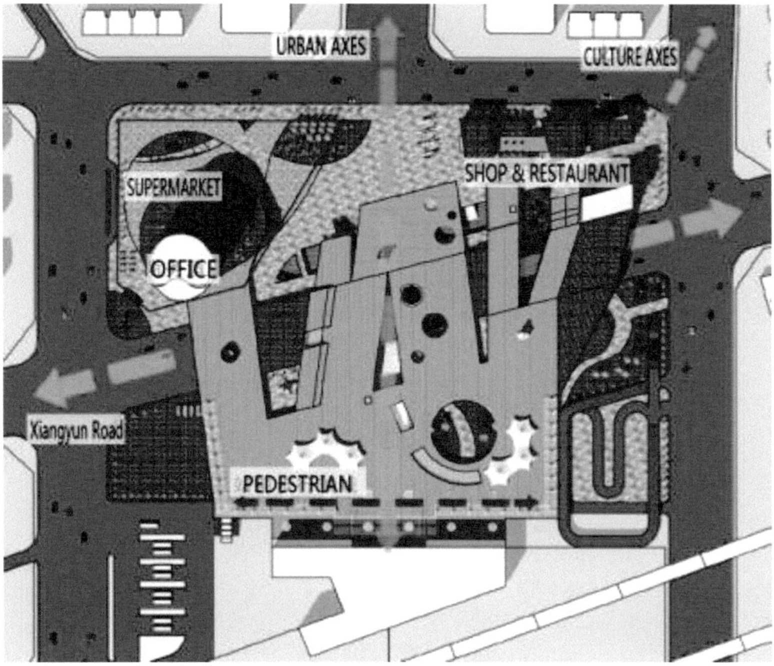


Fig. 35.5 Site plan of the station square

#### 35.4.2.2 Square Internal Traffic Space Layout and Transfer Organization

Station square internal traffic space layout follows the stereo development, efficient transfer, the idea of compact layout with reasonable layout of bus station, and taxi station. Traffic logic can follow "traffic space scale", "traffic space layout", and "streamline organization and the passenger transfer order" (Fig. 35.6).

#### 35.4.2.3 Folding Platform Forms a Pedestrian Line

From city streets into the square on the ground, the folded platform, until the overhead square platform, achieve multi-directional, multi-angle, easy and convenient traffic connection.

#### 35.4.2.4 Parking

Parking facilities adopt the combination of ground and underground way. Combined with the construction development capacity, in the North Station area with underground parking facilities, include public parking and all types of facilities of

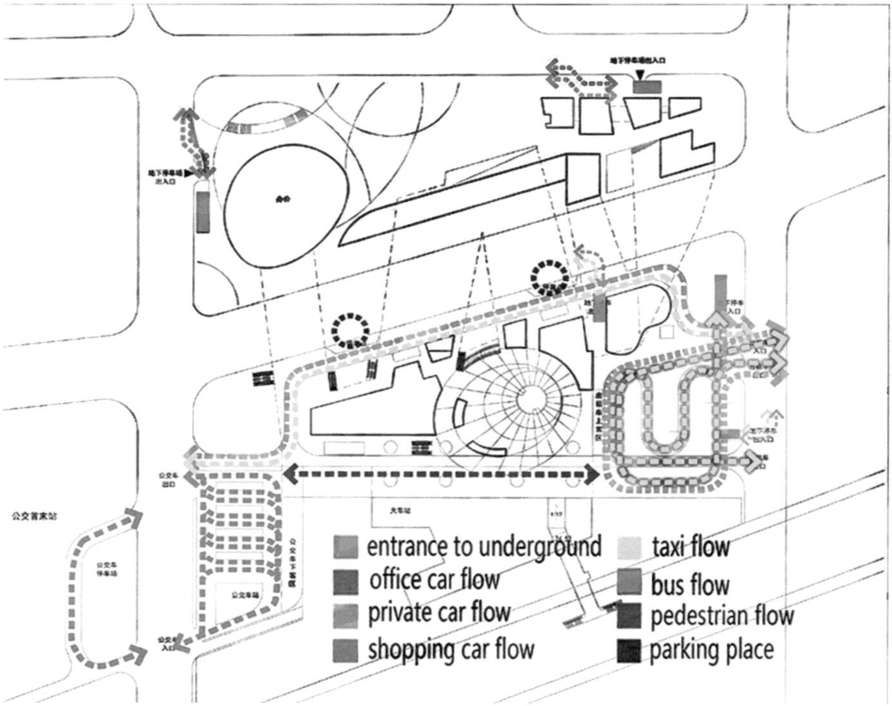


**Fig. 35.6** Traffic organization in the station square

parking lots (garages). Public parking is mainly concentrated in the underground part of D-1 block; total parking spaces for 720, where on the ground 50 and 670 underground.

### *35.4.3 Open Space and Landscape*

The plot between Guanghua Road and Xiangyun Road, is defined as active open space. The effect as follows:

1. Draw back the building line and overhead construction to release the ground space of square.
2. Make station open square be congruency, contrapuntal with end of city main Road axis.
3. Control the height of middle part, let two sides higher, to limit a relatively open and rich open space (Fig. 35.7).

**Fig. 35.7** City open space in the new plan

### 35.4.4 *Building Height Control*

The office building at the north end of Station Road has the height control of 150 m, the rest buildings are residential towers with height control of 100 m. The commercial belt between Guanghua Road and Xiangyun Road is controlled by 90 m. Therefore, we suggest that the height control of west side of D-1 block is 90 m, the height control of east side of D-1 block is 24 m.

## 35.5 Architecture and Landscape Design

### 35.5.1 *Commercial Interface*

Commercial boundary space should open its interface to adjacent city and form natural flexible good communication space, making the business better attracting people stop, suitable for entry. Put the original ground floor supermarket to underground exploitation with high-rise commercial complex on the ground. Set group commercial pedestrian streets on another side, streets freely intersperse, and echo with city road and Temple Pagoda landscape, the layout which breaks up the whole into parts can form a wide variety of small shops (Fig. 35.8).

### 35.5.2 *Soften Platform Edge and Greens Stereo*

Draw back overhead interface edges of square platform, forming a buffer ground square next to the adjacent streets. Take more effective ramps and steps, escalator and elevator, in order to generate short path, to provide users with smooth mobile function selection. Set greens combined with above of Xiangyun Road and square, sculptures enforce interest and openness of square space.

**Fig. 35.8** Commercial interface in the new plan

## 35.6 Summary

Comparing to other areas in the city, station squares face more complicated problems. Design is always a learning of problem solving, planners and architects in facing these problems, must try to make the most possible optimization, which can create livable environment. Finally, through our efforts, this research results have been adopted and implemented by the relevant departments.

**Acknowledgment** This research is supported by: Technical consulting projects (A12X00040); longitudinal research projects funded projects of Beijing JiaoTong University (2011JBM180, 2010RC030, 2012JBM120); Ministry of Education "New Century Excellent Talents" (NCET-10-0217).

## Further Reading

1. Liu Cong (2009) The research to the south square of Shanghai Railway Station. Archit J, 93–98
2. Zhu Yandong (2007) Influence to area re-development by reconstruction of railway station at urban core areas: case of Zhengzhou railway station. Modern Urban Res 10:45–51
3. Zhang Quan (2008) Study on the problems in squares before railway station in China and the countermeasures. Anhui Archit 1:28
4. Xu Yulian (2010) Transport system planning of railway station hub area alteration – take Xiamen Railway Station as an example. Transport Standardization. 2 Half of July 2010, No. 225, pp 112–115

# Part III
# Others

# Chapter 36
# The Optimization Model of Urban Transit Departure Frequency

**Yuan Li and Bingfeng Si**

**Abstract** Transit operator is the key of public transport management. To enhance the level of public transport, departure frequency management is an important means of improving the service quality of urban public transport and attracting passengers. This paper analyzes urban transit departure frequency in terms of service quality and efficiency then builds transit service frequency optimal model, with multitarget optimization model to passengers′ travel cost as well as for the benefit of the transit company. The genetic algorithms is used to solve this model. Based on realistic transit service survey data, the model is applied in transit route and proved to be useful.

**Keywords** Transit operator • Departure frequency • Optimization model • Genetic algorithms

## 36.1 Introduction

Departure frequency optimization is an important part of the transit operators. Domestic and foreign scholars have conducted a lot of research, in which the majority through a supply or demand model to determine the departure frequency of the line [1–3]. Furth and Wilson [4] establish the model for assigning the frequencies in the given set of transit lines by optimizing the sum of the waiting time and travel time. Hasselstrom sought to determine simultaneously the itinerary and the frequency of transit lines on a general network. Stephanedes and Kwon consider the fare as a decision variable, as well as the line frequencies.

The departure frequency optimization directly affects the condition of urban transit. Because t frequency optimization is an important research subject that can

Y. Li (✉) • B. Si
School of Traffic and Transportation, Beijing Jiaotong University,
100044 Beijing, China
e-mail: 11120963@bjtu.edu.cn

H. Xia and Y. Zhang (eds.), *The 2nd International Symposium on Rail Transit Comprehensive Development (ISRTCD) Proceedings*,
DOI 10.1007/978-3-642-37589-7_36, © Springer-Verlag Berlin Heidelberg 2014

solve practical transit problems and can produce high benefits, it is very necessary to formulate more practical models and algorithms.

This paper aims at minimum passenger travel time costs and the economic benefits of transit corporation, establishes the transit frequency optimization model. Because the model is a transit scheduling problem, the practical application of the mode will be complexity with the increase of travel times, sections and site, which means very difficult to solve the model accurately. In this paper, an improved genetic algorithm is used to solve this problem.

## 36.2 Model Optimization

### *36.2.1 Assumptions*

All vehicles are running along lines, not allowing more stations and overtaking [5]; The vehicles are running with the scheduling timetable, traffic jams and other accidents are not considered [6]; The parking time of the vehicle at each station is a fixed value; In the period of optimization, the demand flow is independent from departure time instant; The waiting time passengers accept is a fixed value, over the value passengers will generate dissatisfaction.

### *36.2.2 Variable Definition*

f—departure frequency; $\lambda_i(t)$—passenger arrival rate of the i-th station at time t, people/min; $t_{ij}$—the time of the j-th vehicle arriving the i-th station; $S_j^i$—the parking time of the j-th vehicle at the i-th station; $M_p^w$—the time costs of passengers waiting at the station; $M_p^t$—the time costs of passengers running in the vehicle; l—passengers order; H—the waiting time passengers accept; $\omega$—the number of seats in the vehicle; $\rho_0$—the satisfaction membership value of standing passengers when not crowded; $x_1$—the number of standing passengers in the vehicle; $G_j^i$—due to the limitations of the vehicle capacity, the number of passengers not boarding the vehicle and continue waiting when the j-th vehicle arrives the i-th station; $h_{\max}h_{\min}$—maximum/minimum departure interval of adjacent cars (min).

### *36.2.3 Total Passenger Costs*

In addition to the fare, the main source of the passengers cost can be converted into the waiting time in the station and the in-vehicle time, multiplied with the corresponding time value, we can obtained the corresponding total passengers time cost.

#### 36.2.3.1 The Waiting Time of Passengers at a Station

The waiting time of passengers at a station $C_h$ includes the cost passengers waiting vehicle arriving $c_\rho$ the cost passengers waiting boarding $c_u$ and the remaining passengers extra time cost $c_\varphi$

$$C_h = c_p + c_u + c_\varphi. \tag{36.1}$$

Where

$$c_\rho\left(f, w_j^i\right) = \zeta_h M_p^w \sum_j \sum_i \sum_{i=1}^{w_j^i} \left(1/f + t + S_j^{i-1} - l/\int_{t_{ij}-H}^{t_{ij}} \lambda_i(t)dt\right). \tag{36.2}$$

$\zeta_h$ is the satisfaction coefficient of the passenger waiting, which is the satisfied passengers proportion of the total number. Considering the number of overtime passengers $u_j^i$ corresponding to in the j-th vehicle in the i-th station, divided into three cases;

$$u_j^i = \begin{cases} 0 & P_j^i - \int_{t_{ij-H}}^{t_{ij}} \lambda_i(t)dt \le 0 \\ P_j^i - \int_{t_{ij-H}}^{t_{ij}} \lambda_i(t)dt & P_j^i - \int_{t_{ij-H}}^{t_{ij}} \lambda_i(t)dt > 0 \text{ and } P_{j-1}^i - \int_{t_{ij-H}}^{t_{ij}} \lambda_i(t)dt - U_{j-1}^i \le 0 \\ P_j^i - \int_{t_{ij-H}}^{t_{ij}} \lambda_i(t)dt - \left(P_{j-1}^i - \int_{t_{ij-H}}^{t_{ij}} \lambda_i(t)dt - W_{j-1}^i\right) & else \end{cases} \tag{36.3}$$

Given an Gi initial state, we can calculate the number of unsatisfied passengers vehicle-by-station are not. Therefore, the satisfaction of the passengers waiting meets

$$\zeta_h = \frac{\sum_{i=1}^m \int_{T_1}^{T_2} \lambda_i(t)dt - \sum_{i=1}^m \sum_{j=1}^n u_j^i}{\sum_{i=1}^m \int_{T_1}^{T_2} \lambda_i(t)dt} \tag{36.4}$$

The cost passengers waiting boarding $c_u$ is the waiting time before vehicle arriving and the time passengers spent during the parking time,

$$c_u\left(W_j^i\right) = M_p^w \sum_j \sum_i \sum_{i=1}^{w_j^i} (l-1)\overline{u} \tag{36.5}$$

The remaining passengers extra time cost $c_\varphi$ is the sum of the waiting time for the former vehicle, the waiting time for the present vehicle and the parking time of the present vehicle, i.e.

$$c_\varphi\left(f, G_j^i\right) = M_p^w \varphi \sum_j \sum_i \left( \sum_{i=w_j^i+1}^{w_j^i+G_j^i} (1/f + S_j^{i-1} - (l-1)/\int_{t_{ij-H}}^{t_{ij}} \lambda_i(t)dt \right) + G_j^i/f \tag{36.6}$$

#### 36.2.3.2 In-Vehicle Time Cost

The in-vehicle time cost $C_c^{'}$ including two parts: the in-vehicle time of the in vehicle passengers $c_{t1}^{'}$ and the in-vehicle time of the passengers getting off at the i-th station $c_{t2}$, i.e.

$$C_c^{'} = c_{t1}^{'} + c_{t2} \tag{36.7}$$

Introducing the passenger comfort $\zeta_c$ adjust the in-vehicle time proportion in the total cost, i.e.

$$c_{t1}^{'} = \zeta_c M_p^t \sum_j \sum_i \left( A_j^i t_0 + S_j^i A_j^i (1 - k_i) \right) \tag{36.8}$$

The density of the passengers in-vehicle can effectively reflect $\zeta_c$, expressed as a function of the degree of membership. The maximum of membership is 1, becoming smaller with the increasing in-vehicle congestion. The satisfaction membership of the standing passengers in-vehicle is expressed by $\zeta(x_1)(0, V_j - w)$, i.e.

$$\zeta(x_1) = \begin{cases} \rho_0 & 0 < B_j^i - w < \delta_0 \\ \dfrac{V_j - w - x_1}{V_j - w - \delta_0}\rho_0 & \text{else} \end{cases} \tag{36.9}$$

Thus

$$\zeta_c = \frac{\sum_{j=1} \sum_{i=1} b_j^i}{\sum_{j=1}^{n} \sum_{i=1}^{m-1} B_j^i} \tag{36.10}$$

The passengers getting off the vehicle at the i-th station time cost is

$$c_{t2} = M_p^t \sum_j \sum_i \left( \sum_{i=1}^{w_j^i} \left( S_j^i - (l-1)\overline{u} \right) + \sum_{l=1}^{A_j^i q_i} \overline{d}n \right) \tag{36.11}$$

#### 36.2.3.3 Total Passenger Costs

The Total passenger costs is the sum of waiting time and in-vehicle time, i.e.

$$R_p^z = w_h C_h + w_z C_c^{'} \tag{36.12}$$

### 36.2.4 Transit Company Income

#### 36.2.4.1 Operating Income

Operating income main depends on the flow, that the product of the total passenger and fare, i.e.

$$N_s = C_p^r \sum_j \sum_i w_j^i \tag{36.13}$$

#### 36.2.4.2 Operating Cost

The operating cost includes the vehicle cost and the variable cost, i.e.

$$N_c = \varepsilon_v + \varepsilon_0 \tag{36.14}$$

the vehicle cost

$$\varepsilon_v = C_c^v T/f \tag{36.15}$$

the variable cost

$$\varepsilon_0 = C_c^0 \sum_i D_i T/f \tag{36.16}$$

#### 36.2.4.3 Total Income

The total income equals the operating income minusing the operating cost, i.e.

$$R_c^z = N_s - N_c \tag{36.17}$$

### 36.2.5 The Departure Frequency Optimization Model

In summary, the departure frequency optimization model is as following:

$$\max R = R_p^z - w_s R_C^Z \tag{L1}$$

Subject to

$$\left|\left(t_{j+1} - t_j\right) - \left(t_j - t_{j-1}\right)\right| \le \sigma \quad (j = 2, 3, \ldots, n-1)$$
$$h_{\min} \le t_j - t_{j-1} \le h_{\max} \quad (j = 2, 3, \ldots, n)$$
$$N_s \ge R_{\min}$$

## 36.3 Model Algorithm

Based on the Genetic Algorithm, the solution is expressed using One-dimensional string structure data. The repeatedly genetic manipulations bring the constantly evolutionary processes; finally it would converge to the best solution.

**Step 1**: The alleles are constituted by {0, 1}; we set the length of the encoded string as the scheduling period represented as letter T, the accuracy of the scheduled time is 1 min, each of the encoded bits is corresponded to 1 min in the scheduled time. The value of the encoded bit represents the selection of docking site, '1' means pit stop, and '0' means pass by [7].

**Step 2**: Find and note the position of the '1' in the encoded string in proper order, and then note the number of the '1'.

**Step 3**: First decoding the individual the code string getting the individual phenotype. Than calculate the objective function value. At last, according to the conversion calculate the individual fitness.

**Step 4**: Using penalty function to express the constraints.

**Step 5**: Using heuristic methods to generate the initial population of each individual gene.

**Step 6**: The length of the individual l is T, the size of the population is 100, the crossover probability is 0.6, the mutation probability is 0.005, the terminate algebra corpse is 500.

**Step 7**: Iterative to P(500), the algorithm terminates; if 10 consecutive generations keeps the same, the algorithm terminates.

## 36.4 Example

The parameter value is shown in Table 36.1.

Through the matlab simulation, the most suitable departure interval is 1 min in peak and 10 min in Non-peak.

**Table 36.1** The parameter value

| Parameter | Value | Parameter | Value |
|---|---|---|---|
| Operation time | 6:00–22:00 | Fare | 2 yuan |
| Peak: | 7:00–9:00 | $h_{\min}$ | 1 |
| Second peak | 9:00–10:00,<br>16:00–17:00,<br>19:00–20:00 | $h_{\max}$ | 10 |
| Waiting time (peak) | 3 min | $\sigma$ | 3 |
| waiting time (Non-peak) | 6 min | $R_{\min}$ | 120 yuan |

## 36.5 Conclusion

The model describe the process of transit operators accurately, such as the use of the passenger arriving function, the proportion of the passengers to alight function and the remaining passengers in the station.

But in the specific use, it is pending further investigation and analysis whether the comfort of the membership function can reflect the actual situation.

**Acknowledgment** This work is supported by the National Natural Science Foundation of China (71071016).

## References

1. Ceder A (1984) Bus frequency determination using passenger count data. Transp Res A 36:439–453. Elsevier Science
2. Ceder A (1989) Optimal design of transit short-turn trips. Transp Res Rec 175:8–22. Elsevier Science
3. Shih MC, Mahmassani HS (1995) Vehicle sizing model for bus transit networks. Transp Res Rec 23:35–41. Elsevier Science
4. Furth PG, Wilson NHM (1982) Setting frequencies on bus routes. Theory Pract Transp Res Rec 156:1–7. Elsevier Science
5. Zhou M, Sun SD (1999) Principle and application of genetic algorithms. J Syst Sci Math Sci 26:352–361. The Chinese Academy of Science Institute of Mathematics and System Science
6. Xuan GN, Cheng RW (2004) Genetic algorithms and engineering optimization. J Manag Sci China 11:12–18. The National Natural Science Fund Committee Management Science
7. Niu XQ, Chen Q (2003) Optimal model of urban bus frequency determination. Traffic Transp Eng 110:68–72. Elsevier Science

# Chapter 37
# Long Time Forecasting of Rail Transit Passenger Volume

**Haifeng Song, Tao Tang, Chenling Li, and Yi Ding**

**Abstract** The theory and application of railway passenger traffic volume (RPTV) forecasting is one of the focuses in the research areas of traffic transportation. RPTV forecasting is an important responsibility of the railway transportation management departments, RPTV is influenced by multiple factors, and the action mechanisms of these factors are usually unable to be described by accurate mathematical linguistic forms, so the theory and method of RPTV forecasting remains a focus in research all the time. And RPTV has a great important in the choosing of route, the construction of station before a new line was built, and the operation after the line was using.

**Keywords** Railway • Passenger traffic volume • Forecasting • Long time

## 37.1 Introduction

With the reforming of China's railway passenger market, especially in recent years, China's highway, civil aviation has been greatly developed through the open management, while railway transportation has been challenged powerfully, the competition between different modes of transportation will become increasingly intense. Highway, civil aviation, pipeline transportation and other transportations, are intended to be further extended to the advantaged area of traditional railway, and the quality, efficiency of transportation will be the focus of competition in the future.

---

H. Song (✉) • T. Tang • C. Li • Y. Ding
State Key Laboratory of Rail Traffic Control and Safety, Beijing Jiaotong University, No. 3 Shang Yuan Cun, Beijing, Hai Dian District, China
e-mail: songhaifeng2011@gmail.com; ttang@bjtu.edu.cn; 12111047@bjtu.edu.cn; 1125061@bjtu.edu.cn

H. Xia and Y. Zhang (eds.), *The 2nd International Symposium on Rail Transit Comprehensive Development (ISRTCD) Proceedings*, DOI 10.1007/978-3-642-37589-7_37, © Springer-Verlag Berlin Heidelberg 2014

In order to further improve the competitiveness and economic efficiency of railway, full system analysis research need to be done to make decisions for market operating.

## 37.2 The Important of Railway Passenger Traffic Volume

According to China medium and long-time railway network plan, in order to solve the problem of insufficient trunk railway transport capacity, the national will build Passenger Dedicated Lines (PDL), which will be more than 12,000 km, between capital cities and large and medium cities before 2020. The project of the PDL will be involved with operation, organization, station layout and built, passenger transfer and relief and so on inevitably, all the aspects referred above have a relation to the passenger forecasts. Therefore, the passenger forecast for PDL is becoming more and more important nowadays.

The RPTV forecasting can offer scientific basis for the establishment of policy and making of transportation development plan.

## 37.3 Long Time Foresting for the Train Passenger

According to foreign experiences, after the PDL was built, the passenger volume will increase dramatically in short time, this kind of phenomenon is different from the flow increased year by year pattern. For example, Tokaido Shinkansen in Japan after 1 year opening, the passenger volume dramatically increased 52 % between Shizuoka and Hamamatsu, compare the condition before opening, Tokaido Shinkansen traffic soared 66 % 2 years after opening [1]. The remarkable change should be taking into full account when make long-term forecasting in PDL.

Considering the passenger flow's structure of the PDL, the passenger flow can be divided into three parts: trends passenger flow, induced broker flow and transfer passenger flow. The influence factors of these three flows are not the same, the characteristics of the passenger flows change also have difference, and in the operation of the PDL in different periods three constitutions ratio will change. If the three parts were taken as a whole to analysis in the study of the passenger flow forecast, it will lead to a large deviation to prediction, so in my paper, I make different models apply to the three different flows, so that the result could be more close to the real situation.

### *37.3.1 Trends Passengers Flow*

Trends passenger flow is a flow that with the increasing of population and economy in a country or a region, more and more passengers choose railway to travel.

At present, because of overload in transportation capacity, the main railway lines of China, such as Beijing-Guangzhou, Beijing-Shanghai, Beijing-Harbin and Longhai lines, all of the main lines cannot afford to meet the growth of the passenger travel demand [2]. On this basis, the government of China decided to build PDL to separate the freight transport and passenger transportation, and with the building of PDL, the flow which belong to the existing lines will shift to the PDL, and this trends passenger flow shows a continuous gradual increase, and in the model choosing I thought the grey system theory is good.

#### 37.3.1.1 Grey System Theory

Grey System Theory is a study on indeterminate systems based on small samples and little information [3]. It deals with the original data and creates Grey Model to discover the development rules of the systems. The merits of this method are its easy use and high precision. Especially, it can assure that the number of samples and the volume of calculation will not increase with the time. Whereas its shortcoming lies in its unreliable forecasting result, especially when data samples are few and the fluctuation of data is obvious.

#### 37.3.1.2 Modeling and Procedures

Trends passenger flow is an order bounded set along with the development of economy, there are some unknown rules between it, considering the passenger transportation system is a huge Grey System, Grey Model is very suitable to predict transport volume [2–4].

Step 1: Collect the passenger traffic volume original data (OD matrix), and get the original series:

$$q^{(0)} = \left\{q^{(0)}(1), q^{(0)}(2), q^{(0)}(3), \ldots, q^{(0)}(t)\right\},$$

Where $q^{(0)}(t)$ ($t = 1, 2, 3, \ldots, t$) is respectively the passenger traffic volume of existing lines, which collected from the historical data of the first n years.

Step 2: Applying AGO (Accumulated generating operation) on $q^{(0)}$, provided that:

$$q^{(1)} = \left\{q^{(1)}(1), q^{(1)}(2), q^{(1)}(3), \ldots, q^{(1)}(t)\right\},$$

Where $q^{(1)}(t) = \sum_{k=1}^{t} q^{(0)}(k), \quad t = 1, 2, 3, 4, \ldots, n$

Step 3: Provided sum matrix $B$ and constant vector $Y_T$:

$$B = \begin{bmatrix} -\frac{1}{2}\left(q^{(1)}(1) + q^{(1)}(2)\right) & 1 \\ -\frac{1}{2}\left(q^{(2)}(1) + q^{(1)}(3)\right) & 1 \\ \cdots & \cdots \\ -\frac{1}{2}\left(q^{(1)}(t-1) + q^{(1)}(t)\right) & 1 \end{bmatrix};$$

$$Y_t = \left\{q^{(0)}(2), q^{(0)}(3) \ldots q^{(0)}(t)\right\}^T$$

Step 4: Perform a least squares estimate for the parameters, the simulated parameters can be obtained that

$$a \wedge = [a, u]^T = \left(B^T B\right)^{-1} B^T Y_t$$

Where $a$ and $u$ are parameters.

Step 5: figure out $q^{(0)}(t+1) = -a\left(q^{(0)}(1) - \frac{u}{a}\right)e^{-at}$

Now we get the trends passenger flow.

### *37.3.2 Induced Broker Flow*

Induced broker flow means as a result of the traffic facilities and other hardware change such as high speed railways and PDL build, or software conditions change such as improvement of traffic control measures and policy guidance, people produce more wishes to travel, then a kind of flow is formed. For example, the Beijing-Tianjin Express Railway makes people travel between Tianjin and Beijing more frequent. This kind of passenger flow express three steps: gradually formed, rapid growth, gradually stable.

In the paper, the road traffic of the gravity model is applied to forecast the induced broker flow, all the influence factors are called potential factors with same parameters in the paper. There are many affecting factors in the induced broker flow forecasting, such as resident population, industrial production, cultural exchange, and so on.

#### 37.3.2.1 The Road Traffic of the Gravity Model

Gravity models can be used in cases such as new rail station or construction of a new railway line, for which there are obviously no statistical data.

According Gravitation traffic accessibility model (Casey 1995) [5–6], we get the original model $q_{ij} = k\frac{P_i A_j}{R_{ij}^2}$ [5–6], in the days coming some scholars have further researched on the model and get a modificatory one $q_{ij} = k\frac{P_i^{\alpha} A_j^{\beta}}{R_{ij}^{\gamma}}$.

Where $P_i$, $A_j$, $R_{ij}$ have great influences on the $q_{ij}$, with the same theory, we can apply this model to the forecast of induced broker flow of railway, while equation ** presents a too simplistic analogy with gravity law, for this reason, we make a little change in the former model and get a new one as below [2]:

Suppose city *i* and city *j*, and the passenger flow between these two cities is $P_{ij}$, the formula as follow:

$$P_{ij} = k\frac{(E_i \times E_j)^{\alpha} + (F_i \times F_j)^{\beta}}{R_{ij}^{\gamma}},$$

Where:

$E_i$ and $E_j$ is the GDP of city *i* and city *j*,
$F_i$: population of city *i*,
$F_j$: population of city *j*,
$\alpha$, $\beta$, $\gamma$, $k$: proportionality factor,
$\gamma$: parameter of calibration. Various studies estimated values of the parameter $\gamma$ to be between 0.6 and 3.5.
$R_{ij}$: generalized cost of rail transport between cities *i* and city *j*, it is the comprehensive reflection of transportation service.

$$R_{ij} = C_{ij} + T_{ij}W + S_{ij}$$

Where $C_{ij}$: fee cost of rail transport between cities *i* and city *j*,
$T_{ij}$: time cost of rail transport between cities *i* and city *j*,
$W$: time value (dollar per hour),
$S_{ij}$: passengers fatigue.

#### 37.3.2.2 Modeling and Procedures

Step 1: Get the population data and its GDP value between each section of the line.
Step 2: Use the questionnaire method etc. to get the transport impedance between each section before a PDL was built.
Step 3: Make log conversion for improved gravity model, with the multiple regression method to
Step 4: The Step 3 can be worked out the passenger flow for existing lines.
Step 5: Obtained the transportation impedance for the PDL in planning, from the model built result to obtain PDL flow after it in operation.
Step 6: The difference in value between Step 5 and Step 4 can be concluded that the proposed increase induced broker flow for passenger.

### 37.3.3 Transfer Passenger Flow

The transfer passenger flow means with the promotion of the railway system, some passengers who used to take the other travel ways tend to railway system, so we should also take this passenger flow in account [1].

#### 37.3.3.1 Sharing Rate Model

First, the most important part is we should ascertaining the sharing rate of the passenger flow of the railway in the research of forecast model of the passenger flow volume which involves all the available traffic model. Using other city's experience is the most common way to get the original data, and the influence scope of railway should be taken in account [7]. Here we build a model which reflects the probability of choosing the certain traffic modes. The model is as follow:

$$w_i = \frac{\exp(v_i)}{\sum_j \exp(v_i)}$$

Where $w_i$ is the probability of choosing this traffic modes; i, j is the kinds of traffic modes; $V_i$, $V_j$ is the utility function of taking the traffic modes i, j;

In the forecast model, the main task is to distinguish the passenger flow of choosing the railway and other modes. Whether selecting the railway is related to the accessibility of the railway station and the time of resident's trip. So its utility function is as follow:

$$v_i = x_1 C_i + x_2 T_i + \varepsilon$$

Where $C_i$, $T_i$ is respectively the convenient and the time cost of traffic model; $x_1$, $x_2$ is the variable of the utility function; $\varepsilon$ submits to normal distribution.

This model takes the accessibility and trip time & distance as the variable of the utility function, and reflects the sharing rate of railway passenger flow volume will change with the layout of the station and the distance of railway station. The distance to the station is shorter; the accessibility is better; the sharing rate is bigger. This meets the characteristic of railway passenger flow distribution.

#### 37.3.3.2 Modeling and Procedures

Proposed there were $n$ different modes of transportation in a transportation line from $i$ to $j$, and the total traffic volume was $A_{ij}$.

Step 1: Get the effectiveness $v_i$ from the survey of different modes of transportation

Step 2: According to the model, work out the share rate $w_i$ of different modes of transportation.

Step 3: Figure out passenger flow $A_{ij}^{i}$ based on $A_{ij}$ and $w_i$.

Step 4: Obtain the new effectiveness $v_i^{'}$ when mode $i$, one of the transportation modes, had improved its service.

Step 5: According to the model, figure out the share rate of $w_i^{'}$ different modes of transportation.

Step 6: Suppose the situation that the total traffic volume has no change, we can figure out the passenger flow $A_{ij}^{i'}$ in mode $i$ after its service had improved.

Step 7: And now the difference between step 6 and step 3 is the transfer passenger flow.

## 37.4 Conclusion

With the results of Sect. 37.3.1.2 step 5, Sect. 37.3.2.2 step 6 and Sect. 37.3.3.2 step 7, we can get the forecasting of long-term passenger flow of PDL. At the different period operation of PDL, the passenger flow is different. In detail, at the beginning the main part is trends passenger flow; when the PDL has operated for a time, the induced broker flow and transfer passenger flow will take more weight, while the trends passenger flow may not change a lot.

So I would like to say the model built to forecasting long-term passenger flow can reflect the real situation in some degree.

Rail transit is a huge, complex system which has the feature of irreversibility, once the line was built on change was available.

Above all, China's high-speed railway and the construction of the rail transit development of China will occupy an important position in the future, this part will occupy a great share of the market, and the government will donate huge financial support into the research, as a result academic research in the area of rail transit will also have a great prospect.

**Acknowledgment** We wish to acknowledge the support of Beijing Laboratory For Mass Transit, the National High-Technology Research and Development Program ("863"Program) of China No. 2012AA112801, The Project of the Fundamental Research Funds for Central Universities No. 2011JB2004.

## References

1. Chen CL, Huang ZT (2008) Study on forecast of the passenger traffic volume of passenger-dedicated line[J]. Technol Econ Areas Commun 10(1):75–76
2. Wang W (2009) The research of passenger volume in passenger dedicated lines[J]. Railway Oper Technol 15(2):43–45

3. Xu W, Qin Y, Huang HK (2004) A new method of railway passenger flow forecasting based on spatio-temporal data mining. In: 2004 IEEE intelligent transportation systems conference, October 2004, London, pp 402–405
4. Qi F, Liu XY, Ma YH (2009) Prediction of railway passenger traffic volume based on neural tree model. In: 2009 second international conference on intelligent computation technology and automation, October 2009, Beijing, pp 370–373
5. Liu CQ (2001) Advanced traffic planning[M]. China Communications Press
6. Mao BH, Zeng HX, Yuan ZZ (1999) Transportation planning model and application[M]. China Railway Publishing House, Beijing
7. Xie H, Yan KF, Wen Y (2008) A forecast model of urban passenger flow containing new railway project. In: 2008 workshop on power electronics and intelligent transportation system, August 2008, Guangzhou, pp 435–439

# Chapter 38
# A Study on City Logistics Design Planning Based on Logistics Requirements Potential Model

**Fang Lu and Bo Lu**

**Abstract** It is extremely important to predict the logistics requirements in a scientific and rational way for the city logistics design planning. However, in recent years, the improvement effect on the prediction method is not very significant and the traditional statistical prediction method has the defects of low precision and poor interpretation of the prediction model, which cannot guarantee the generalization ability of the prediction model theoretically, but also cannot explain the models effectively. Therefore, in combination with the theories of the spatial economics, industrial economics, and neo-classical economics, taking city of Baotou as the research object, the study identifies the leading industry that can produce a large number of cargoes, and further predicts the static logistics generation of the Baotou and its hinterlands. By integrating various factors that can affect the regional logistics requirements, this study has established a logistics requirements potential model from the aspect of spatial economic principles, and expanded the way of logistics requirements prediction from the single statistical principles to a new area of special and regional economics for the city logistics design planning.

**Keywords** City logistics • Design planning • Logistics requirements potential model • Prediction

F. Lu (✉)
College of Fine Arts, Dalian University, Dalian, China
e-mail: shining_112@live.cn

B. Lu
Institute of Electronic Commerce and Modern Logistics,
Dalian University, Dalian, China
e-mail: lubo_documents@hotmail.com

H. Xia and Y. Zhang (eds.), *The 2nd International Symposium on Rail Transit Comprehensive Development (ISRTCD) Proceedings*,
DOI 10.1007/978-3-642-37589-7_38, 

## 38.1 Introduction

The regional logistics requirement which was firstly proposed in the 1990s [5], is very important for formulating city logistics design planning. After nearly 20 years of development, it has made great progress. By sorting out comprehensive overview about existing method, most researchers deal with them as a regression problem. According to the development process and the level of intelligence, it can be roughly divided into three stages:

The prediction method based on statistics in the first stage. Main methods include regression analysis [1], elasticity coefficient method, freight intensity method, clustering method, gray theory model, Markov chain, input-output model [3], space-time multi-term probability model and decision support system. The major features of this type of methods can process the sequencing and linear data, and explain the construct models effectively. The prediction method based on artificial intelligence in the second stage. Later on, to enhance the prediction accuracy, the researcher [6] adopted the artificial intelligence method, such as the artificial neural network (ANN) and its improvement model. The prediction methods based on statistical learning theory is the third stage.

Although certain achievements have been made, some urgent problems still exist in the regional logistics requirements prediction [4]. Therefore, the generalization capability of the prediction model cannot be guaranteed from theory, which makes the predictive models after training to have no stable prediction effect for the new logistics requirements data set. Third, among these methods, although the artificial intelligence methods and support vector machine method have certain advantage in the prediction of accuracy, it still cannot explain the models effectively.

In fact, regional logistics requirements is a derived requirements determined by the level of regional economic development, and the indicators that affect logistics requirements prediction such as economic, logistics industry, environmental factors will directly or indirectly affect the growth or reduction. Therefore, this study has explored the logical relationship between the regional logistics requirements and the relevant factors based on the spatial economics and new economic geography. From the perspective of logistics market potential, this study has established the logistics requirements potential model and expanded logistics requirements prediction way from the single statistical principles to new area of space economics, to enrich research method of the regional logistics requirements.

The paper is structured as follows: after introductory section of Chap. 1, there will be followed by description of logistics requirements potential model. In so doing, the research procedure is included in Chap. 2. Moreover, research scope and logistics requirements prediction are derived in Chap. 3. Finally, conclusions are drawn in Chap. 4.

## 38.2 Research Method

### *38.2.1 Logistics Requirement Potential Model*

The concept of market potential was firstly proposed [7]

$$M_j = \sum_k Y_k g(D_{jk}). \tag{38.1}$$

where, $M_j$ is the market potential of City j; $Y_k$ is the income of every area; $g(D_{jk})$ is the attenuation function of distance; $D_{jk}$ is the distance between City j and City k. However, there is no micro-structure, which only showed that the market potential is proportional to market purchasing power of various places (i.e. the market scale), and inversely proportional to the distance from this place to the market. The theoretical innovation on this issue was originated from the early 1990s when Krugman had built the New Economic Geography [2, 8, 9]. Krugman claimed that two forces determine whether the economic behavior and economic factors are convergent or diffusive in spaces in the model: centripetal force and centrifugal force. The centripetal force makes the economic activities and economic elements to reflect the industrial agglomeration in the spatial distribution in the industry level, which is why the new economic geography and spatial economics attaches great importance to industrial agglomeration.

Later, Fujita established his market potential index formula:

$$\begin{aligned} D_x(j) &= \sum_{h=1}^{l} c_x(h|j) = P_x(j)^{-(\mu+1)} \sum_{h=1}^{l} \frac{\alpha E_h T_X^{-(\mu+1)}}{\sum_{k=1}^{l} n_k \{P_x(k) T_x(k,h)\}^{-\mu}} \\ &= \left\{\Lambda^{-1} W_j^{1-\beta} \Gamma(j)^{\beta}\right\}^{-(\mu+1)} \Omega(j). \end{aligned} \tag{38.2}$$

where, $\Omega(j) = \sum_{h=1}^{l} \left[ \frac{\alpha E_h T_x(j,k)^{-(\mu+1)}}{\sum_{k=1}^{l} n_k \left\{ W_k^{1-\beta} \Gamma(k)^{\beta} T_x(k,h) \right\}^{-\mu}} \right]$ is market potential of City j, where has been set the producing area; $\alpha$ is market distribution share; $E_h$ is market capacity of City h; $T_x(j, h)$ is transport factor from City j to City h; $n_k$ is differentiated products quantity of City k; $W_k$ is wages of workers in City k; $\Gamma$ is price index of City k; $\mu$ is redefinition of substitute parameter; $1 - \beta$ is investment share of labor in the process of intermediate product. The form of market potential is more clearly, and is also convenient for specific calculation. The numerator is prototype of Harris' market potential. Differing from Harris, the denominator of Fujita's market potential includes the spatial competition factors.

However, Krugman and Fujita proposed the potential model was just from the perspective of economic scale and regional economic agglomeration. On the basis of that, this paper has introduced the related logistics factors to construct the logistics requirements potential model:

$$\ln p_{rt} = \alpha_0 + \alpha_1 \ln E_{rt} + \alpha_2 \ln l_{rt} + \alpha_3 \ln e_{rt} + \alpha_4 \ln f_{rt} + \alpha_5 \ln w_{rt} + \alpha_6 \ln d_{rt} + \mu_{rt}. \quad (38.3)$$

where, r indicate the city, t indicate a particular year, $\mu_{rt}$ is random error.

For simplifying the calculation difficulty, this study has set the all power exponent of factors and random error equal 1. Therefore, according to the positive/negative correlation, the logistics requirements potential model can been re-constructed as:

$$p_i = \sum \frac{l_i E_i e_i f_i w_{1i} w_{2i}}{d_{ij} d_{ip}}. \quad (38.4)$$

where, $p_i$ is the potential value of region i, E is the GDP, e is human capital (employee of logistics/population of the city), f is the proportion of investment in fixed assets (logistics investments/the whole investments of society), w is highway freight transport, $d_{ij}$ is the Euclidean distance between region i and j, $d_{ip} = 2/3 \sqrt{S_i/\pi}$ which is the inner region distance, where S is the region area.

### 38.2.2 Research Procedure

By constructing the model of the logistics requirements potential, one hand, this study need identify the cities which have high relevance with logistics of researched city to jointly constitute a logistics system. Then, based on the leading industry of researched city, this study could predict the cargo quantity of each city according to these industries by data fitting. Another hand, by applied with potential model to predict the logistics requirements, this study need calculate the parameters within the potential model. Then, the study could respectively identify the logistics requirements potential value of each city. Furthermore, the potential value of each city has been calculated independently, and the research scope of this study, which is centering on city of Baotou, has formed a system composed of 20 cities surrounding Baotou. Therefore, the study should apply the normalization processing to the independent potential values of each city: $\bar{p}_i = \frac{p_i}{\sum_i pi}$.

And then, by multiplying the processed potential value by the total cargo production of each industry within the system, the logistics requirements of each industry for each city could be obtained. The formula is: $\bar{n}_i = \bar{p}_i \sum_i n_i$, where, n is the total cargo production of each industry, and $\bar{n}$ is the logistics requirements of each city (Fig. 38.1).

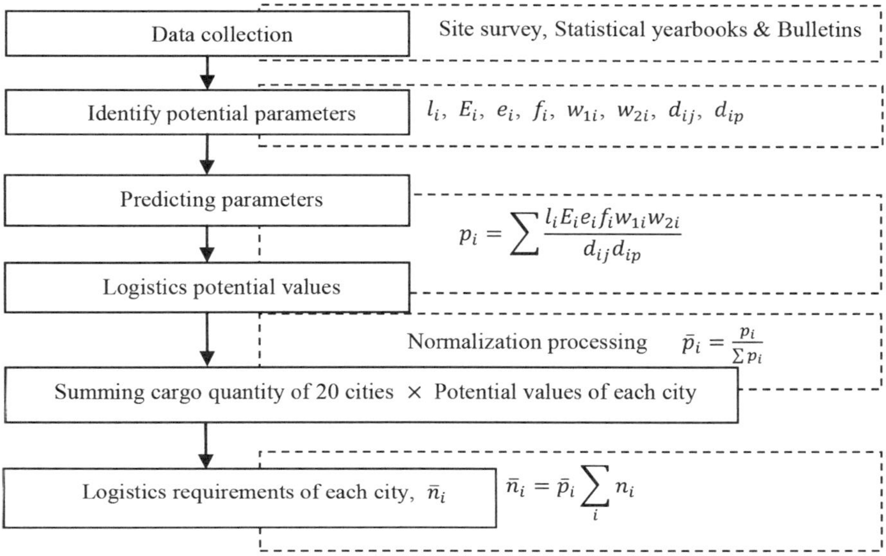


**Fig. 38.1** Research procedure

## 38.3 Result Analysis

### *38.3.1 Research Scope*

The study is centering on city of Baotou, scope of I, II, III area surrounding, I area is as per radius of 100 km, II area is as per radius of 100–300 km, III area is as per radius of 300–500 km. Definition of I, II, III surrounding areas is as following: central city: Baotou; I area: Erdos; II area: Hohhot, Shuozhou, Yulin, Wuhai and Bayan Nur; III area: Shijiazhuang, Taiyuan, Xi'an, Yinchuan, Erenhot, Ulan Tsab, Zhangjiakou, Datong, Xinzhou, Lvliang, Alashan, Shizuishan and Wuzhong.

### *38.3.2 Prediction for Static Cargo Quantity*

Each city has its independent industrial type, and the production and consumption have different features. Therefore, judging the leading industry based on Baotou, this study has predicted the logistics requirements of these industries in each city.

The rapid economic development of Baotou has direct correlation with its industrial structure. The major industry in Baotou is the second industry with relatively complete category, by combining with the status of economic development, the following industries, such as: grain, vegetables, fruit, meat production,

**Table 38.1** Logistics requirements potential value of each city in 2015

| Cities \ Items | Baotou | Hohhot | Shi J.Z. | Taiyuan | Erdos | Shuo Z. |
|---|---|---|---|---|---|---|
| Logistics requirement | 17,232.7 | 6,356.53 | 4,758.31 | 2,036.71 | 1,601.94 | 388.45 |
| Potential value | 961,216 | 354,558 | 265,411 | 113,605 | 89,354.1 | 21,666.9 |
| Normalization processing | 52.42 % | 19.34 % | 14.47 % | 6.20 % | 4.87 % | 1.18 % |
| Xi'an | Zhang J.K. | Ulan Tsab | Datong | Wuhai | Lvliang | Yulin |
| 339.96 | 55.93 | 24.87 | 22.19 | 20.49 | 13.01 | 12.54 |
| 18,962.23 | 3,119.80 | 1,387.40 | 1,237.71 | 1,143.09 | 725.56 | 699.24 |
| 1.03 % | 0.17 % | 0.08 % | 0.07 % | 0.06 % | 0.04 % | 0.04 % |
| Xinzhou | Yin C. | Alashan | Bayan Nur | Erenhot | Shi Z.S. | Wu Z. |
| 7.38 | 2.07 | 1.06 | 0.96 | 0.11 | 0.08 | 0.004 |
| 411.56 | 115.45 | 59.12 | 53.54 | 5.82 | 4.16 | 0.246 |
| 0.02 % | 0.01 % | 0.00 % | 0.00 % | 0.00 % | 0.00 % | 0.00 % |

Unit: Hundred thousand tons

milk, metallurgical industry, non-metallic mineral products, coal mining and washing, textile, beverage manufacturing, chemical materials and chemical products, food processing and the flow of commerce will bring a large quantity of logistics requirements, have been set as the research centre.

### *38.3.3 Prediction for Dynamic Logistics Requirements*

The predicting results of logistics requirement for each city in 2015 have been summarized in Table 38.1, and the following observations can be made. The column and row totals represent, respectively, the logistics requirements and potential values of each city in 2015. It is clear that, Baotou has the best performance among the twenty cities. Hohhot and Shijiazhuang ranked as the second and third best in this model, respectively. The values were more than 10 %, with these potential values far exceeding that of other cities. According to potential values, although Baotou, Hohhot and Erdos have more potentiality than others, there is a big distance, however, among them. With the similar economic scale and location, the huge gap of logistics requirements and potential values which are depended on the logistics potential parameters among three cities, can be explained that Baotou has most convenient logistics infrastructures, therefore it is possible to become the biggest logistics centre and the most important node in Inner Mongolia, firstly.

In the contrary, logistics potential values of most others are less than 1 %, thus indicating that they would need to improve their investment of resources to enhance the logistics competitive power. Among them, Wuzhong has the least potential value.

## 38.4 Conclusions

Implementation of West Development Strategy has laid a policy advantage for development of city logistics industry in Inner Mongolia. The state has intensified support for logistics infrastructures, such as logistics parks. However, the vacancy rate of logistic park is as high as 60 %, resulting in huge wasting of resources. Therefore, construction of logistics parks should follow a uniform design planning and should not be repeated. Because of that, a scientific prediction is a basic work for proper planning. Prediction of regional logistics requirements can provide a basis in theory for national and regional economic administrations. Moreover, according to predicting results, the government could evaluate the contribution of the logistics industry to local economic development which is useful in helping the government to formulate for the logistics industry and guide optimum configuration of resources in the logistics market. From a micro perspective, logistics enterprises should properly configure limited resources according to the prediction of logistics requirements in order to minimize investment risks and maximize benefits.

There are also some limitations in this study. In the construction of model, this study has set the all power exponent of factors and random error equal 1. However, the weight of each factor is inequality, obviously. Moreover, rationality of research scope also needs further explaining. Therefore, in the future research, the research will concentrate on the two aspects.

## References

1. Chu LY, Tian ZG, Xie XL (2010) Application of an combination forecasting modal in logistics demand. J Dalian Marit Univ 30(4):43–46
2. Fujita M, Thisse JF (2002) Economics of agglomeration: cities, industrial location and regional growth. Cambridge University Press, Cambridge, MA
3. Guo XC, Xie SH, Hu B (2009) Regional logistics demand analysis model and solution. J Southeast Univ 31(3):24–31
4. Luo SG, Ye S, Hu R (2010) A research of forecasting the logistics amount based on multi-output support vector regression. J East China Jiao tong Univ 27(5):67–71
5. Xia GE (2010) A study on the development of regional logistics requirements prediction. China Logist Purch 4:68–69
6. Yin YL (2010) Logistics demand forecast based on adaptive neural network. J Henan Polytech Univ 29(5):700–704
7. Harris CD (1954) The market as a factor in the localization of industry in the United States. Ann Assoc Am Geogr 44:315–348
8. Krugman P (1991) Increasing returns and economic geography. J Poetical Econ 99:483–499
9. Acs Z, Anselin L, Varga A (2002) Patents and innovation counts as measures of regional production of new knowledge. Res Policy 31(7):1069–1085

# The Editor's Note

The development of China rail transit has entered into an unprecedented era. The construction of rail transit interacts with the urban development. The construction scale and rate of rail transit are astonishing, which has brought about unprecedented impact on the urban development. Under such circumstances, there are many problems need to be solved, including the technical issues of rail transit, management issues of urban planning, and investment and operation issues. All these issues are interweaved together, which need to be handled by many departments. Hence, in order to carry out the research on such complex situation, broader and more comprehensive perspectives are highly required. This is also the purpose of holding this conference by Beijing Jiaotong University, as well as calling for the publication of collected papers. Before holding the conference, we also carried out particular investigations on the complex development of China urban rail transit construction. Through the investigation, we collected information, as well as found problems and proposed suggestions, in order to summarise some useful conclusions into this assembly documents. We do hope these efforts could arouse attention from relevant departments, in order to promote the complex healthy development of China rail transit.

This assembly documents includes 40 papers, which are divided into three parts. All these academic papers reflect relevant problems and research progress of China rail transit construction and development. Embodying strong academic significance, we do hope readers will be benefit thorough reading it.

Mentioning the seminar and the publication of this assembly documents, special thanks will goes to honorable academician Shi, Zhongheng, academician Zou, Deci, for their supportive and enthusiastic speaking before the seminar. We are urged and encouraged by their speakings, and we are encourage to constantly make progress for China urban development and urban rail transit. Meanwhile, we would

H. Xia and Y. Zhang (eds.), *The 2nd International Symposium on Rail Transit Comprehensive Development (ISRTCD) Proceedings*, DOI 10.1007/978-3-642-37589-7, © Springer-Verlag Berlin Heidelberg 2014

like to present our gratitude to China Urban Planning Academy, CCDI, Springer, and all the members from the School of Architecture and Design who made great contributions to the publication of this assembly documents.

Haishan Xia
Dean/Professor/Ph.D.
The School of Architecture and Design, Beijing Jiaotong University
20th March 2013

MIX
Papier aus verantwortungsvollen Quellen
Paper from responsible sources
FSC® C105338

If you have any concerns about our products,
you can contact us on
**ProductSafety@springernature.com**

In case Publisher is established outside the EU,
the EU authorized representative is:
**Springer Nature Customer Service Center GmbH**
**Europaplatz 3, 69115 Heidelberg, Germany**

Printed by Libri Plureos GmbH
in Hamburg, Germany